World Nuclear Energy

This book has been prepared under the auspices of The Royal Institute of International Affairs.

Edited by Ian Smart

WORLD NUCLEAR ENERGY

Toward a Bargain of Confidence

THE JOHNS HOPKINS UNIVERSITY PRESS
Baltimore and London

Printed in the United States of America

The Johns Hopkins University Press, Baltimore, Maryland 21218
The Johns Hopkins Press Ltd., London

Illustrations by George Hartfield

Library of Congress Cataloging in Publication Data
Main entry under title:

World nuclear energy.

Includes index.
1. Atomic energy industries—Addresses, essays, lectures. 2. Nuclear nonproliferation—Addresses, essays, lectures. I. Smart, Ian.
HD9698.A2W59 333.79'24 82-179
ISBN 0-8018-2652-7 AACR2

Contents

List of Figures

List of Tables

World Nuclear Energy

1

Introduction

Ian Smart

An introduction to this somewhat unusual book has to explain not only what the papers in it contain but also how and why they came to be written, for this particular collection of papers about international nuclear policy issues cannot be fully understood outside the context of its origin.

No one has ever doubted the technical difficulty of exploiting nuclear energy. No one has ever been unaware of the challenge its use presents to those concerned with policy at the local or national level: to men and women in government, but also to ordinary citizens. By 1976, however, some of us who had been involved in one aspect or another of those issues found ourselves increasingly worried also about what we saw as a growing danger of international friction over the ways nuclear energy was being developed for civil use, and especially over the alleged relationship between civil nuclear power and nuclear weapons.

The Non-Proliferation Treaty (NPT), which came into force in 1970, had suggested to many that governments in the West, the East, and the Third World were on the way to establishing an agreed framework of effective rules for separating the atoms of peace from the atoms of war. It seemed that the group of only five countries that had already proved their possession of nuclear explosives might, after all, grow no larger, and that civil nuclear cooperation might meanwhile, on that basis, flourish in an atmosphere of international harmony. Sadly, the hope was short-lived. The problems that, in fact, remained in 1970 turned out to be of daunting proportions. They were increasingly complicated, moreover, by the progressive international diffusion of nuclear technology capable of serving either civil or military purposes. Finally, in 1974, the possible implication of failure to solve those problems was forcefully signaled when a sixth country, India, made and exploded a nuclear device.

Faced with a rising tide of evidence pointing to the enormous difficulty of regulating the spread of inherently ambiguous technology and materials, many governments, in countries at very different levels of nuclear development, began to question the adequacy or acceptability of existing international arrangements. "Supplier" countries in the industrial world, newly alarmed and suspicious about the use to which such technologies or materials might be put, sought, unilaterally or in concert, to reinforce the barriers to their uncontrolled dissemination. "Recipient" countries, not least in the Third World, found much to resent or fear in what they sometimes saw as an unjustifiable attempt to hinder their civil nuclear development. Thus, by 1976, when the issues of nuclear energy use and nuclear export policy emerged prominently in an American presidential election, the relations between countries engaged in nuclear trade or cooperation had already become increasingly fractious, and sometimes embittered.

The first of the conversations that led eventually to the establishment of the International Consultative Group on Nuclear Energy (ICGNE) took place at about that time between Mason Willrich, while he was still professor of law at the University of Virginia, and myself, while I was still director of studies at the Royal Institute of International Affairs. We both felt that the situation that had come about since 1974 entailed two risks of a kind that might not easily be addressed through formal intergovernmental channels. One was the obvious risk of growing political, as well as commercial, friction between nuclear exporters and importers – suppliers and recipients – especially where the division between them coincided with that between the industrial world of the "North" and the Third World of the "South." The other risk was the more general one that, because of the inhibitions that naturally characterize formal communication between governments, and because so many of the issues likely to affect future nuclear power use were sufficiently delicate or speculative to fall foul of those inhibitions, some prospective sources of serious international friction were unlikely to be explored quickly or effectively enough through official channels.

With the latter risk particularly in mind, we were persuaded that it might be helpful if there could be brought together, under nongovernmental auspices, a small group of people experienced in international nuclear relations, including both senior policymakers and independent observers, to address longer-term issues in that field in a less inhibited manner. Since it would necessarily draw upon experts from importing countries, including those in the Third World, as well as on nuclear suppliers, the group might also have some value as a vehicle for the frank discussion of reasons for, and possible responses to, the growing friction between those groups.

Our interest in the idea of such a group grew stronger as international nuclear relations deteriorated further. The adoption of a more restrictive policy on nuclear exports by the United States, first under President Ford and then under President Carter, was matched by tighter regulations on the part of other exporters of uranium or nuclear equipment and services. Importing countries were audibly unhappy about the effects. Moreover, exporting countries were by no means in complete agreement as to what should or should not be done. When President Carter, in April 1977, gave notice that his administration would seek to revise bilateral cooperation agreements to reflect its own opposition to separating and using plutonium, the critical international reaction was no more than the predictable expression of a wider malaise.

The great majority of governments continued to share a desire to prevent control of nuclear weapons from spreading to additional countries, as they also shared a general interest in securing for the world the peaceful benefits of nuclear power. Palpably, however, nations were increasingly at odds about the appropriate means to those separate ends and about the way they were to be reconciled. There was every reason, in those circumstances, to persist in the attempt to investigate those problems through the informal medium of a new international group of experts. Our plans were therefore pressed to fruition. The Rockefeller Foundation (to which Mason Willrich had moved to direct the International Relations Division) was willing to consider an application for funds, and the Royal Institute of International Affairs (from which I was preparing to depart) agreed in principle to act as the proposed group's cosponsor. The way was thus clear to issue preliminary invitations, in June 1977, to those who might form the ICGNE.

It required no great perspicacity in 1977 to see the need for a larger effort to understand and resolve international discord over nuclear energy development. Certainly the governments concerned were well aware of the costs and risks involved. It came as no surprise, therefore, when the United States government took the lead in launching, from October 1977, a broad intergovernmental assessment of the technical relationship between civil nuclear programs and nuclear weapons proliferation, in what was entitled the International Nuclear Fuel Cycle Evaluation (INFCE). By that time, plans for establishing the ICGNE had of course reached maturity. Clearly, however, the fact that governments had agreed to set up INFCE at about the same time raised a question about the sense of international deficiency that underlay the ICGNE concept.

On reflection, those in New York and London who had been planning the ICGNE enterprise were firm in the view that the proposal had lost none of its force as a result of government plans. Although motives for the two initiatives overlapped, and though some individuals might be involved in both exercises, their expected results were quite different. Three things stood out. First, the ICGNE would be predominantly concerned with political, rather than technical, issues, whereas INFCE was expressly to limit itself to technical assessment. Second, the ICGNE, partly because of its bias toward politics, would have a broader and longer perspective. Third, the ICGNE would be a small group of not more than twenty persons, meeting privately and speaking exclusively in their personal capacities, without any duty to represent states or national policy positions. Their deliberations could therefore expect to be less inhibited than those in a larger and more open government forum.

A measure of incidental complementarity between the roles of INFCE and the ICGNE could not be excluded, but no conflict was seen. Nevertheless, it was prudent to submit that view to the judgment of those who would actually constitute the ICGNE. The formal invitations spoke, therefore, of an initial meeting "to evaluate our tentative plans for the Group's activity, before committing ourselves and those we are inviting to a longer programme of meetings." In the event, the unanimous opinion was that the ICGNE should be constituted for at least two years. On that basis, the Rockefeller Foundation agreed to provide the additional funds needed and to act formally, with the Royal Institute of International Affairs, as the group's sponsor.

The group invited to meet for the first time in London at the end of October 1977 had a broad but not unlimited purpose and a diverse but not unconstrained membership. The cosponsors had agreed at an early stage that, if useful results were to be obtained in a reasonable time, some limit would have to be placed on the ICGNE's agenda. The most important limit was that the group should not be expected to debate the intrinsic virtues and vices of nuclear power. Given that the aim was to focus on questions of international cooperation and conflict, it should rather assume that attempts to exploit nuclear power would occur in a growing number of countries and should concentrate on how such an expansion might be made internationally tolerable and harmonious. From that decision there flowed conclusions not only about the agenda but also about the group's composition. It became unnecessary, for instance, to include a complete spectrum of opinion on the inherent value of nuclear fission as a means of supplying energy. Arguments for and against nuclear power, in terms of cost, safety, environmental impact, and social effect, were familiar ground, and no one involved in the ICGNE was insensitive to their significance or to the sometimes delicate balance between them. But it was never intended that the ICGNE should conduct, still less resolve, arguments of that kind. Nor, therefore, was it necessary that its membership should embrace the diverse strands of opinion on those issues.

What was necessary was that the ICGNE's membership should be balanced in at least three other respects, all bearing on the central theme of international nuclear relations. It should include a range of views on the rate at which civil use of nuclear power could realistically be expanded in different parts of the world. It must also include a diversity of interests in civil nuclear activities and the nuclear fuel cycle: the supply and processing of nuclear raw materials, the design and operation of nuclear power plants, the regulation as well as the management of nuclear industries, and above all the conduct of international negotiations on nuclear energy and its control. Finally, it was essential that the ICGNE should include personally authoritative voices from a representative selection of regions and nations, covering suppliers and recipients and developing as well as industrialized states. When that last point was weighed against the earlier conclusion that not more than twenty persons should be involved, it led to the decision that, excepting the chairman and rapporteur, no more than one ICGNE member should be of any one nationality. Within those limits, the criteria for selecting members had no reference to whether they shared the policy views of incumbent governments. In the event, some did but some clearly did not. What mattered, since all would take part in their personal capacities, was that those chosen should be highly qualified as individuals and that together they should make up a group that was balanced on the bases mentioned.

The list of members thus selected to take part in ICGNE activities, as it appears later in this volume, is a striking one. It includes distinguished personalities from seven developing and nine developed countries, as well as from the International Atomic Energy Agency (IAEA). Three members were forced by circumstances to leave the group before its work was completed. Two others joined the group after its inception. The great majority, however, remained throughout the term of the seven plenary ICGNE meetings, from October 1977 to January 1980. Their continuing interest was the best vindication of the premises on which they had been selected.

The second product of the premises underlying the ICGNE at its inception was the group's agenda. No single encapsulation of that could be adequate, partly because the group revised its plans as it went along, and partly because discussion constantly strayed beyond the boundaries of any written agenda. At each plenary meeting, for example, it became customary to devote an opening session to a free-ranging exchange of comments, and sometimes complaints, about recent national and international events. However, one way to give the flavor of the ICGNE's preoccupation is to quote from the schematic outline of a work program that was broadly agreed on at the second plenary conference in May 1978. Additional topics were imported later, and others were discussed only briefly. The outline does, however, suggest the scope of the group's concern.

A. Circumstantial Background
  1. *Historical Experience*
    a. *Government Confidence*
       Changes affecting the confidence placed by the governments of developed and developing countries in international agreements, relationships, and standards in regard to nuclear energy.
    b. *Social and Political Confidence*
       Changes affecting social and popular confidence in advanced technology relevant to nuclear energy, in nuclear energy itself, in nuclear industries, and in nuclear policies.
  2. *Future Assumptions*
    a. *Alternative Futures*
       Alternative future energy supply outcomes in regard to

nuclear power, and their geopolitical implications, including the implied need for the provision and/or movement of materials, equipment, and technology.

B. Policy

1. *Phase I*

   a. *Nuclear Proliferation Issues*

   Re-creating an international consensus on the institutional and political means, including global, regional, and multinational arrangements, appropriate to the acceptable management of risks to physical security associated with the distribution of nuclear weapons capability.

   b. *Economic and Social Viability of the Nuclear Energy Option*

   Identifying minimum operational standards and economic, structural, and institutional conditions necessary to the maintenance of an adequate, timely, and generally available industrial capacity to provide nuclear power.

2. *Phase II*

   a. *World Energy Interdependence*

   Establishing necessary and generally acceptable terms and conditions for future international cooperation on the exchange of nuclear materials, equipment, and technology, in a manner capable of contributing rationally to national and social requirements for energy security in all countries.

It is illuminating to compare that outline of the ICGNE agenda with an extract from the summary report by the Technical Coordinating Committee of INFCE, defining its accepted field of interest.

> The Organizing Conference agreed that INFCE was to be a technical and analytical study and not a negotiation. . . . It was recognized that a decision by a government to construct nuclear weapons is essentially a political decision motivated by political and national security considerations, among which is the relationship between vertical proliferation and horizontal proliferation and the existing and undiminished arms race. But it was not the task of INFCE to discuss these political matters. In this respect, the task of the study was to consider the possibility that the know-how, facilities or materials used in the nuclear fuel cycle might be misused for the purposes of constructing nuclear weapons—although it was recognized that nuclear energy for peaceful purposes is not the only or perhaps even the preferred route to that end—and also to consider whether any technical, institutional or improved safeguards measures might be introduced to discourage such misuse. Although possibilities for diversion by sub-national groups were also considered to some extent, it was recognized that the physical protection of nuclear materials and facilities is primarily a national responsibility. [INFCE 1980*b*, pp. 1–2]

Although the ICGNE's primary purpose was to permit informal discussion of unusually large policy issues within an unusually small group, it was envisaged from the outset that members would want to draw not only on statements and commentaries by their own number but also on a variety of analytical work carried out on the group's behalf. During the early months of its existence, the ICGNE devoted a good part of its time to debating the limits of what it might usefully address. During the last part of its life, it concentrated on the text of a report to be made collectively. The remainder of its debates focused, however, on analyses and working papers specifically prepared for its consideration. A number of those papers were written by members of the ICGNE or its supporting staff, sometimes with the assistance of colleagues outside. Others were drafted, at the group's request, by experts having no other connection with the ICGNE enterprise. None was initially intended for publication; their essential purpose was to inform discussion within the group. As time passed, however, it was recognized that some of these internal working papers had more general relevance and could serve an additional purpose if they were adapted for wider distribution. In particular, certain papers, if published, seemed likely to promote the wider and better-informed public discussion of international nuclear policy issues that it was one aim of the ICGNE to encourage.

The group itself was moved to select working papers for individual publication from those submitted to it. In the end, eight such papers were informally printed and distributed by the sponsoring institutions, the first in November 1978 and the last in March 1980. All are reproduced in this volume, essentially as originally published and with a minimum of editing. In each case the date of original publication is indicated. One purpose of this is to emphasize that each represents a view of its particular topic as seen from the vantage point of a particular historical moment. Another purpose is to make it clear that the order in which the working papers appear here is not chronological. Instead, they have been rearranged thematically. That said, it must be admitted that the order remains somewhat arbitrary. On balance, however, it seemed appropriate to begin with the broad historical review of international nuclear relations since 1954 by Bertrand Goldschmidt and Myron Kratzer that was also, by chance, one of the first ICGNE papers to be published individually. It was, after all, central to the conception of the ICGNE in 1976–77 that relations between governments in regard to nuclear power had become sadly strained, and the group was immediately at pains to understand how that had come about. Goldschmidt and Kratzer, the one a member of the ICGNE and the other its rapporteur, responded with a strikingly lucid account, given added force by the fact that both had held senior official positions in their own countries during much of the period they described.

The Goldschmidt and Kratzer paper has the ring of undoubted authority, sharpened by its Franco-American authorship. At the same time, its authors inevitably reflect perspectives from the industrial world, albeit from two parts of that world whose respective governments have not always seen eye to eye on nuclear policy issues. With that in mind, the ICGNE felt that their analysis might usefully be complemented by a view of the same historical context from the Third World. The resulting working paper by Munir Khan of Pakistan, another ICGNE member, stands second in this collection. It was not designed as a rebuttal of the analysis by Goldschmidt and Kratzer, and it has a good deal in common with their characterization of past events. It nevertheless demonstrates that any attempt to understand international policy disagreements in the nuclear field, or a fortiori to resolve them, must start by recognizing the extent to which interpretation both of disagreement and of the responsibility for it is a function of the angles of vision imposed on different governments and individuals by their particular circumstances.

Those two opening papers go a long way toward explaining the problems and conflicts that prompted the ICGNE's establishment and dominated many of its debates. They also refer extensively to national and international efforts to solve the problems or contain the conflicts, especially in regard to nuclear proliferation, through treaties or less formal arrangements. At the heart of many such efforts, since 1959, has stood the international safeguards system of the IAEA, coupled more recently with the safeguards provisions of the NPT. The paper on international safeguards by David Fischer of the IAEA, again an ICGNE member, forms a logical bridge, therefore, between the definition of the problem and the fuller discussion of possible solutions. It serves also to recall both the extent to which international efforts to erect a convincing barrier to nuclear proliferation – buttressed by IAEA safeguards – have, in fact, been successful, and the importance of persisting in the attempt to close the remaining gaps in that barrier through further agreement.

Fischer assesses the position on participation in the NPT and the application of IAEA safeguards as it was in 1979. Two more ICGNE members, Ryukichi Imai of Japan and Robert Press of Britain, look further into the future in the paper that follows. Their joint interpretation of the nuclear proliferation issue and of policy options in that connection is cast in political and psychological, rather than technical, terms. It thus echoes the view widely held in the ICGNE that, while technology must play a part in dissipating, as in prompting, fears of proliferation, the central issues involved are inescapably political and demand a political address.

The same political realism informs the following paper, by Russell Fox of Australia and Mason Willrich of the United States, both ICGNE members, on the international management of plutonium stocks. Here, however, the authors sought to give an institutional form to political will. In doing so, they anticipated more recent action in the IAEA. They also provided an example of the way ICGNE discussions, while often starting from broad principles or political generalities, looked toward specific and concrete action that might be taken internationally. In one sense, since related proposals are now before governments in the IAEA context, their paper has been overtaken by events. In another sense their perception of the problems to be solved by a practical arrangement for managing plutonium stocks internationally has become even more topical.

As all the five papers so far mentioned demonstrate, problems and fears of nuclear proliferation and of friction in international nuclear commerce figured largely in the ICGNE's work. They did not, however, monopolize it. It had been decided at the outset that another of the group's principal concerns should be with the structure and international distribution of technical and industrial capacity in the nuclear power field, and considerable time was in fact devoted to comparing the performance and prospects of nuclear industries in different countries. The sixth and seventh essays printed here, one by Richard Lester and the other by Måns Lönnroth and William Walker, bear testimony to that theme.

Lester's paper on nuclear power plant lead times had a particular background. During the first plenary ICGNE meeting, participants commented forcefully on the problems created in their own countries by the time taken to license, build, and commission a commercial nuclear reactor. Two separate propositions emerged: that this lead time was everywhere becoming longer, especially as a result of regulatory delays, and that reactor lead times in different industrial countries were becoming more various, with consequently divergent effects on the relative economics of nuclear power. To test those propositions, data on reactor lead times in a number of industrial countries were collected and analyzed. Special studies were commissioned in the United States and West Germany, and comparable information was gathered, through ICGNE members, from Japan and Canada. Thereafter Richard Lester, the group's research associate, was asked to collate and compare the results of those analyses in the single working paper reprinted here. It constitutes only a preliminary essay, based on a small sample of national cases. Its measured conclusions nevertheless underline the value of systematic international comparisons of nuclear planning and construction experience, too few of which have yet been attempted.

The next paper, by Måns Lönnroth, an ICGNE member from Sweden, and William Walker, the group's secretary, takes off, in one sense, from Lester's conclusions. The question they addressed concerned the future economic viability of the nuclear industry in different countries, with particular reference to reactor manufacture. Throughout the world, investments have been made and industrial capacity created on the basis of certain assumptions about the rate at which nuclear power production and associated fuel cycle operations will expand. Some of those assumptions have already been invalidated. Others seem increasingly vulnerable as economic recession moderates the growth of electricity consumption. Lönnroth and Walker set out to examine the potential effect of that slackening demand on a heavily capitalized industry. Their pointed and sometimes alarming conclusions deservedly attracted wide attention when their paper first appeared.

In their assessment of industrial viability, Lönnroth and Walker drew in part on preliminary results from an ambitious investigation of future nuclear demand and supply conducted for the ICGNE by a group at the West German nuclear research establishment in Jülich, under the overall direction of another ICGNE member, Karl-Heinz Beckurts. The main report of that study appears as the last of the working papers in this volume. In it, Thomas Connolly, Ulf Hansen, and Wolfgang Jaek, with the support of Beckurts, describe the elaborate model they constructed of nuclear power growth up to 2020. The ICGNE itself decided against attempting any new forecast of the rate at which world demand for nuclear power would increase. At an early stage, however, it became apparent that it would be valuable to have a reasonably detailed scenario of future demand, less for its own sake than in order to identify implications for the various parts of the nuclear fuel cycle, from uranium mining to spent fuel reprocessing and breeder reactor development. In particular, the group wanted a plausible and internally consistent projection of supply and demand in all those areas so that it might reflect upon the consequential international flows of nuclear materials, equipment, services, and technology. A notable feature of the Jülich study is thus the disaggregation of projected demand not just to the regional level but to the level of individual countries. Not surprisingly, much turns out to depend not only on the rate at which nuclear electricity use increases but also on the choice of fuel cycles, and especially on the role of plutonium and breeder reactors. The Jülich model was designed, therefore, to study three alternative fuel cycles in relation to fifty-nine (later sixty-nine) different countries. The resulting scenarios reported here remain just that: founded as they are on simplifying, and even arbitrary, criteria, they are certainly not intended as forecasts in the full sense. Nevertheless, while its main value is heuristic, the Jülich study, in its detail and its degree of disaggregation, is arguably unique.

It should be stressed that the eight working papers presented here were never conceived as the components of a single volume, and still less as complementary chapters in a consciously integrated survey. That they represent views from different temporal vantage points has already been mentioned. Just as obviously, their scope and content are not closely coordinated. In general, moreover, they have deliberately not been revised since their first appearance. Terminology has been standardized (and a single glossary added). The opportunity has also been taken to correct minor errors or to clarify particular obscurities. In addition, as editor, I have inserted a limited amount of new material into some of the bibliographical notes that appear at the end of each paper, where that seemed likely to benefit a wider audience. In only one paper, however, has any substantive amendment been made since first publication: that is, in the final working paper "World Nuclear Energy

Paths," where subsequent work with the model constructed by the group at Jülich has yielded supplementary details of their scenarios that it seemed wrong to withhold. In all other respects the substance of the papers has been "frozen" as it was when they were initially distributed outside the ICGNE. It follows that they do not pretend to be other than contributions to an analytical scrapbook, varying in date, method, and style, sometimes overlapping in subject, and occasionally revealing gaps in their collective coverage.

The working papers also reveal divergences of interpretation or opinion. That is not surprising. Indeed, the most important thing to recognize about these essays is that they are the expressions of individual views. Earlier drafts of each were discussed at plenary ICGNE meetings. It came to be a rule of practice, in fact, that no paper should be eligible for publication until it had been debated during at least two such sessions. The authors revised their texts in the light of those debates and often as a result of additional private discussion and correspondence with ICGNE members. Throughout that process, however, it was emphasized that a decision to publish would rest not on whether the ICGNE as a whole agreed with what was said in a working paper, but only on whether members agreed that the paper was one that could contribute to informed discussion beyond the group's own limits. No one except the named authors, therefore, bears the responsibility for whatever is said in the papers themselves. That was the essence of the rubric the ICGNE decided to have printed in front of each published paper, and that should equally be seen to stand in front of them here:

> The following paper was originally drafted for internal discussion by the International Consultative Group on Nuclear Energy (ICGNE), which is sponsored jointly by The Rockefeller Foundation (New York) and The Royal Institute of International Affairs (London). It does not express the opinion of the joint sponsors, which are unofficial bodies that do not, as such, take positions on policy. Nor does its publication imply that ICGNE, any of its individual members or any organization with which those members are associated necessarily agrees with the views here presented. It is, in fact, to be assumed that particular members do not concur in at least some of the premises, data or interpretations in such a paper. Where they and the sponsoring institutions agree is in believing that its wider circulation will help to stimulate a responsible public debate on the international issues relating to the civil use of nuclear energy. Responsibility for the contents of the paper rests exclusively, however, with the authors in their personal capacities.

It remains to add that the same caveat is to be read as standing in front of this introduction, the text of which has not been discussed, let alone agreed on, by the ICGNE membership in general or the authors of working papers in particular.

Given that the working papers reflect only the views of their individual authors, for any expression of a collective view one must look not to those essays but to the ICGNE report, first issued in January 1980 and reprinted here. Even then, one must note that the report committed not the ICGNE as such, but rather the fourteen members who signed it. It is, however, the only document reproduced here that can claim some collective authority. A preliminary statement on future demand for nuclear power is followed by sections on three broad themes: the need to develop technical options in an ordered manner, the need to respect public interest in nuclear safety and otherwise, and the need to reconcile growing international trade in nuclear goods and services with effective checks on the spread of nuclear weapons. Finally, the lessons drawn from those themes are distilled into an assertion of five conditions that must be satisfied if nuclear power is to meet future energy needs.

The ICGNE did not start from the premise that it should issue a collective report. Its first aim was rather to serve as an instrument of informal communication in an area increasingly shadowed by international friction. As its debates proceeded, however, a large majority of members came to feel that their frequent disagreements were balanced by the discovery of a sometimes surprising extent of common ground, and that some simple statement of their joint views might be both possible and desirable. Eventually the task of drafting the report on the group's behalf fell on the reluctant shoulders of its chairman, together with a unanimous injunction that the ICGNE should be the first body to report on nuclear power issues in not more than twenty pages. There followed a long negotiation, during which the exact terms of the report were chiseled out in debate and correspondence. Although much common ground evidently existed, drawing up a specific statement of joint views turned out to involve a great deal of argument and compromise. In the end, however, almost all the participants in the group's later sessions felt able to add their signatures to the result. The text of the report is thus not only unusually succinct but also unusually delicately balanced. Achieving that balance was a significant event. Indeed, I shall repeat here what I said publicly when the report was first issued: that it must be seen not as the outcome of an academic inquiry but as a notable example of willingness to seek and achieve exactly the sort of international compromise, albeit at a personal level, that the ICGNE had singled out as the most important of all conditions for a globally acceptable nuclear future. Views within the group on the ideal nature of policy might, and commonly did, diverge. Ultimately, however, all those involved came to be dominated by the conviction that there was no sensible alternative to international cooperation on the basis of mutual compromise, since the pursuit of disagreement to its logical conclusion could only, in the real world, do harm to the interests of all. It was in the spirit of their personal experience within the group, therefore, that ICGNE members called so strongly in their report for a new international "bargain of confidence," as the only escape from universally costly and finally unproductive confrontation.

The ICGNE and its product cannot be divorced from their international context, any more than this volume can be divorced from the genesis and purposes of the ICGNE. It is difficult, however, to compare the work of the ICGNE with that of any other body, since none has recently resembled it in all important respects. That it was so widely international distinguishes it, for example, from a number of privately sponsored groups that have considered nuclear energy policy from the perspective of a single nation or a homogeneous community: groups such as that which produced the Ford/MITRE report of 1977 in the United States (Keeny et al. 1977). At the same time, the breadth and political emphasis of the ICGNE's agenda differentiate it from the smaller number of groups that have privately investigated particular international aspects of nuclear energy policy, such as the group the Rockefeller Foundation had previously invited to study international cooperation on breeder reactors (Gray et al. 1978). Conversely, that the ICGNE was assembled under nongovernment auspices and that its members served only in their personal capacities separates it from all the official bodies that have considered international nuclear policy issues, ranging from the INFCE working groups to the various assemblies of the IAEA or to the 1975 and 1980 NPT review conferences. Despite all of this, it is natural to ask how the outcome of the ICGNE's efforts can be related to other exercises within the same broad field.

The first answer must be that comparisons are strictly inadmissible, because the ICGNE, as such, had no outcome. All that emerged was produced by its individual members: by the authors of working papers or by the signatories of the ICGNE report. Nevertheless, on a less formal level, it is reasonable to look for respects in which approaches characteristic within the ICGNE resembled or differed from those discernible elsewhere. On occasion, for example, they tended to reflect some skepticism concerning the merit or utility of particular aspects of the national or international policies adopted by particular nuclear suppliers. To illustrate that, those who signed the ICGNE report clearly felt that an uninterrupted effort to develop viable fast breeder reactors (FBRs) and associated reprocessing techniques, albeit without immediate prejudice to commercial deployment, was internationally imperative. Many would contrast that conviction with the April 1977 decision by President Carter in the United States to defer indefinitely commercial reprocessing development and construction of a prototype commercial FBR, even if United States policy subsequently evolved slightly in the direction preferred by many ICGNE members, to the extent of conceding that some countries have a stronger case for pursuing those options. Equally, however, the several references in the ICGNE report to the desirability of limiting the multiplication of enrichment and reprocessing plants, in proportion to demonstrated energy needs, might be held to contrast with the assertion by some traditional recipient countries of their unqualified national right to establish such facilities.

In a broader context, the ICGNE report implicitly invited comparison with some of the concerted, as well as several, actions of governments. Its repeated insistence, for instance, on the need to base non-proliferation policy on free cooperation between suppliers and recipients represented at least a different emphasis from the collective attempts of supplier governments alone, especially in the Nuclear Suppliers Group, to regulate the international flow of nuclear materials, equipment, and technology. However, that is only one of the implications of what can certainly be identified as a general theme in much of the volume—in the working papers by Goldschmidt and Kratzer, Fischer, Imai and Press, or Khan, as much as in the ICGNE report itself. The theme is one of voluntary and reciprocal international restraint as the only possible source of effective policies to limit both the probability and the fear of proliferation. Separate exploitation of particular leverage by any country or group of countries, whether supplier or recipient, in contravention of that principle is equally, says the report, to be deplored.

> Whatever is done to combat fears of proliferation must be done in a spirit of international cooperation rather than confrontation. . . . Cooperation, in turn, must be founded on reciprocal constraint. Any attempt to engage in political or economic extortion by exploiting particular strength—whether strength derived from ownership of resources, control of technology, or a capacity to disturb international stability—is thus strongly antithetical to it.

The conclusion drawn by ICGNE members from their wide experience is not so much that efforts by individual governments to impose their will on others in this field are somehow morally or legally reprehensible as that they are fated to be unsuccessful and may even produce the opposite effect from that intended.

The stress in the ICGNE papers on the need for an international approach reaches beyond the question of reciprocal restraint in non-proliferation policy, to the question of how nuclear power is to be further developed for energy purposes. The report called for wider international collaboration in developing commercially applicable fast breeder reactors. Such collaboration already occurs, within limits, between continental West European countries, but the report's signatories indicated that more than that would be needed, especially on the part of those other countries—Britain, Japan, the Soviet Union, and the United States—that have also made significant national advances in FBR development. At the same time, they saw a specific need to intensify international cooperation in three other fields, loosely related to each other: nuclear safety, the management of radioactive waste, and the handling of nuclear accidents. While the ICGNE made no attempt to reassess the technical issues involved, its members were clearly convinced that governments and industries share a duty to demonstrate to the public at large that safe provision can and will be made to dispose finally of nuclear waste and, more generally, that the inevitable risks of exploiting nuclear power can and will be held, in practice, at a publicly acceptable level. In addition, they pointed to the need for governments, in appropriate concert with the IAEA, to assist each other more actively in resolving those issues, by sharing information and experience about safety and waste management and by working together to deal with any nuclear accident that may nevertheless occur. All of those are areas in which some steps have been taken toward international cooperation. All, it was nevertheless suggested, are areas in which more — and more effective — effort is needed.

One form of international collaboration that touches both issues, nuclear power development and non-proliferation policy, is an attempt to manage "sensitive" aspects of the nuclear fuel cycle on a multinational, rather than national, basis. The ICGNE report was suitably cautious in endorsing that course. Recognizing its difficulties and limitations, it asked only, at the general level, for fuller and more urgent international consideration of possible multinational arrangements. In the particular case of controlling separated plutonium, however, it revealed a conviction that international custody of surplus stocks, given agreed conditions, would indeed be a positive step forward. One way that step might be taken had, moreover, been spelled out earlier in the working paper by Fox and Willrich. It was very much in accord with views expressed in the ICGNE, therefore, that government experts began in December 1978, under IAEA auspices, to study proposals for managing plutonium stocks internationally. Similarly, on a wider front, there were echoes of the ICGNE report in the recommendation by several INFCE working groups that further attention should be given to a variety of possible multinational fuel cycle institutions (INFCE 1980*b,* pp. 44–48). Since that time, official discussions about an international plutonium storage (IPS) scheme, together with parallel exchanges from June 1979 on international spent fuel management (ISFM), have moved ahead only slowly and have yet to produce agreed proposals. The ICGNE report suggests that many members had hoped for swifter progress.

In partial contrast, the progress made between October 1977 and February 1980 in INFCE itself probably matched the hopes of most ICGNE members and exceeded some of their fears. INFCE, for the reasons explained earlier, was in no way directly comparable to the ICGNE. The former was intergovernmental in form and predominantly technical in substance; the latter was both unofficial and predominantly political in its concerns. Frequently, moreover, the ICGNE's time horizons were even more extensive than those in INFCE, looking forward far into the twenty-first century, but, even more important, looking backward into the period since the 1940s to understand the origins of contemporary difficulty or friction. In addition, there turned out to be more subtle differences between the two assemblies, and thus between their products. The ICGNE working paper contributed by Connolly, Hansen, Jaek, and Beckurts, writing from Jülich, suggests, for example, that at least some of those associated with the ICGNE had somewhat more modest expectations concerning the rate at which nuclear power use would plausibly expand internationally in the future. For example, INFCE participants, in their final reports, looked to the deployment of 833–1,209 gigawatts (GW) of nuclear generating capacity in the non-Communist world by 2000 (INFCE 1980*a*, p. 53), compared with the 695 GW foreseen in the Jülich scenario. Moreover, the ICGNE paper by Lönnroth and Walker reflected even greater skepticism, implying that installed nuclear capacity in those same countries at the end of the century would be only about half that predicted by INFCE. Conversely, exactly because of their preoccupation with politics, those who signed the ICGNE report found, if anything, even stronger reasons than INFCE participants for making all prudent use of nuclear power, not only in the interest of those countries that might themselves exploit it, but also to reduce the overall pressure on future world energy supplies in the interest of smaller or poorer developing countries that were most unlikely to have their own nuclear plants in the near future. Indeed, in their emphasis on the long-term danger of human suffering and international conflict on account of energy scarcity, some passages in the ICGNE report find their recent echo more in the report of the Brandt Commission (ICIDI 1980) than in the results of specifically nuclear studies such as INFCE.

International events since the ICGNE's work ended both justify and fall short of what its members might have hoped. INFCE aside, the Second Review Conference of the Non-Proliferation Treaty in August–September 1980 was obviously, for instance, a disappointment, in that its participants failed to agree on a substantive final document. Significantly, however, their failure came less from an inability to find sufficient common ground in regard to reconciling nuclear development, cooperation, and trade with the constraint of horizontal proliferation—the area with which the ICGNE had been much concerned—than from unresolved disagreements about checking the vertical proliferation of nuclear arms in the hands of existing nuclear-weapon states—an area largely outside the ICGNE's agenda. The ICGNE report acknowledged that either horizontal or vertical proliferation would make the world more dangerous and noted the risks to international security in the present deployment of nuclear weapons. But it concentrated much more on the problems of civil nuclear energy expansion, as affected by possibilities and fears that additional countries might acquire nuclear weapons as a by-product of that process. In doing so it may have avoided some of the disagreements about other aspects and versions of proliferation that so heavily handicapped the subsequent NPT Review Conference.

The work of the ICGNE is to be seen, in fact, as standing in the line not so much of debates among governments about nuclear weapons and their control as of attempts by them, in and after INFCE, to build an internationally acceptable basis for future nuclear trade and nuclear power use. To that extent one should look less to the 1980 NPT Review Conference as a test than to intergovernmental efforts since 1980, in the IAEA and elsewhere, to exploit the technical foundation laid in INFCE. The IAEA working groups on plutonium storage and spent fuel management, established during the ICGNE's term, have already been mentioned. Potentially, however, the newer Committee on Assurances of Supply (CAS), set up under the IAEA Board of Governors in June 1980, is even more clearly relevant to the ICGNE's wider interests. The CAS is intended to examine and, if possible, agree on ways international nuclear trade, covering services as well as goods, "can be assured on a more predictable and long-term basis in accordance with mutually acceptable considerations of non-proliferation." Unusually explicitly, therefore, it offers an opportunity to pursue major components of that new "bargain of confidence" between states, involving credible assurances "that the international supply of nuclear materials, services and technology will be maintained on agreed and predictable terms," of which the ICGNE report spoke. Whether governments, including both suppliers and recipients, will fully seize that opportunity remains to be seen. What is already apparent is that persuasive evidence of their inclination to follow what was identified in the ICGNE report as the only profitable way forward can now come most conveniently from a convincing demonstration of their willingness and ability to seek a mutually acceptable formula within the CAS.

Formulas are one thing; performance is sometimes another. ICGNE members showed, in all the group's products, that they were well aware of the preeminent need for practice to match agreed principles: in the application of measures to establish the exclusively peaceful purpose of nuclear programs, in the strict observance of international contracts to supply nuclear goods and services, or in the safe pursuit of nuclear energy activities. Indeed, the awareness that desire – or design – has a tendency to outrun performance is characteristic of much in this volume, in regard not only to relations between states but also to the evolution of nuclear power programs. Although much continues to be said, in a number of countries, about the need to facilitate the licensing, construction, and commissioning of nuclear plants, little has occurred in practice to modify the picture painted in the working paper by Lester of lengthening lead times. Nor have the conclusions reached by Lönnroth and Walker concerning the serious problems faced in the real world by major reactor manufacturers yet been invalidated. All the evidence of performance continues to suggest, in fact, that the studiedly restrained projection of worldwide nuclear power growth constructed by Connolly, Hansen, Jaek, and Beckurts is likely to prove nearer to reality than are many more ambitious forecasts. None of that, however, diminishes the gravity of the political issues posed at the international level by the civil use of nuclear power. There too, the message in these ICGNE papers is that it is performance, rather than pronouncement, that constitutes the only proper test of political will.

Having reached a formal conclusion with the publication of its report, the ICGNE no longer exists. It is not for me, as its chairman, to pass any final judgment on the utility of its work. Nor is it for me to state or interpret its collective view of events. All I have offered here is my personal assessment, which may very well be regarded by other ICGNE members as eccentric. In any case, though it is proper to have said something about the ICGNE and its work, this is not only a book about an international group. It is also a collection of notable papers on particular issues in international nuclear policy. They should be, and no doubt will be, read in that light, as contributions to the understanding of their specific topics. In the final analysis, it is nevertheless germane to each of them that each survived the furnace of ICGNE activity, to which the authors would be the first to acknowledge some debt. The group, in turn, must be grateful to all those who prepared working papers, whether published or not. In addition, it must be grateful to individuals who made its work possible: to Myron Kratzer, the group's rapporteur; to Richard Lester, its research associate, who shared the task of acting as the group's collective memory during and after its meetings;* to William Walker, its secretary, who not only carried a part of the same burden but also prepared the working papers for original publication and labored full time for two years as the group's chief administrator; to Mary Ryan, who represented the Rockefeller Foundation in organizing material for publication; to Elizabeth Watson, who bore the lion's share of responsibility for managing the ICGNE's documentation, and sometimes its members as well. Above all, the ICGNE must acknowledge its great debt to the two institutions that brought it into being and sustained it: the Rockefeller Foundation in New York, which provided the necessary funds, and the Royal Institute of International Affairs in London, its cosponsor and on several occasions its host. In return, it must hope that this volume represents a legacy of some continuing value, to its sponsors, to any group that seeks to go farther along the road the ICGNE formerly trod, but more widely to all of us who necessarily share a responsibility for the future.

REFERENCES

Gray, John E., et al. 1978. *International cooperation on breeder reactors.* New York: Rockefeller Foundation.

Independent Commission on International Development Issues. 1980. *North-South: A programme for survival.* Cambridge, Mass.: MIT Press.

International Nuclear Fuel Cycle Evaluation. 1980*a*. Vol. 1. *Fuel and heavy water availability.* Vienna: International Atomic Energy Agency.

______. 1980*b*. Vol. 9. *Summary volume.* Vienna: International Atomic Energy Agency.

Keeny, Spurgeon M., Jr., et al. 1977. *Nuclear power issues and choices: Report of the Nuclear Energy Policy Study Group.* Cambridge, Mass.: Ballinger.

* During the first plenary meeting, that role was also played by Dr. Gene I. Rochlin.

2

Peaceful Nuclear Relations: A Study of the Creation and the Erosion of Confidence

Bertrand Goldschmidt and Myron B. Kratzer

INTRODUCTION

Nuclear power and its relation to the proliferation of nuclear weapons has been the subject of numerous reviews in recent years, including several by the present authors. In this study the issue is examined from the viewpoint of the historical forces that shaped the arrangements that prevailed for some twenty years and the dramatic changes set in motion by various events and trends in the early 1970s.

Two major speeches by American presidents twenty-three years apart have marked this history: the Eisenhower Atoms for Peace declaration of December 1953 and the April 1977 Carter pronouncement (Carter 1977*a*). The different philosophies that inspired these two declarations are at the base of the two schools of thought that prevail today. One school holds that proliferation in the future, as in the past, depends principally upon conscious political decisions and that efforts to restrain the application of nuclear power will not materially lessen proliferation risks but could conceivably increase them by stimulating a new round of independent efforts. The other school, which has emerged more recently, introduces the new element of "temptation" and holds that the widespread dispersion of sensitive facilities and materials implicit in earlier policies would create grave risks of diversion and proliferation even on the part of countries with no premeditated design to go nuclear.

For two full decades following the inauguration of international nuclear relations, a broad consensus prevailed on the fundamental acceptability of the policy of nuclear supply under international controls. In 1974 this consensus weakened with some suddenness. This process, designated in this paper as "the erosion of confidence," was by no means unilateral. Critics of past nuclear policy, largely centered in the United States and other supplier

Date of origin, November 1978.

countries, lost much of their earlier faith in the efficacy of the existing nonproliferation regime to contain proliferation in the future. Concurrently, supporters of the policies of the past, especially in recipient nations, lost confidence in the ability or will of their traditional suppliers to meet their growing needs for nuclear energy resources on reasonable and nondiscriminatory terms.

Regardless of individual stands on the merits of the issues, there is widespread agreement that this erosion of confidence, coming as it does at a crucial juncture in the formulation of world energy policy, is a disturbing and even dangerous development in its own right. Confidence between nations, like trust between persons, is the result of a delicate balance and deteriorates very quickly if this balance is upset. It has been said that the surest way to make somebody untrustworthy is to distrust him and show him your distrust. The same applies to nations. The past steps taken to avoid the emergence of independent and uncontrolled national nuclear programs suffer the risk of serious and perhaps irreversible setbacks if confidence on either side falls to the level where an independent course is viewed as offering greater benefits than continued dependence on outside supplies.

In this essay we have tried to identify the underlying factors that led to a period of relative stability and trust, and the events or processes by which this trust was eroded. We hope this will prove a helpful tool in the search for a new consensus. We believe too that our joint effort, undertaken against two different backgrounds of lengthy involvement in the development and implementation of policies that were sometimes in opposition, adds a dimension to the results that might not be achieved in a study reflecting only a single viewpoint.

Among the conclusions reached in this review is that, despite the comparative suddenness with which the pattern of stability of 1954–74 was weakened, some of the roots of this development can be traced surprisingly far back into the era of "consensus." Neither the assurance against further proliferation, on the one hand, nor the assurances of cooperation and supply, on the other, were as effective and accepted during that period as popularly supposed then or now. The search for consensus in which the world is now engaged will become more and more difficult as time goes on and as irreversible patterns are adopted. It is therefore of the utmost importance that the new consensus should derive maximum benefit from the lessons of the past.

In developing this review of the erosion of confidence in the international nuclear environment, we have attempted to the best of our abilities to act as reporters — or, since most of the events dealt with are in the past, as historians — and not as commentators or critics. In identifying causes for the erosion, we have sought to avoid value judgments either on the merits of the event or on the merits of the reaction — the decrease in the mutual trust — it triggered. Nevertheless, it is our thesis, as we suggested earlier, that the erosion of confidence itself, in a field where trust had been so painstakingly built up through unique measures, is an unfortunate result that should be reversed even at the expense of some compromise in other objectives.

THE STABLE YEARS: 1954–74

The period of international cooperation and commerce inaugurated by the Eisenhower proposal of December 1953 followed eight years characterized by the Anglo-American policy of secrecy and monopoly of non-Communist uranium resources: that is, a policy of denial of transfer of nuclear materials, equipment, and technology. This postwar period had also been marked by the United States proposal for the internationalization of all nuclear activities all over the world as conceived by the Acheson-Lilienthal plan (U.S. Department of State 1946). The failure of the Soviet Union to accept the loss of national sovereignty implied in such a plan jeopardized the last chance for humanity to live in a world free from atomic weapons. The Eisenhower proposals and the subsequent changes in United States law saw the development of new policies, programs, and institutions that transformed worldwide perceptions of nuclear energy.

The Eisenhower speech came against the background of a growing spread of nuclear weapons. Among the first five countries to initiate significant nuclear programs – the United States, the Soviet Union, Britain, France, and Canada – only Canada chose to forgo nuclear weapons production. The Soviet Union had entered the ranks of the nuclear weapons powers in 1949, and Britain followed in 1952. France had begun a substantial program of plutonium production whose military end, even though not announced, was predictable.

The only countries that officially renounced nuclear weapons in this early period had been forced to do so as a consequence of their defeat in World War II. Some, such as Bulgaria, Finland, Hungary, and Italy, had been obliged to take this step as early as 1947 in their peace treaties. West Germany in 1954 renounced the manufacture of nuclear weapons on its territory as a condition of conventional rearmament and membership in NATO, and Japan's constitutional renunciation of war as an instrument of national policy and its national atomic energy law stand in the way of any acquisition of nuclear weapons.

The cornerstone of the new Atoms for Peace concept was the policy of offering assistance in peaceful uses in exchange for verified undertakings that the assistance would not be diverted to military purposes. This verification was to be accomplished through a system of safeguards, including on-site inspection of peaceful nuclear activities by inspectors from outside the host country. The International Atomic Energy Agency (IAEA) was created in 1957, empowered both to promote international cooperation in peaceful uses and to implement the safeguards arrangements noted above. A central feature of the cooperation was the supply, principally by the United States at the outset, of nuclear materials under favorable terms and political conditions, and the transfer of reactors and other nuclear equipment by several suppliers.

An important and often overlooked aspect of the Atoms for Peace approach is that it was not explicitly a non-proliferation policy. (The term "non-proliferation" is itself of relatively recent origin, having appeared only in the mid-1960s.) There was, in this approach, an intent and hope that the offer of peaceful nuclear assistance under safeguards would direct nations away from independent and uncontrolled nuclear programs and thus delay the emergence of additional nuclear powers. Nevertheless, there was no prohibition in the Atoms for Peace program against assistance to nations that concurrently were developing independent efforts even of a military nature. Thus France, which openly broadened its program to include both civilian and military objectives in the late 1950s, continued to receive United States assistance after this time for its peaceful activities, though this assistance, which took the form mainly of enriched uranium and later plutonium for its research reactors, did not contribute to France's military effort.

The progress of the Atoms for Peace program was rapid. By 1955 "agreements for cooperation" between the United States and several other countries had been concluded. Major international conferences on the peaceful uses of atomic energy were held in Geneva in 1955 and 1958, serving as the mechanism for the declassification and dissemination of a vast body of technical information on peaceful nuclear applications developed by all the leading nuclear nations, including the Soviet Union. Dozens of countries initiated significant peaceful nuclear programs, and the first sales of research and power reactors took place. The first power reactor sales as well as the first large research reactor sales, contrary to the common assumption, were of natural uranium units of British, French, and Canadian manufacture, not of the United States enriched uranium type.

As time went on there were further developments that seemed to confirm and strengthen worldwide interest in, and the acceptability of, nuclear power. By the mid-1960s the IAEA had established an inspection capability and, as originally intended, had virtually displaced the bilateral safeguards arrangements relied on initially. The conversion of the Soviet Union to advocacy of IAEA safeguards was an indispensable ingredient in this development.

It was only at this stage that the first real policy of non-proliferation could be discussed and translated into a general agreement. The Non-Proliferation Treaty (NPT), by which nonnuclear powers of their own will agreed to renounce independent uncontrolled efforts, was opened for signature in 1968, attracting a large and steadily growing number of adherents. Throughout the 1960s, only France and China were added to the list of nations that had achieved a nuclear explosive capability. In both these cases, as for the earlier nuclear powers, the development had clearly not been dependent on the nuclear power fuel cycle.

The application of nuclear power grew steadily, and forecasts indicated that nuclear generation would provide on the order of a third of the electricity consumed by many industrialized countries by 1985. Independent capabilities for reactor and fuel element manufacture were achieved by a number of industrialized countries, through licensing arrangements and other forms of technology transfer, as well as by indigenous national developments.

From 1954 onward no aspect of peaceful uses, including the sensitive steps of the fuel cycle, was prohibited. Although reprocessing was singled out for special treatment in light of the recognized difficulties of safeguarding this activity, nuclear planning from the outset of the Atoms for Peace program envisaged reprocessing and the recycle of plutonium, as well as the development of the plutonium fueled breeder reactor.

In today's world of declining expectations and seemingly intractable problems, the two decades between 1954 and 1974 are often viewed by nuclear advocates as the golden era of nuclear cooperation, where everything seemed possible and the few problems were rapidly overcome by goodwill and common sense. The reality, however, was rather different. While the achievements were real and have by and large survived the critical review that began in earnest in 1974, the problems and tensions of the earlier era were appreciable and set the stage for the erosion of confidence that is the focal point of this review.

In this analysis, the various elements that contributed to the stability and predictability of the era of cooperation are identified and reviewed, to isolate both their strengths and their weaknesses. The general picture is then examined during the period of reassessment, with a view toward identifying changed perceptions and their role in the erosion of confidence.

SAFEGUARDS

The central feature of the new cooperative regime initiated in 1954 was, and remains, safeguards. They represented a major and innovative departure in international relations – the first successful major effort to subject undertakings among nations to verification by outside authorities in place of the historical reliance on the good faith of each nation.

The decision to insist upon safeguards as a condition of nuclear commercial assistance was an important break with the past reliance on sovereign undertakings. As such, safeguards are per se an expression, if not of distrust, then at least of unwillingness to accept fully the assurances of another sovereign government. In discussions with recipient countries, safeguards were usually described as a means by which both supplier and recipient could reassure the world at large of the peaceful character of their arrangements. While this characterization was not entirely a euphemism, recipients had no doubt that the safeguards they were being asked to accept were intended in large measure to satisfy the suppliers' concern that their assistance would not be diverted to military purposes.

Although the decision to impose such safeguards as a condition of nuclear assistance was in many respects a bold one, it fell short of other possible approaches to the control problem. The Acheson-Lilienthal plan of 1946 specifically rejected international inspection, as insufficient to control peaceful nuclear activities, and called for the more extreme approach of international ownership and management of all dangerous nuclear activities. The selection of the more limited approach of the Eisenhower plan may have been, in part, the product of the search for a solution that would be acceptable to the Soviet Union for application to its exports, given its strong opposition to the international management principle of the Acheson-Lilienthal plan.

The other important respect in which the safeguards approach was limited was the decision to confine its application to directly related assistance, leaving recipients free to pursue independent nuclear programs or to seek additional assistance from suppliers that did not insist on safeguards. However, Mexico decided voluntarily to place all its nuclear activities under IAEA safeguards in the mid-1960s.

The original safeguards arrangements negotiated by the United States and other suppliers called for bilateral safeguards, a euphemism for unilateral inspection by the supplier. These arrangements were made necessary, despite a preference for international inspection from the outset, by reason of the delay first in establishing the IAEA and thereafter in the development of the IAEA safeguards system and staff. The IAEA system was initiated in 1959 on a limited basis, over strong Soviet opposition. It was not until 1963 that a system applicable to full-scale power reactors received approval, this time with Soviet support.

By this time unilateral inspections, principally by the United States, in view of its dominant supply role, had become generally accepted, and the effort of the United States and other suppliers to transfer safeguards responsibility to the IAEA met strong resistance. With the exception of Euratom, whose regional safeguards system was accepted as adequate by the principal suppliers, transfer of safeguards responsibility to the IAEA was accomplished in the early 1960s, but not without considerable resistance from a number of countries that had become accustomed to bilateral links and safeguards arrangements. Only the implied threat of nonrenewal of supply arrangements was able in some cases to achieve acceptance of IAEA safeguards.

The primacy of the IAEA in the implementation of safeguards ultimately gained general recognition, and the earlier preference for bilateral safeguards ceased to be a problem by the time NPT negotiations were under way. However, the special status of the Euratom safeguards system has continued to be an issue to the present day, in view of the insistence of countries outside the European Community on nondiscriminatory treatment in the safeguards area. This has complicated the development and implementation of IAEA safeguards arrangements with Euratom as well as with individual countries outside the European Community.

In the mid-1960s attention began to be given to the problem of indigenous or other unsafeguarded paths to the acquisition of nuclear weapons, leading to both United States and Soviet proposals for the Non-Proliferation Treaty that was eventually opened for signature in 1968. Since that time, virtually all attention has been focused on the treaty's effectiveness in avoiding the acquisition of nuclear weapons by additional countries through independent nuclear programs.

A central feature of the NPT was its requirement for IAEA safeguards on all of the peaceful nuclear programs of its nonnuclear parties. This fact broadened still further the implication of distrust inherent in the safeguards principle, although here again the argument was made that it was in each country's self-interest to reassure its neighbors of its peaceful intent in nuclear matters. Through its safeguards requirements, the treaty overcame in part the original limitation of the safeguards approach to assisted nuclear activities. However, the change was not complete, since adherence to the treaty was naturally voluntary, and the treaty allowed suppliers that were treaty parties to continue to make nuclear sales to non-parties that did not accept safeguards on their whole programs, so long as the assistance itself was safeguarded. This feature of the treaty has been subject to debate and criticism to the present time by those who consider that countries not party to the NPT should not benefit from outside nuclear supplies.

While steady progress was made by the IAEA in the development and improvement of its safeguards capabilities, concern continued to exist, especially within supplier countries, that the technical effectiveness of the system—that is, its ability to detect diversion—though adequate, was less than desirable. The technical effectiveness of IAEA safeguards, including the provision of sufficient financial support for them, has accordingly been a continuing problem over the period.

Notwithstanding the limitations described above, safeguards played a central and indispensable role in the creation and maintenance of confidence in the non-proliferation regime during the period of stability. Of particular importance, safeguards were the sine qua non for most of the major suppliers to provide assistance in the peaceful uses of nuclear energy.

NATURAL URANIUM

As American military needs for uranium declined sharply in the late 1950s, the United States decided against renewal of its uranium purchase contracts with Canada and South Africa, thus creating a glut of uranium available at low prices that lasted until the early 1970s. It was a buyer's market. Thus, throughout most of the period of stability, uranium supplies seemed assured, and there was little concern over long-term availability. Moreover, the supply of enriched uranium, which accounted for an increasing proportion of natural uranium demand as the adoption of light water reactors (LWRs) spread, was initially accomplished by direct sale, which placed responsibility for uranium procurement on the United States as the enrichment supplier.

Such pressures as existed were largely in the direction of allowing the use of indigenous uranium for the preparation of enriched uranium ordered by non–United States customers. Partly in response to these pressures, the United States decided in 1964 to adopt toll enrichment, which placed responsibility for the procurement of uranium feed on the enrichment customer and thus allowed clients in the non–United States market to make use of uranium from any source.

Concurrently, the Congress insisted that the internal enrichment market be confined to United States natural uranium until the viability of the domestic uranium industry was assured by adequate market growth. The resultant embargo on the use of foreign uranium in the United States market contributed to a depression of world uranium prices to the level of five dollars a pound or less: too low to cover production costs and, even more, to give producers the necessary cash flow for the exploration needed to assure future supplies. The outcome of this was swamping of the market and a slump in prices at the very time that increased prospecting effort was necessary to cope with the increasing demand of the ambitious nuclear power plant programs foreseen for the 1970s and beginning of the 1980s.

Elsewhere, in 1965 Canada, having refused to sell France uranium for peaceful uses without safeguards, as it had done up to then with the United States and Britain, resolved the issue of discrimination by deciding to impose safeguards on all future sales to nuclear-weapon as well as non-nuclear-weapon states.

The chief United States manufacturer and exporter of power plants helped, at the beginning of the 1970s, to perpetuate the stagnation of uranium demand by capturing and neutralizing part of the domestic and foreign market in the United States. The manufacturer offered along with the sale of power plants all the uranium needed to fuel them for their thirty years of operation, without covering itself by buying the requisite stocks and thereby artificially reduced demand. The efforts of producer countries other than the United States, acting in concert to remedy the situation, were slowly beginning to produce results when suddenly, in 1974, matters were reversed and a shortage again emerged, accompanied by its inevitable counterpart – a considerable increase in the price of uranium.

A more or less simultaneous occurrence of several factors was responsible for this reversal: the oil crisis; the decision taken by some electricity companies to cover themselves on a long-term basis and even to stock up with uranium, which they had never done before; the persistent failure of

Australian producers to come on the market, owing to the influence of trade unions hostile to nuclear energy, at a time when large deposits had just been discovered there; the slowdown, and later discontinuation, of Canadian exports as a result of constant doubt as to the political terms of sale; the emergence of the United States as a buyer on the foreign market as a consequence of the import embargo's gradually being lifted; and finally the revelation that many sales thought to be firm were not backed by contracts with uranium producers.

Despite these complications, which disturbed the commercial market, the generally adequate availability of uranium for civil purposes was an important positive element in the stability of the period.

ENRICHMENT

In contrast to uranium, the availability of enrichment was a cause of concern from the outset of the cooperative period, and the United States, as the only commercial supplier, engaged in a continuing effort to reassure users as to the future availability of this resource. These efforts took the form of long-term "agreements for cooperation" and subsidiary long-term supply contracts, with identical and attractive prices, terms, and conditions for United States and non-United States users, accompanied by frequent high-level pronouncements on the ability and readiness of the United States to supply all the foreseeable needs of its partners.

The change to toll enrichment, referred to earlier, was partly a United States effort to respond to the interest of foreign enrichment customers by allowing them to procure uranium feed material indigenously or on the world market. In a similar vein, the conditions of supply were progressively liberalized to allow foreign users to convert and fabricate United States enriched uranium, first for themselves and later for third country or United States customers as well. A key early decision was that the availability of United States enriched uranium would not be limited to use in United States manufactured reactors. All these steps were a deliberate part of the effort to reassure foreign users that, aside from safeguards and related non-proliferation restraints, United States enrichment supply arrangements were both dependable and without restriction of a commercial nature.

The results were mixed. United States supply assurances and arrangements were regarded with sufficient favor to allow light water reactors to come to dominate the world reactor market. At the same time, the natural instinct to avoid dependence on a single source of supply led to continuing concern and consideration by other nations of measures to establish alternative supply possibilities. These concerns were heightened by several features of the United States supply arrangements, including:

- the inclusion of ceiling quantities and the consequent need for periodic extension of the agreements, which involved both executive branch and congressional review;

- requirements for special approval of highly enriched uranium supply, which limited consideration of advanced reactors such as the HTR and worked as a reminder of the United States supply monopoly;
- prohibition against retransfer of United States supplied material except to countries with which the United States had nuclear arrangements, and requirements for case-by-case United States approval, which gave rise to particular concern on the part of countries interested in exporting reactors themselves.

As a consequence of these and similar concerns, a consensus was finally reached in Europe that an independent European enrichment capability was required. This consensus was reflected in the 1968 decision of Great Britain, the Netherlands, and West Germany to establish the tripartite Urenco enrichment consortium, and in the 1973 decision by France to construct a commercial diffusion plant with multinational ownership through Eurodif. European moves toward the creation of independent enrichment capacity stimulated the 1971 United States offer to share diffusion technology for use in a multinational plant. The negative reaction to the restrictive conditions proposed by the United States contributed to the European determination to proceed independently.

In 1973 the United States announced major changes in future enrichment contracts that made them significantly less attractive than in the past. Under these new arrangements, customers were obliged to commit themselves for long periods in advance at prices in effect at the time of delivery, which were under unilateral American control. Although the new conditions were applied uniformly to both domestic and foreign customers, the switch to less attractive conditions was a forceful reminder of the unilateral United States ability to dictate the terms of enrichment availability in a noncompetitive market, and it further strengthened the European resolve to develop independent capability.

Concurrently, the Soviet Union, which had made an open offer to supply enrichment to the free world in 1969, entered the market actively, contracting initially with France and later with West Germany, Sweden, and additional countries.

In summary, enrichment supply was a fundamental building block in the development of international nuclear commerce. Although the distribution policies adopted were generally successful in meeting world needs, continued concern over reliance on a sole source of supply grew and eventually led to decisions by some European countries to diversify sources of enrichment as well as to decisions by some Asian and Latin American nations to select natural uranium heavy water reactors. The enrichment problem was heavily influenced by a unique feature: enrichment technology alone among the technologies applicable to civil nuclear power remained classified. This fact, coupled with the high investment costs and economies of scale of the gaseous diffusion process, was an important element in limiting the ability of additional nations to enter the field, while increasing their incentives to do so. As the only commercial supplier during the greater part of the period, the United States openly followed the policy of dominating the supply role for as long as possible and to the maximum possible extent. Despite the tensions inherent in this situation, enrichment supply was, on balance, a very positive element in the nuclear power field.

REACTORS

The supply of reactors was another element that contributed to the environment of stability and predictability. Reactor availability benefited from the outset from a competitive situation that, while involving only a small number of suppliers, provided strong assurances of availability along with favorable terms and prices.

The initial sales of nuclear power reactors, as well as large research reactors, were of natural uranium fueled units. The early transfer of a Canadian research reactor (employing United States heavy water) to India in the 1950s, under arrangements that included no safeguards, was the action that ultimately led to the most significant single event in the history of nonproliferation policy: the Indian nuclear explosion of 1974. The first sales of power reactors were Magnox units sold by Britain to Italy and Japan in the late 1950s, but Britain was unable to follow up on this early start and ultimately abandoned the Magnox line itself. Nevertheless, the controversy between proponents of natural uranium and proponents of enriched uranium power reactors dominated the early reactor market and served to give it the competitive characteristics of a conventional commercial market.

France competed in some early reactor sales efforts and provided one unit of its gas-graphite line to a joint Franco-Spanish venture in Vandellos, Spain. Although IAEA safeguards were not applied to this project, it is in practice a multinational enterprise with all spent fuel to date reprocessed in France, which has purchased the resultant plutonium.

Additional natural uranium heavy water reactor sales have been made by Canada (to Argentina, India, Pakistan, and South Korea, and also a large research reactor to Taiwan) and by West Germany (to Argentina). With these important exceptions, the international reactor market came to be dominated by the light water reactor. In addition to competition between United States light water reactor manufacturers and natural uranium reactor vendors in Britain, France, and Canada, intense competition among the United States suppliers themselves helped ensure favorable terms for buyers and contributed to the commercial normality of the reactor market.

Export financing was another important element in stimulating reactor sales, especially to developing countries. Loans on clearly concessionary terms were an important element in a United States sale to India and in a Canadian sale to Pakistan. Elsewhere, export financing arrangements, even though on nonconcessionary terms, were an important catalyst of sales.

In addition to commercial considerations, other factors helped stimulate reactor sales. The United States–Euratom cooperative program to build 1,000,000 kilowatts (kW) of reactor capacity in the European Community reflected both American political support for the Community and the desire of the United States to increase its export sales, while acquiring useful development experience for good measure. Although criticized by Britain as an unfair export subsidy and though falling short of its quantitative goal, the cooperative arrangement was successful in helping to create a light water reactor industry in Europe, especially in West Germany, and in stimulating interest in nuclear power.

An important feature of the international reactor market was the willingness of United States vendors to license their technology to foreign manufacturers, a step that also required and received the approval of United States authorities. As a consequence, independent light water reactor manufacturing capacity now exists in at least France, West Germany, Italy, and Japan, while a number of other countries can produce a large proportion of the necessary components. Sweden independently developed a capability for the manufacture of boiling water reactors (BWRs), which it has supplied to both the Swedish and the Finnish markets.

These facts make it clear that reactor availability has not been a restraining or destabilizing influence in the nuclear power field. On the contrary, overcapacity became an increasing cause of concern toward the end of the period, as reactor construction schedules slipped and nuclear power sales and growth forecasts were progressively reduced. The ready availability of reactors as well as the transfer of reactor technology on commercial terms thus contributed to confidence and stability.

This open market in the transfer of commercial reactor equipment and technology extended also to the fast breeder reactor and other advanced reactors in the process of development. From an early stage, extensive cooperation took place in the breeder reactor field, which represented the area of maximum effort and prestige in national reactor development programs. During the 1960s the United States, Britain, and, later, Canada supplied substantial quantities of plutonium for breeder reactor development programs both in the European Community and in Japan. These transfers were accomplished without undue objections. While technical exchange began to be restricted for commercial reasons as development in some countries—notably France—neared the stage of industrial application, these restrictions were not related to proliferation or other concerns peculiar to nuclear energy. In the early 1970s French technical assistance was extended to India to build a breeder research reactor to be fueled with Indian plutonium and French enriched uranium. There was no public opposition to this arrangement, which incorporated bilateral safeguards.

In only one area – the high temperature gas-cooled reactor (HTR), whose most promising configuration involves the use of highly enriched uranium (HEU) fuel – did proliferation concerns affect international cooperation and availability. United States enrichment distribution policy, as noted earlier, required special case-by-case approval for the distribution of material of greater than 20 percent enrichment, and, while substantial quantities of HEU had been distributed for research purposes, United States policy on the supply of very large amounts of HEU for power reactor purposes remained undecided. In 1973 the United States agreed to provide HEU for the 200 megawatt (MW) West German prototype high temperature reactor, the THTR. This decision, however, was without prejudice to the ultimate policy on supply for commercial HTRs, should this reactor type achieve commercial status. Given the absence of pressures favoring its adoption, uncertainty about the ability to commercialize the HTR, which in any event relates more to enrichment than to reactor policy, was not a serious exception to the general rule of free cooperation and unrestricted availability of reactors.

REPROCESSING AND PLUTONIUM RECYCLE

The earliest concepts of nuclear power exploitation via the light water reactor assumed that reprocessing and recycling of plutonium would take place, and economic projections of nuclear power costs routinely included credits for plutonium and recovered uranium, coupled with a charge for reprocessing, the net effect of which was to reduce estimated overall fuel cycle costs significantly. The early gas-graphite natural uranium reactors also assumed reprocessing and plutonium recovery, although the later Canadian CANDU concept, which used natural uranium more efficiently, was promoted on the basis that reprocessing was unnecessary.

Despite the assumption of reprocessing in the LWR fuel cycle, no capability for actually accomplishing the task was present in the United States at the outset of commercial LWR deployment. To overcome this deficiency, the United States Atomic Energy Commission (AEC) undertook to accept spent fuel and give credit for its residual energy value on the basis of costs of a "conceptual" reprocessing plant, pending the establishment of a commercial capability.

United States policy relating to reprocessing of spent fuel from foreign reactors was reflected in special provisions. The earliest bilateral agreements indicated that reprocessing of United States supplied fuel would occur only in American facilities or in other facilities "acceptable to the United States." The language of this provision makes it clear

- that reprocessing was contemplated;

- that it was expected to take place in the United States;
- that while reprocessing outside the United States was not ruled out, the United States retained an unqualified veto power over it.

An interesting feature of these agreements, which was to assume importance only many years later, was that United States approval rights extended only to United States supplied fuel. No such rights were acquired with respect to fuel of foreign origin irradiated in United States supplied reactors. This distinction had little practical significance so long as the United States remained the sole supplier of enriched uranium.

Although United States agreements appeared to contemplate reprocessing in the United States as the norm, as observed earlier, no capability for accomplishing this operation existed in the United States at the outset; and United States policy on whether it should seek to fulfill foreign reprocessing requirements – as it clearly wished to do with respect to enrichment – was at best ambivalent. This had much to do with the later evolution of the reprocessing provisions incorporated in United States agreements.

An additional influence was the Statute of the International Atomic Energy Agency. Unlike early United States bilateral agreements, which accorded the United States an unqualified right of approval over reprocessing, the Statute gave the IAEA, in arrangements subject to its safeguards, a right of approval "solely to ensure that this chemical processing will not lend itself to diversion of materials for military purposes and will comply with applicable health and safety standards" (article XII.A.5).

In later or amended United States bilateral agreements, a formulation closer to that of the IAEA Statute was adopted, in which United States approval for reprocessing was related to the "safeguardability" of the proposed foreign facility. This change further confirmed the presumption of reprocessing on which the arrangements were based, as well as the acceptability in principle of reprocessing outside the United States.

A special case of reprocessing arrangements in United States agreements for cooperation, which also assumed major importance at a much later date, is that of the United States-Euratom agreement. The United States had decided even before the formal establishment of the European Atomic Energy Community to accord special arrangements to the Community as an expression of United States political support for European integration and of confidence in its regional safeguards system for verifying the use of nuclear materials for their declared purpose. The United States-Euratom agreement concluded in 1958 was therefore unique, containing a number of special benefits not present in any United States agreement with individual nations. Among these was the omission of the usual United States right of approval over reprocessing. Moreover, the United States agreed to reprocess material for Euratom if requested to do so and if it were then providing reprocessing services to its own domestic licensees.

The situation with respect to reuse of plutonium was similar. Plutonium recycle in one form or another was contemplated from the outset of the civilian nuclear power era, and plutonium recycle research and development was an important part of the United States–Euratom joint program. The United States Atomic Energy Act of 1954 required "buy-back" of all domestic plutonium offered to the AEC, with payment based on its intended use. As we noted earlier, nuclear power economic studies consistently included a credit for produced plutonium, which was a significant factor in the close economic balance between nuclear and conventional power and between enriched and natural uranium reactors. United States agreements gave the United States the right to approve the storage facilities for any recovered fissionable material that was in excess of a country's needs, and early agreements included the right of United States buy-back of this excess. In a similar manner, its Statute gave the IAEA the right to store in its own facilities "excess" produced material "to prevent stockpiling."

Although the arrangements outlined above are indicative of an acceptance of the concept that plutonium and other energy values should be recovered from spent fuel and made use of, at the same time they also reflect a recognition of the special proliferation implications of plutonium and reprocessing and of the need for special arrangements to limit plutonium accumulation in national hands. This concern was based on two factors: first, since reprocessing plants were viewed as assemblies of conventional equipment, it was felt that their export could not, as such, be made subject to safeguards; second, it was recognized that it would be technically difficult to apply to such facilities safeguards that could detect diversion promptly and with a high degree of assurance.

National policies relating to the diffusion of reprocessing technology reflected the same ambivalence. Reprocessing technology was not included in the information declassified by the United States and Britain for presentation at the first Geneva Conference in 1955. However, information on solvent extraction reprocessing was presented there by France, and the United States and Britain followed suit at the 1958 Geneva Conference. The Soviet Union, on the other hand, has withheld reprocessing information to the present time, while also requiring its East European partners to return spent fuel elements to the USSR for reprocessing.

Early consideration of the possibilities of controlling the transfer of detailed reprocessing technology, or of restricting the transfer of reprocessing plants or components, was strongly influenced by the view cited above, that such technology and hardware are largely of a conventional nature, making controls difficult or impossible to implement. In the early 1960s French industry, already experienced in the reprocessing field as the result of the French national program, took the leadership in the construction of the Eurochemic reprocessing facility at Mol, Belgium, a joint enterprise of thirteen West European countries under OEEC (later OECD) auspices. The United States cooperated with this endeavor – the first example of multinational operation of a major nuclear facility — at least partly in the belief that it was preferable to steer sensitive activities such as reprocessing into multinational channels.

Despite its technical success in building and operating a semicommercial pilot reprocessing plant, Eurochemic experienced many difficulties, especially of a financial nature, and its plant ceased operation in 1974 without having created the nucleus of a multinational reprocessing undertaking in Europe.

A surplus of reprocessing capacity at the end of the 1960s, which was responsible for some of the Eurochemic difficulties, even led Britain and France, already in possession of large plants, and West Germany, which was about to build one, to join forces. Their 1971 agreement (at first commercial and later technological), giving rise to United Reprocessors, was intended to avoid the risks entailed by an uncontrolled development of reprocessing capacity before existing plants had been fully employed. A sudden reversal of the situation, so typical of the development of nuclear energy, has now shown up in a shortage of available capacity and has led the three partners to reconsider the role of this industrial undertaking.

Reprocessing facilities were also built in a number of other countries, with varying, but usually minor, amounts of outside assistance. These included a plant of significant size in India, with a plutonium recovery capability, completed in 1965 and built without significant outside help. They also included smaller facilities for research reactor fuel reprocessing in Spain, Argentina, and Italy.

A small pilot reprocessing plant for power reactor fuel was completed in West Germany in 1970, on the basis of its Eurochemic experience, and Japan entered into a contract with French industry, against British competition and at least some United States industrial interest, for a sizable pilot plant at Tokai-mura.

By the late 1960s United States attitudes toward assisting in the construction of reprocessing plants began to undergo a significant change. The United States had always been concerned about the possible construction of unsafeguarded reprocessing facilities, which could in principle be available for the reprocessing of fuel from unsafeguarded reactors. With the approaching conclusion of the NPT and the growing likelihood that there would be significant sources of unsafeguarded fuel and reactors, the United States government took informal steps to restrict any American industrial assistance in reprocessing.

These steps were formalized by a 1973 amendment to the United States regulations that made any such assistance subject to specific government approval. The criteria for consideration included, among other items, the recipient country's NPT status and whether the facility would be under multinational auspices. The new policy was not a prohibitive one that ruled out all United States help for reprocessing. It was a policy designed to allow such assistance under carefully controlled conditions and providing one means of according advantageous treatment to NPT parties: a strategy that had considerable support in the United States administration at the time.

Toward the end of the period 1954–74, it became increasingly apparent that reprocessing and the subsequent steps of high-level waste disposal and plutonium recycle were likely to become the first of the major nuclear goods and services to be unavailable in adequate supply. This development was largely the result of domestic factors in a number of countries. A small United States commercial reprocessing plant, placed in operation in 1966, was shut down in 1972 for an expansion that never came about. A second United States plant was essentially completed in 1973, but technical difficulties prevented its being placed in operation. British and French capacity, based on modification of military plants, was available to a limited degree. Planned facilities elsewhere, such as in Japan and Germany, were experiencing delays of various kinds, including growing public opposition. An especially significant and precedent-making challenge was mounted in the United States, where environmental interests called in 1973 for a generic environmental impact statement on the use of mixed oxide fuel (GESMO), which would subject the entire concept of reprocessing and plutonium recycle, including its safeguards and non-proliferation implications, to scrutiny.

In summary, while reprocessing and plutonium use were always viewed as a problem requiring special treatment from the non-proliferation and safeguards viewpoint, there was no thought that these operations should be eliminated from the power reactor fuel cycle. There was little to suggest the dramatic changes in attitudes toward reprocessing and recycle that were to take place later.

THE EROSION OF CONFIDENCE

Confidence, in the international nuclear environment, is a two-way street. It has involved, on the one hand, confidence that non-proliferation undertakings will be met and, in a broader sense, that nuclear exchange will not foster proliferation. While confidence in this sense has been of particular importance to the policies and actions of suppliers, all nations, including recipients and their neighbors, acknowledge an important stake in its maintenance. On the other hand, the term also embraces the confidence of recipient nations in the dependability of the supply undertakings of suppliers, and in a broader sense, the willingness of the advanced nations to share the benefits of peaceful nuclear technology in a nondiscriminatory manner. An important factor affecting national attitudes on both sides of this equation is that many nations, including most of the principal suppliers, occupy the roles of both supplier and recipient. Their attitudes therefore tend to reflect complex mixtures of these two roles.

The two faces of confidence described above are not in conflict. On the contrary, they are opposite sides of the same coin. The binding force between them has been the international consensus in favor of non-proliferation. Confidence, in both the directions described above, is an important part of the overall non-proliferation regime through which this consensus is preserved and proliferating actions are avoided.

In the following sections the discussion is divided into two periods; the first, from 1974 to 1977, was the period of transition; the second, beginning with the Carter administration, saw the United States attempt to introduce a substantially new policy.

THE TRANSITION YEARS: 1974–77

The main event triggering the erosion of confidence in the effectiveness of the non-proliferation regime was the Indian nuclear explosion of May 1974. It was accomplished by the nation that was the first among the developing countries to embark on the peaceful application of nuclear energy and had therefore at one time been a field of competition among some of the main exporters. India had also refrained from signing the NPT and had both unsafeguarded research facilities, including a small reprocessing plant built without direct external help, and an unsafeguarded power reactor program under construction, including heavy water power reactors, heavy water production plants, and a larger reprocessing plant. India also had adequate indigenous sources of natural uranium that, of course, were not subject to safeguards.

In addition to these unsafeguarded activities, India had purchased from the United States a nuclear power station containing two BWRs, together with the necessary enriched fuel under a long-term supply contract. It also was constructing a heavy water reactor power station, with Canadian help and thus subject to IAEA safeguards, under arrangements that, however, allowed India to employ the transferred technology in later stations not subject to safeguards.

The Indian nuclear explosion employed plutonium, not from any of its safeguarded activities but from the large heavy water research reactor built with Canadian assistance in the 1950s. That earlier transaction had taken place under a pledge of peaceful uses, with bilateral safeguards on the plutonium produced by use of nuclear materials of Canadian origin, but with no safeguards on plutonium that would be produced by the use of Indian nuclear material. At the time of the sale of this research reactor (1955), international safeguards in the present form did not exist. Thus Canada, one of the countries most devoted to non-proliferation, found itself an active contributor to allowing India to become the first additional country in a decade, and the first outside the five permanent members of the United Nations Security Council, to produce a nuclear explosion. The Canadian dismay was not appreciably lessened by the fact that the initial heavy water required for the operation of the reactor was provided by the United States in an arrangement similarly lacking in safeguards.

India took the position that its peaceful use pledges remained unbroken because the explosion was carried out for peaceful purposes, and it cited the peaceful nuclear explosive programs of the United States and the Soviet Union in support of this position. Canada, on the other hand, argued that the clear intent of these pledges was to preclude nuclear explosions of any kind and that, in any event, India could not unilaterally decide upon the interpretation of such an important point.

The Indian nuclear explosion constitutes a discontinuity in international reaction to proliferation, attracting far more criticism than the first explosions by the preceding nations. This was partly because for the first time outside assistance made available for peaceful uses enabled a country to make a nuclear test. However, the reaction would most probably have been radically stronger in the Indian case even if outside assistance had not been involved, owing to changing world attitudes toward the danger of proliferation.

Although no safeguards arrangements or NPT commitments were broken by the Indian nuclear explosion, the event cast doubt on the adequacy of the entire structure of non-proliferation restraints and especially on the liberal and relatively stable export policies that had evolved in the safeguards and NPT environment. This doubt extended not only to policies on export to the few countries that had refrained from adhering to the NPT but also, somewhat surprisingly, to the policy on exports to non-nuclear-weapon states that were parties to the NPT.

Despite the importance of the Indian nuclear explosion in triggering distrust, a number of other events and circumstances, some of them building up for some time previously, were also important. Most of these had their origin in the United States and first affected United States attitudes toward the adequacy of the non-proliferation regime.

Also in 1974, in an unfortunate juxtaposition of timing with the Indian nuclear explosion, the Nixon administration offered nuclear power assistance to Egypt and Israel. While both of these possibilities had been considered and publicly discussed without any violent reaction during an earlier and calmer epoch of United States history, the offer, coming as it did during an intense political crisis in the United States, was the subject of immediate and severe criticism, leading to the adoption of the first new restrictive legislation on United States peaceful nuclear cooperation in the twenty years since the adoption of the Atomic Energy Act of 1954.

At an even earlier date, American opinion on the hazards of nuclear energy had taken a sharp turn toward increasing criticism. Among the hazards that received particular attention was the possibility that terrorists might acquire plutonium and construct a crude nuclear bomb. Concern over this threat was an important factor in the 1974 decision by the Atomic Energy

Commission to stop further movement toward the commercialization of plutonium reprocessing and recycle until completion of the GESMO study, in which analysis of the security threat of plutonium would be a key feature.

It is not without significance that these specifically nuclear events—the Indian nuclear explosion, the Egyptian and Israeli nuclear offers, and the GESMO study—took place against a backdrop of the growing strength in the United States of the environmental movement and of decreasing confidence in the "establishment," particularly the federal government, as well as in science and technology. Although these attitudes tended to appear first in the United States, they occurred independently, with varying degrees of promptness and strength, in other industrialized countries.

Another major stimulus toward the declining faith in the existing nonproliferation regime was the development of arrangements for the transfer of reprocessing and enrichment technology to developing nations. This took place at a time when there was a perception, largely inaccurate, that the use of nuclear energy itself, under the impetus of the oil crisis, was growing much more quickly than anticipated earlier and was spreading rapidly to the developing nations. In fact, application of nuclear power by developing countries had been anticipated virtually since the outset of the peaceful nuclear program, and it had been encouraged by both the IAEA Statute and the NPT, both of which were ratified by the United States Senate. Moreover, global nuclear power estimates reached their peak in 1973, before the oil crisis, and at most one additional country, Iran, taking advantage of its new wealth, turned to nuclear power following the oil price increases, while Brazil expanded the size of its program considerably.

The disclosure of the sale of reprocessing and enrichment facilities and technology, even though under safeguards, had a major, and quite possibly crucial, impact on confidence in the United States. Whatever their justification, United States congressional, press, and academic attitudes, undoubtedly stimulated by the example of the Indian explosion, clearly regarded developing nations as less likely to show stability and exercise restraint in face of pressures to turn their sensitive nuclear facilities and materials to developing nuclear explosives. The relatively latent fears of an increasingly nuclear-armed world were reawakened.

Contributing to the overall sense of frustration in the United States was a newfound concern over the effectiveness of the central feature of the nonproliferation regime: international safeguards. Despite the circumstances that this institution had been strongly encouraged and closely monitored by the Congress from 1954 onward and that its limitations had never been concealed, a newly concerned group became aware that safeguards were a mechanism for the detection of diversion, not for its prevention, and that even the IAEA's capability of detecting unauthorized removals by individuals was far from foolproof. This group made no real distinction between prevention of terrorism, which is the responsibility of national governments, and safeguards against diversion by governments, which are different in their objectives and are alone the task of the IAEA.

The events described above came at a time when a number of important changes were taking place in past international nuclear patterns. In particular, the United States monopoly was broken by the emergence of the Soviet Union and the two European multinational enterprises, Urenco and Eurodif, as enrichment suppliers. Concurrently, the ability of European manufacturers, especially the dynamic West German industry, to respond to the demand for the preferred light water reactors was established. These developments helped make possible the diffusion of nuclear technology that had caused so much concern in the United States and set the stage for the next events.

THE REAPPRAISAL

The first major reaction to the concerns noted above were United States and Canadian efforts to seek strengthened undertakings among the principal suppliers on new non-proliferation ground rules to govern nuclear exports. Consultation and cooperation among exporters was not new, and the principle that export competition should not extend to weakening or undermining safeguards requirements had acquired fairly general support among suppliers relatively early in the cooperative era. In 1974 the United States sought to bring the suppliers together once again. A key event was the decision of France, at the December 1974 summit meeting of Presidents Ford and Giscard d'Estaing, to abandon its free-lance role of the past and to join in a concerted effort by the main suppliers, meeting in London, to avoid competition in safeguards and non-proliferation arrangements.

Also in December 1974, the Canadian government announced its decision to strengthen its control over exports by renegotiating its bilateral agreements, obliging its customers to accept new and more stringent safeguards or face an interruption of Canadian supplies. This introduced for the first time the concept of nonfulfillment of existing supply commitments as a negotiating tool.

The London meetings showed that there existed among the supplier countries two different schools of thought on the possible means of reinforcing non-proliferation through some kind of selective embargo. One group, lacking confidence in IAEA safeguards and in the NPT pledge (which can be revoked in exceptional circumstances on three months' notice), wanted to revert to technological barriers to proliferation. They felt that the greatest restraint should be shown with regard to materials that can be used for weapons or to installations such as reprocessing or enrichment plants that can produce materials of that kind. They wanted to prohibit any foreign assistance to national reprocessing, enrichment, or heavy water plants (because of the proliferation danger of the heavy water natural uranium reactor) and to provide such assistance only to multinational plants managed with the participation of the supplier country.

Another group still had confidence in legal commitments and wanted only to prohibit the sale of material or equipment on an established "sensitive" list to any country that would not submit to verification by the IAEA of the peaceful and nonexplosive character of its entire fuel cycle. The strongest supporters of non-proliferation policy favored combining both groups' viewpoints. France remained opposed to any form of nuclear embargo aimed at putting pressure on the energy policy of a country that refused to submit its full fuel cycle to safeguards, feeling in particular that such pressures could encourage independent programs.

Nonetheless, by the end of 1975 the London group agreed on minimum guidelines that they would in any case follow in their nuclear exports, with the understanding that a country was always free to apply stricter rules. An important precondition of achieving this agreement was the attitude of the Soviet Union. The Soviet Union had, in its own sphere, supplied nuclear assistance only to NPT parties (a condition fulfilled by all the East European countries) and had precluded participation by its Comecon partners in any part of the fuel cycle (except the mining and milling of uranium concentrates to be sold to the USSR). Thus the Soviet Union could welcome more restrictive policies without any inconvenience to its own export program.

The London guidelines were disclosed in general terms in mid-1976. The precise text, only slightly amended as a result of the 1976–77 meetings of an enlarged group of suppliers, was finally made public in early 1978, when fifteen supplier nations officially informed the IAEA of the rules and their individual decisions to adopt them.

The Canadian call for renegotiation of its supply agreements, the concerted action of the suppliers, and the new United States attitude all contributed importantly to the erosion of confidence on the part of recipient nations in the continued availability of nuclear power benefits. That the London negotiations and their results were secret (although the existence of the group and the occurrence of its meetings became known to the press on every occasion) contributed to the creation of distrust by generating suspicion that the group was conspiring to maintain its position of dominance by preventing the transfer of technology, whereas in fact the group itself included recipient countries already threatened with possible embargoes by some of the suppliers. The ultimate publication of the guidelines can thus be viewed as a step contributing to the restoration of confidence by dispelling exaggerated fears as to the severity of the measures agreed upon.

Long before publication of the guidelines, recipient nations, especially the developing countries, expressed strong opposition to the suppliers' cooperation and its result. Article IV of the NPT had assured the non-nuclear-weapon states that were parties to the treaty of "the inalienable right . . . to develop research, production and use of nuclear energy for peaceful purposes," and "the right to participate in . . . the fullest possible exchange of equipment, materials and scientific and technological information." These countries viewed the guidelines, which call for "restraint in the transfer of sensitive facilities, technology and weapons-usable materials," as an infringement of the provisions of article IV.

Except for the period of denial that extended from 1946 to 1954, efforts to avoid the spread of nuclear weapons through the abuse of peaceful nuclear activities have, in the main, depended on political and institutional restraints, including safeguards, the NPT, and bilateral non-proliferation assurances. The Acheson-Lilienthal plan itself, with its far-reaching proposals, did not rule out any activity so long as it was conducted under international management and ownership. Under the guidelines, certain steps of the fuel cycle were for the first time being considered dangerous even under safeguards and were singled out as calling for special restraint in trade with non-nuclear-weapon states, even though there was no effort at this stage to outlaw these activities.

The restrictions against reprocessing and enrichment were reflected not only in the requirement for "restraint" in the transfer of these technologies but also in the noncompulsory recommendation that suppliers secure rights of approval over reprocessing—a condition that previously had appeared only in United States agreements. An early application of this recommendation was the Canadian decision to press for reprocessing approval rights in the renegotiation of agreements with Euratom and Japan that it had called for earlier. By the end of 1976, Canada's threat to cut off supplies if renegotiation were not successfully concluded was placed into effect through the imposition of embargoes, even with respect to existing contracts for natural uranium shipments to these two clients. The embargo was lifted at the end of 1977, following an interim agreement to be reviewed on conclusion of the International Nuclear Fuel Cycle Evaluation (INFCE), but not without a residue of dissatisfaction.

Another important feature of the guidelines was the adoption for the first time of a requirement that is tantamount to a kind of safeguard on the transfer of sensitive technology itself—as distinct from the attachment of safeguards to sensitive facilities, equipment, or materials. Although its Statute authorized the IAEA to apply safeguards on information, this provision had never been implemented on technology. The new provision of the guidelines was directed toward preventing a country that acquired sensitive technology through the transfer of a sensitive but safeguarded facility from using this technology to build its own additional unsafeguarded facilities of the same type.

In effect, this new prohibition against "replication" reflected the response of particular suppliers—especially West Germany and France—to criticism of their arrangements for the transfer of reprocessing and enrichment technology: technology that, in the view of the United States and some other suppliers, was better controlled by avoiding any further dissemination.

At the time the London meetings were begun in 1975, France had nearly completed two prolonged series of negotiations for French assistance in the sensitive field of reprocessing with South Korea and Pakistan. During the same period the West German government was negotiating a major agreement with Brazil, covering the construction of a number of large power reactors as well as an enrichment plant and a small reprocessing facility. These three projected agreements were modified, with some difficulty, at an advanced stage of the negotiations to conform to the new and still-unpublished guidelines by including provisions for the application of safeguards to any replication. They were presented to and approved by the IAEA Board of Governors in 1975 and 1976.

While South Korea canceled its projected reprocessing plant under external pressure, Pakistan refused to yield, even though the United States intervened at the highest level and Canada decided to interrupt the uranium and heavy water supplies indispensable for the functioning of the only Pakistani power reactor (a Canadian-built unit that supplies the city of Karachi with half its electricity). The Canadians also cut short their long-standing agreement with India, following lengthy negotiations in which agreement provisionally reached at the negotiating level was not finally confirmed by the Canadian government.

France, preferring to avoid the path leading to the abrogation of existing agreements, did not cancel the freshly signed Pakistani contract. The building of the plant, which was scheduled to be completed in 1981, is under the sole responsibility of the Pakistanis, with French industry assisting only as a general consultant. Pakistan could probably "go it alone" at this time and assemble a less sophisticated unit over a longer period.

The United States also expressed strong opposition to the West German–Brazilian agreement, which, in light of the current slackening in nuclear power plant orders, is of particular importance for the West German industry because of the magnitude of the Brazilian plans that include reprocessing and enrichment. West Germany refused to cancel its contractual commitment. Neither the West German–Brazilian arrangement nor the French arrangement described above were in conflict with the recently adopted London guidelines.

The French government, in December 1976, stated publicly that it would not enter into new arrangements, at least for the time being, for the bilateral transfer of its industrial reprocessing technologies, and thus it joined for the first time the path of denial of technology. The following year the West German government took a similar position.

Concurrent with the development of the guidelines, other developments were occurring in the area of bilateral export policy. Although its enrichment monopoly had ended in the early 1970s, the United States remained an extremely important supplier of enrichment services, particularly since the new facilities of Urenco and Eurodif would not be available for large-scale production for some time. By mid-1974 the United States had completely committed its existing and authorized enrichment capacity, without, however, having decided upon the construction of new facilities. As a result, it imposed a moratorium on new contracting. Unfortunately, a number of contracts that had been completely negotiated at that time (and on which down payments had been made) were caught by this moratorium and were not executed.

After a period of examination, the United States concluded all the outstanding domestic contracts and several of the foreign contracts, while offering a number of foreign customers "conditional" contracts that would become firm if and when it decided to proceed with plutonium recycle. This action served to remind other nations in a dramatic fashion of the perils of relying on a single source of supply and further fortified their determination to seek independent or alternative sources of supply, including indigenous ones wherever possible. Brazil was one of the countries whose pending enrichment contracts became "conditional," and the West German–Brazilian arrangement that would provide Brazil with independent enrichment and reprocessing capabilities has been attributed at least in part to this episode.

In 1975 responsibility for the issuance of United States nuclear export licenses was transferred by law to the newly created Nuclear Regulatory Commission (NRC). While the NRC did not purport to set new United States nuclear cooperation or non-proliferation policy, it had the statutory responsibility to determine that each nuclear export was "not inimical to the common defense and security." Until this time, nuclear export licensing had been an essentially formal step that followed as a matter of course for exports that were within established United States policy and authorized by agreements for cooperation. This circumstance was changed by assignment of the responsibility to the NRC, which, as an independent regulatory body, was not subject to presidential direction or review and which saw its role as one of scrupulous independence of executive branch supervision.

While the actions of the suppliers group had their effect principally on the developing nations, the first impact of the new United States export licensing regime was on Euratom, a number of whose imports of highly enriched uranium were delayed for extended periods as the NRC and the executive branch sought to develop a modus vivendi. After some early and considerable delays, license issuance once again became fairly regular, but the potential for substantive disagreement between the executive branch and the NRC remained.

THE NEW PROPOSALS

The new policy reflected in the suppliers' guidelines agreed upon prior to 1976, even though addressed to nuclear exports to all non-nuclear-weapon states, seemed mainly directed against developing countries, a circumstance that was regarded as unfortunate but unavoidable. It tended to divide the world between supplier countries and recipient countries, and between countries that already possessed sensitive fuel cycle facilities and those that could build them without outside help only with great difficulty. For this latter category of countries, the guidelines left little hope that they could purchase these sensitive facilities, but plutonium utilization itself remained as an acceptable part of the fuel cycle.

But this situation did not last, and industrialized countries were soon to feel the effects of a new policy that sought to outlaw, at least for an extended period, reprocessing, plutonium recycle, and plutonium breeding. The first major evidence of this came in October 1976, when the Ford administration in the United States, in response to initiatives by the Democratic candidate in the presidential election campaign, announced that it would henceforth not regard reprocessing and plutonium recycle as necessary and inevitable parts of the fuel cycle, and that commercialization of those activities in the United States would be deferred until the government was satisfied that the attendant proliferation hazards could be dealt with (Ford 1976).

This statement, coming only a few days before President Ford's defeat in the election, attracted relatively little attention. Nevertheless, it had its effect. Approvals for the transfer of spent fuel from several nations to Britain and France for reprocessing, which had been routinely granted in the past, began to be held up. Japan was given no response to its request for United States approval to start up its French-built reprocessing pilot plant on United States supplied fuel. Even earlier, problems were encountered by applications for the export of highly enriched uranium that had been routinely approved in the past, as the Ford administration began to reconsider its policy in this area as well. The administration came to an end with these issues unresolved.

The new Carter administration made a review of non-proliferation policy one of its priority concerns and began consultations on the issue even before entering office. In two statements in April 1977, President Carter spelled out his policy, calling for indefinite deferral of commercial reprocessing and plutonium recycle and a slowdown and redirection of breeder development away from the plutonium fuel cycle (Carter 1977*a, b*). He underscored the nondiscriminatory nature of the policy by announcing that the private Barnwell reprocessing plant would not be encouraged to operate and that the government's demonstration breeder reactor project at Clinch River would not go forward. While acknowledging that the United States could not impose this policy on other nations, he left no doubt of his hope that it would be followed universally. To counter foreign criticism of the restriction on reprocessing and deemphasis of the breeder, he offered assurances that the United States would again fulfill its role as a reliable supplier of nuclear fuel. But this assurance was given against the background of the lack of United States natural uranium export capability, of continued United States delays in expanding its enrichment capacity, and of export licensing obstacles to deliveries under existing United States fuel supply contracts. The proposal to renegotiate existing cooperative agreements while commitments under these agreements were not being fulfilled was viewed as contrary to the assurances of reliability of supply and further contributed to the erosion of confidence.

The Carter administration's policy statement was presaged by a long awaited report on nuclear policy by the Ford Foundation, which reached similar conclusions (Keeny et al. 1977). Like the Carter statement, the Ford/MITRE report was criticized, particularly because of its optimistic estimates of the future availability of uranium and its failure to take into account the much more limited energy options of countries outside the United States.

The new policy attracted widespread criticism in both Europe and Japan, which had advanced programs on reprocessing, plutonium recycle, and breeder reactors, and among the developing nations, which saw these activities as important to the achievement of greater nuclear energy independence in the future. A coincidence in timing made the new policy a principal topic of discussion at the Teheran Conference on Nuclear Technology Transfer, held in April 1977, and a critical statement on the policy was adopted informally.

The consensus outside the United States was that the new policy failed to take adequate account of the differing energy circumstances of the United States, with its still vast energy resources, and of most other nations, which have limited indigenous fuel reserves. Thus the new policy, particularly in what seemed to be its opposition to the breeder reactor, which was widely viewed as fundamental to the long-term economic and resource rationale of nuclear power, was perceived as not only "antiplutonium" but also opposed to nuclear energy itself. The fear – which was borne out in some measure – was that the policy, however laudable its motives, would be seized upon by the opponents of nuclear power as one more proof that this energy option was too risky to pursue.

Further annoyance arose when the United States announced its choice of the centrifuge process for the expansion of United States enrichment capacity, notwithstanding its greater proliferation risks. There was criticism that a double standard was involved in the selection of this process for reasons of energy conservation in the United States at the same time that the United States opposed the adoption of reprocessing, plutonium recycle, and the introduction of the breeder reactor, whose energy conservation features were deemed important by others.

The Carter policy recognized the need for international cooperation and proposed an International Nuclear Fuel Cycle Evaluation (INFCE), focused on alternative fuel cycles. At the London economic summit conference of May 1977, the participating nations (Canada, West Germany, France, Italy, Japan, Britain, and the United States) agreed to make a preliminary study of such an evaluation, in which all countries would be able to participate and which would seek to reconcile future nuclear energy needs with non-proliferation concerns. In the ensuing discussions, the United States agreed that INFCE should examine not only alternative proliferation-resistant fuel cycles, but the conventional plutonium fuel cycle as well.

As time went on, the United States administration also took pains to clarify that it was not opposed to the breeder reactor per se but only to the conventional plutonium fuel cycle. The emphasis of United States policy pronouncements shifted from the importance of deferral to the importance of review before irrevocable decisions were made on the future nature of the fuel cycle. INFCE itself was carefully restricted to an evaluation, with no responsibility for negotiation on new institutional arrangements, and assurances were given that "the evaluation will be carried out in a spirit of objectivity, with mutual respect for each country's choices and decisions in this field, without jeopardizing their respective fuel cycle policies or international cooperation, agreements, and contracts for the peaceful use of nuclear energy, provided that agreed safeguards measures are applied" (INFCE 1980, p. 55).

Among the most disturbing of the developments that arose in the post-1974 period was the tendency on the part of some suppliers to judge existing arrangements on the basis of unilaterally conceived new rules. This retroactivity, fostered in the United States by congressional attitudes and the grant of licensing authority to the independent Nuclear Regulatory Commission (NRC), has been one of the strongest factors in the erosion of confidence. The Carter administration was at least as successful as its predecessor in resisting attempts at retroactive application of its policies in concrete cases, and existing United States fuel supply commitments, with certain important exceptions, have continued to be fulfilled.

A more recent and complex development is the passage of new United States legislation: the Nuclear Non-Proliferation Act of 1978. This highly complicated law contains both provisions that are likely to further erode confidence and provisions that are designed to restore confidence and that, if effectively administered, might help to do so. In setting new export criteria—in particular, the requirement for United States rights of approval over reprocessing and for full-scope safeguards after an eighteen-month transition period—it conflicts with United States assurances that the results of INFCE would not be prejudged and, in any case, gives legislative endorsement to the recent practice of applying new standards retroactively to existing commitments. At the same time, the new legislation restores to the president (although in a complex manner and subject to congressional veto) the ultimate responsibility for issuing export licenses. It also reconfirms in strong, though not functional, terms the long-standing United States commitment to international nuclear cooperation, to the concept of reliable supply, and to prompt action on export and similar requests. The final consequences of this legislation cannot be predicted at present, and much will depend on the way it is administered both by the administration and by the NRC.

In the short run, the provision insisting, as a condition of continued United States supply, on consent to renegotiation of agreements for cooperation that do not grant the United States the right of approval of reprocessing has already had its effect on the erosion of confidence. The thirty-day grace period for application of this requirement passed without Euratom's consenting to renegotiate, and United States export licensing to the European Community was accordingly suspended until an accommodation was reached a few months later. Although the immediate problem of renegotiation was resolved by a compromise formula delaying consideration of major questions until INFCE is completed, the substantive differences of opinion on whether the United States should be granted reprocessing approval rights represent a major issue.

Canadian-Euratom relationships have followed a somewhat similar course. Although interim agreement was reached for the time period of INFCE, allowing resumption of Canadian uranium exports to Euratom after a one-year embargo, the substantive differences on the reprocessing approval issue have not been resolved, and they could lead to a renewed delivery moratorium if negotiations are unsuccessful.

Reliability of supply thus emerges as one of the critical ingredients in the maintenance of confidence. Reliability has been described as involving not only the fulfillment of existing commitments but also the expansion of capacity as needed, stability of conditions of supply without unilateral change, and fairness in pricing.

There seems to be a growing realization of the close connection between assurances of supply and successful non-proliferation policy, or, put conversely, of the fact that the failure to offer reliable supply arrangements will lead to the development of multiple independent sources of supply, with a consequent loss of control. The new United States legislation accepts this principle and calls for efforts to seek agreement on the creation of an international fuel bank or fuel authority that could provide assurances of supply independent of national interference. But it is the rigidity of the conditions attached to such assurances that will decide their degree of attractiveness to recipient countries.

CONCLUSION

International nuclear cooperation and non-proliferation policy are in a state of flux. There are positive influences and opportunities that, if exploited, could lead to a new consensus and the revival of confidence. INFCE, in particular, provides a forum in which differing views can be exposed not only to technical examination but to the test of international acceptability. There are a few encouraging signs that the more extreme positions on every side are being discarded in favor of more moderate ones. Any new consensus is likely to involve restrictions on the accumulation of plutonium in national hands that did not exist before. At the same time, it is likely to involve recognition that the energy values of spent fuel must be put to use in one form or another. Above all, a new consensus probably depends upon forthright acknowledgment of the importance of nuclear energy itself in meeting world energy needs.

Difficult problems also remain, along with influences that could frustrate a new consensus. Among these are the continuing application of retroactivity to implementation of existing agreements. Mr. Justice Parker, in his opinion on the Windscale matter in Britain, referred to this problem in these terms: "I do not accept that the best way to achieve a new bargain is to break an existing one" (Parker 1978, p. 21).

Confidence is a two-way street. In nuclear affairs, it means that, for cooperation to flourish, there must be confidence both in the efficacy of the non-proliferation regime and in the reliability of cooperative undertakings. The failure to accept this fundamental principle could still lead to a breakdown of the system, with less cooperation and more risk of proliferation than before. We believe this can and must be avoided.

REFERENCES

Carter, James. 1977*a*. Nuclear power policy: Statement by the president on his decisions following a review of U.S. policy, 7 April. Washington, D.C.: White House. Mimeographed.

______. 1977*b*. Message to the Congress transmitting the nuclear non-proliferation bill, 27 April. Washington, D.C.: White House. Mimeographed.

Ford, Gerald R. 1976. Nuclear policy: A statement by President Gerald R. Ford, 28 October. Washington, D.C.: White House. Mimeographed.

Goldschmidt, Bertrand. 1980. *Le complexe atomique: Histoire politique de l'énergie nucléaire*. Paris: Fayard.

International Nuclear Fuel Cycle Evaluation. 1980. Vol. 9. *Summary volume*. Vienna: International Atomic Energy Agency.

Keeny, Spurgeon M., et al. 1977. *Nuclear power issues and choices: Report of the Nuclear Energy Policy Study Group*. Cambridge, Mass.: Ballinger.

Multilateral agreements. 1959. Legal Series no. 1. Vienna: International Atomic Energy Agency.

Nuclear proliferation factbook. 1980. Prepared for the Committee on Governmental Affairs, U.S. Senate, and the Committee on Foreign Affairs, U.S. House of Representatives, by the Congressional Research Service, Library of Congress. Washington, D.C.: Government Printing Office.

Parker, Justice. 1978. *The Windscale inquiry: Report by the Hon. Mr Justice Parker*. Vol. 1. London: Her Majesty's Stationery Office.

United States Department of State. 1946. *A report on the international control of atomic energy*. Publication no. 2498. Washington, D.C.: Government Printing Office.

3

Nuclear Energy and International Cooperation: A Third World Perception of the Erosion of Confidence

Munir Ahmad Khan

ENERGY NEEDS AND DEVELOPING COUNTRIES

The secure availability of adequate supplies of energy is essential to the economic development of any country, particularly a developing country. A review of the world's commercial energy production and consumption indicates that there are serious imbalances in the utilization of energy in different parts of the world. The developing countries of the Third World, which contain more than two-thirds of the world's population, account for only 16 percent of commercial energy consumption, while 84 percent of world energy is consumed by the advanced countries with less than one-third of the world's population. The United States alone, with only 6 percent of the population, consumes 36 percent of world energy.

The world average annual per capita consumption of electricity and commercial energy stands at 1,600 kilowatt hours (kWh) and 2 tons of coal equivalent (tce) respectively. In Asia, per capita consumption is only about one-fifth of these figures. In many of the poorer, more populous developing countries, particularly those in South Asia, as little as 150 kWh and 0.2 tce is consumed by the average inhabitant each year. The corresponding figures in the advanced industrialized countries have reached 10,000 kWh and 12 tce. There is thus a highly uneven energy consumption pattern between industrialized and Third World countries.

The less developed countries (LDCs) need more energy inputs for accelerating their development programs, and some means will have to be found to ensure that adequate energy is made available to them at reasonable cost. The industrialized countries developed their basic industrial infrastructure at a time when energy costs were very low: oil was priced at one to two dollars a barrel and coal at two to three dollars a ton as recently as the mid-1960s. But

Date of origin, September 1979.

now, when the LDCs have started building their industrial infrastructure and require enormous inputs of energy, the cost of energy has risen by an order of magnitude, making it extremely expensive for them to embark upon large-scale development programs. Ironically, until recently, planners in the Third World did not comprehend the crucial significance of long-range planning in the energy sector. It was only after the worldwide 1973 energy crisis that they began to appreciate the importance of energy in relation to economic development.

The decision of the Organization of Petroleum Exporting Countries (OPEC) in 1973 to increase oil prices fourfold caused sudden and serious economic repercussions, partly because it was so unexpected. According to OPEC sources, the upward correction in oil prices was long overdue because they had been kept at very low levels for too long, lulling the world into a false sense of security with regard to the availability and price of oil. Whatever the justification, this sudden and drastic increase in oil prices severely jolted the world economy and forced both the industrialized and the developing countries to make painful economic adjustments. The degree to which individual countries were affected depended very much on relative dependence on imported oil, reliance on foreign trade, and domestic economic policies. It led to a worldwide inflation and a reevaluation of currencies, which still continue. There has been a general slowing of economic growth and increase in unemployment in several industrialized countries. However, the advanced countries, because of their greater technological, economic, and managerial resources, have shown considerable resilience in meeting the challenge posed by increased oil prices. Japan was one of the countries that fared best. In spite of the fact that it imports 99 percent of its energy and its energy bill has soared by a factor of six in the last few years, it has consistently increased its trade surplus, generally because of its strong industrial base, advanced technology, and better economic management.

As for the non-oil-producing developing countries, those with the more advanced infrastructures, skilled labor, and dynamic economic policies could benefit from a boom in OPEC countries by exporting goods and services. Such beneficiaries include Brazil, South Korea, Hong Kong, Singapore, and Taiwan. But the economic performance of the remaining non-oil-producing LDCs has been very disappointing. In some cases, the cost of oil imports has risen to over 50 percent of their visible export earnings, leaving very little for their development programs to meet the needs of their growing populations. Such countries, which may be classified as most seriously affected countries (MSAs), have run into serious economic difficulties, have been obliged to retrench in their economic development programs, and have sought debt relief. According to OECD estimates, their aggregate trade deficit is estimated to be $46 billion.

OPEC LDCs have been more fortunate because, with successive increases in oil prices, their export earnings have risen exceptionally. It has been estimated that, with oil price increases, their export earnings for 1979 may reach $177 billion, of which they are expected to spend about 90 percent on imports of goods and services that are also becoming more expensive (*Newsweek,* 9 April 1979). However, their future economic prosperity is not as assured as is often believed. Their oil is disappearing very rapidly, and they have not been able to put their huge surpluses to good use because of lack of indigenous infrastructure, skilled manpower, and necessary management and technology. In fact, at the energy conference organized by the Organization of Arab Petroleum Exporting Countries (OAPEC) in March 1979, it was widely acknowledged that unless they managed to translate their current wealth into something really tangible and lasting, such as technology and development, their future prosperity would be uncertain. It was pointed out that two-thirds of the OPEC member countries could face energy crises before the turn of the century and that preparation for the postoil era should begin at once. It is true that these OPEC countries have limited ability to absorb huge investments quickly and that their surplus wealth is being eroded by rising inflation. They are far from being economic giants, in spite of impressive statistics regarding their earnings. It is interesting that, according to the figures published by the World Bank, the combined trade surplus of OPEC for the years 1977 and 1978 amounted to $35 billion and $30 billion respectively, whereas the corresponding figures for Japan alone were $17 billion and $25 billion. This once again underlines the fact that LDCs, dependent on exports of precious but dwindling raw material resources, cannot match the economic performance of raw-material-deficient but highly advanced countries that have the necessary technology, trained manpower, and economic and fiscal discipline.

There is no doubt that the fossil fuel reserves of OPEC countries are considerable, but they are far from inexhaustible. The data presented at the World Energy Conference in 1977 (table 3.1) indicate that the proved oil, gas, and coal resources of OPEC countries amount to 292×10^9 tce, which is about 22 percent of known world resources. The developed countries have about 66 percent, and the remaining 12 percent is distributed over non-OPEC developing countries. Even the large discoveries of oil and gas recently reported by Mexico do not appreciably change the overall balance between the developed and developing countries. The industrialized nations expect OPEC countries to produce increasing amounts of oil to help sustain their economies and high standards of living, based on high energy consumption, even though they have been too slow or unwilling to enforce conservation measures at home for political reasons. What is happening, therefore, is that a transfer of nonrenewable energy resources is taking place from OPEC LDCs to the industrialized countries, and these are not being replaced with appropriate technologies that can provide OPEC states with energy alternatives when their oil and gas resources run out. Unfortunately, the advanced countries continue to import more oil, preempting all readily available sup-

TABLE 3.1 Economically Recoverable Fossil Fuel Resources, 1977

	Population, 1975 (10^6)	Coal (10^9 tce)	Oil (10^9 tce)	Gas (10^9 tce)	Total (10^9 tce)	Resources per Capita (tce)
Developed countries	1,160	552	170	209	931	803
Developing countries						
OPEC	300	2	165	125	292	973
Non-OPEC	2,486	81	51	25	157	63
World total	3,946	635	386	359	1,380	350

Sources: Fossil fuels: Data presented to World Energy Conference, 1977.
Population: World Bank 1978.

plies and pressing OPEC members to produce still more. This dangerous spiral of increasing dependence by the industrial world on limited oil reserves is increasing oil prices and placing great strains on the world economy, to an extent that may become unbearable for the non-oil-producing LDCs.

Third World countries will also have to explore alternative sources of energy, such as solar, wind, and biomass, but these will need considerable research and development, and it may take a few decades before they can be harnessed economically on a large scale. The developing countries believe that these nonconventional sources cannot offer near-term solutions to their power needs. They are acutely aware of the difficulties of higher capital cost, safety, environmental impact, and sophistication associated with nuclear energy. But, in spite of all these difficulties, the Third World has a strong economic compulsion to turn toward nuclear energy as a practical and viable source to ensure long-term supply of reasonable cost energy for its development.

THE ATOMS FOR PEACE PLAN AND THE GOLDEN ERA OF COOPERATION

In 1953 the United States took an unprecedented initiative in launching the Atoms for Peace plan to share the benefits of peaceful applications of nuclear energy with the rest of the world.

Several reasons have been advanced to explain this far-reaching decision by the United States. It may have been based on recognition that the United States monopoly in nuclear weapons had been broken by the Soviet Union, and that it was no use denying peaceful nuclear technology to friendly countries in Western Europe and elsewhere. Industrial and commercial considerations may have played a part, since the plan helped create a market for peaceful nuclear technology while the United States still had an undisputed lead in this field. The United States was the first to use nuclear weapons, and serious reaction among its intellectuals and scientists may also have led the government to take measures to restore its image at home and abroad.

Whatever the motives behind the plan, it was without doubt a unique and commendable gesture. The end result of the Atoms for Peace plan was a large-scale worldwide transfer of peaceful nuclear technology. The United States trained thousands of engineers and scientists in its own laboratories and provided a large number of research reactors and other facilities to several advanced as well as developing countries. Many developing countries took a keen interest in this new technology, which offered both prestige and potential long-term economic benefits. The net effect of the Atoms for Peace program was decidedly positive and far-reaching as far as LDCs were concerned. Besides imparting know-how in nuclear research, it acted as a catalyst for initiating quiet but meaningful scientific and technological change. But not all these countries had the necessary leadership or infrastructure to make best use of the opportunities afforded by it.

Among those LDCs that benefited most were India, Argentina, Brazil, Taiwan, South Korea, and Pakistan. India had set up its Atomic Energy Commission as far back as 1948, and it made good use of international and bilateral cooperation in the period between the mid-1950s and 1960s, acquiring the ability to assemble the Apsara research reactor in 1956 and to complete a small reprocessing plant in 1964. According to reports presented at United States congressional hearings, the United States trained more than 1,100 Indian scientists and engineers in its nuclear research laboratories and universities. Canada also extended generous nuclear cooperation to India. Argentina embarked upon a program of acquiring nuclear technology and decided to build indigenously an *Argonaut* type of research reactor in 1959. Pakistan signed an agreement with a United States supplier for a research reactor, which was commissioned in 1963. In fact, scores of research reactors were built during the heyday of the Atoms of Peace plan, and thousands of scientists and engineers from developing countries received training.

The Soviet Union initiated close cooperation with China in the nuclear field by supplying a research reactor and helping to develop China's uranium resources. Considerable know-how was transferred in other areas, and thousands of Chinese scientists received training in the Soviet Union. This cooperation agreement was, however, abrogated in 1959 when the prospects of a nuclear China began to appear on the horizon.

In 1964 China conducted a nuclear weapon test that was criticized in several quarters. However, this showed that the spread of nuclear weapons technology was difficult to contain and that even a poor and developing country like China, beset with numerous problems at home, could, if it made a political decision to do so, go nuclear at a cost by investing its technical and material resources preferentially for this purpose. The worldwide impact of the Chinese nuclear test was muted to some extent by the recognition that China was potentially a major world power and that it had its own political compulsions to go nuclear. China's test did not therefore create much alarm about horizontal proliferation, and no concerted efforts were made to restrict the spread of peaceful nuclear technology, which continued to be shared internationally.

Perceptions in the Soviet Union and India were, however, quite different. The Soviet Union could not ignore the presence of a nuclear adversary next door and began reassessing problems of nuclear weapons proliferation. It realized that it had a community of interest with the United States in preventing more countries from becoming nuclear-weapon states. This helped influence Soviet thinking regarding nuclear disarmament discussions at Geneva and the negotiations leading toward the Treaty on the Non-Proliferation of Nuclear Weapons (NPT).

THE NPT AND THE DEVELOPMENT OF SAFEGUARDS

Until the early 1960s, the Soviet Union had consistently opposed the development of International Atomic Energy Agency (IAEA) safeguards. As the negotiations on the NPT advanced, the Soviet attitude underwent a marked change, paving the way for the adoption of the IAEA safeguards system (INFCIRC/66) and its subsequent revisions.*

A conference of non-nuclear-weapon states under the auspices of the United Nations was organized at Geneva in 1968 to discuss their specific needs, particularly their requirements for peaceful applications of nuclear technology and their security problems. The two superpowers showed unwillingness to extend any tangible security assurances to non-nuclear-weapon states against nuclear threat or blackmail. This, indeed, has remained one of the major reasons why many non-nuclear-weapon states are reluctant to sign the NPT. During this conference it was repeatedly pointed out that the treaty was basically discriminatory and tended to establish the hegemony of nuclear-weapon countries and maintain the status quo in their favor. The numerous resolutions adopted at the conference of non-nuclear-weapon states failed to make any impression on the superpowers, which did not agree to modify the draft text of the NPT.

The negotiations for the NPT were conducted largely between the two superpowers but within the framework of the Eighteen Nation Disarmament Committee (ENDC) at Geneva. The IAEA was kept out of it even though, looking back, it appears that the agency could have made very useful contributions in the drafting of this important treaty. However, India, Brazil, and Mexico took part in the ENDC negotiations, and their contributions were especially reflected in the wording of articles IV, V, and VI.

The final text of the treaty reflected a compromise, largely between the two superpowers, on the one hand, and between the United States and its allies in Western Europe plus Japan, on the other. This compromise, however, intro-

* The basic texts for the two versions of the IAEA safeguards system are the agency's documents INFCIRC/66 and INFCIRC/153, first promulgated in 1965 and in 1971.

duced some inherent weaknesses and deficiencies into the treaty. For instance, under article X the signatories have the right to withdraw from the treaty by giving ninety days' notice if their supreme interests are jeopardized: a provision which has not been invoked so far but which could lead to a serious impasse in the future. Article VI of the treaty envisaged an end to the nuclear arms race and a possible reduction in stockpiles of nuclear weapons, but no time limit was imposed for this purpose, nor were any procedures defined. And article IV provided for the fullest possible exchange of nuclear technology between the signatories, with no exclusions or riders on such as "sensitive" technologies. Yet the provisions of this article have not been faithfully implemented, undermining the implied quid pro quo for the adherence of signatories to the NPT regime.

Soon after the adoption of the NPT by the United Nations General Assembly in 1968, the main sponsors launched a campaign to attract more adherents to the treaty. The IAEA safeguards system was developed further, and bilateral safeguards agreements were transferred to the agency, as were the major safeguards functions relating to the NPT. While the backers of the NPT were persuading nonsignatories to join the fold of the treaty, no serious barriers were placed in the way of transferring nuclear technology, materials, and equipment to nonsignatories. In fact it was during the late 1960s that several important transactions took place between supplier states and developing countries, covering power reactors, heavy water, and fuel fabrication plants and other facilities. In the early 1970s the IAEA concluded safeguards agreements with nonsignatories such as Argentina, India, and Pakistan with regard to the power reactors supplied to them by West Germany and Canada. This was also the era of considerable optimism about the future of nuclear power, and a number of advanced countries were interested in marketing power reactors in Third World countries.

However, 1974 witnessed a major international event when India exploded a nuclear device and became the sixth country to demonstrate nuclear explosive capability. India claimed that this underground experiment was a peaceful explosion, but others viewed it differently. It evoked strong protests from Canada, which had generously shared nuclear technology with India and supplied the Canada-India reactor on the understanding that it would be used for peaceful purposes only. (In 1971 the Canadian prime minister made it clear during a visit to India that nuclear technology supplied by Canada was not to be used for conducting nuclear explosions.) The Indian underground test triggered a serious worldwide reappraisal, particularly by the nuclear powers, of the adequacy of existing measures and undertakings to control nuclear proliferation. France, which had hitherto been only lukewarm toward the NPT, started to review its policy of "free-lancing" in nuclear technology. It had developed its own know-how after having been denied access to nuclear technology by the United States in the 1950s, and it felt psychologically less restrained in sharing it with others who were similarly discriminated against. It had also stayed away from NPT negotiations and discussions on nuclear disarmament and had pursued an independent nuclear policy. In the late 1950s it had provided assistance to Israel for the construction of a reactor to which a reprocessing facility was later added (SIPRI 1978).

In a meeting between the presidents of France and the United States in December 1974, an understanding was reached concerning strengthening the non-proliferation regime. France reiterated that, although it would not sign the NPT, it would behave as if it had signed. Thus four nuclear powers including France decided to work closely to achieve the objective of the Non-Proliferation Treaty and arrest the further spread of nuclear weapons technology.

In 1975 certain important agreements between advanced and developing countries were negotiated that were to be covered by IAEA safeguards. These included the West German–Brazilian agreement for the transfer of complete nuclear fuel cycle and power reactor technologies, the Franco–Pakistani agreement covering the supply of a reprocessing plant, and the Franco–South Korean agreement for the transfer of a small-scale reprocessing facility. Although the suppliers claimed that these agreements were signed with the prior knowledge and acquiescence, if not wholehearted approval, of Washington, they soon became a domestic political issue in the United States. In the case of Pakistan, it was thought that the construction of a reprocessing plant would be potentially proliferating, particularly as seen in the regional context. As regards Brazil, the United States did not favor the acquisition of complete nuclear fuel cycle technology by the largest Latin American country, which could have repercussions in the area. A United States supplier had been negotiating large contracts for the supply of nuclear power plants to Brazil but was preempted by the quick action of suppliers from West Germany, which was quick to conclude a bilateral cooperation agreement with Brazil. The commercial interests of the nuclear power industry in the United States were therefore affected, following closely upon India's nuclear explosion. The combined effect of these three agreements was to focus attention on the proliferation issue in the United States, Canada, and some other nuclear countries.

THE LONDON SUPPLIERS CLUB GUIDELINES

To forestall an uncontrolled spread of nuclear technology and to eliminate commercial competition as a factor in negotiating safeguards, the major nuclear countries, led by the United States, established the Nuclear Suppliers Group, which came to be known as the "London suppliers club." Although its first announced meeting was in London in 1975, consultations among its members had taken place earlier. The initial membership of seven nations later increased to fifteen, including some states from Eastern Europe. The main motivation behind the establishment of this group was to strengthen the non-proliferation regime by controlling the supply of nuclear-related equipment, materials, technology, and services. The membership of the group is, however, heterogeneous. Only three members are really self-sufficient in nuclear technology and materials. Others are suppliers as well as recipients of materials or technology or both. Members' interests are therefore not identical. This is reflected in the varying degrees of conviction with which they have supported the "guidelines" adopted by the group.

The guidelines were first conceived in 1975 and were published in 1978. They incorporated some far-reaching and novel concepts, including restrictions on transfers of basic technology and know-how, particularly in the "sensitive" fields of fuel enrichment, reprocessing, and heavy water production. The use of derived know-how acquired through use of supplied technology and equipment was also subjected to controls, and safeguards on the subsequent generation of fissile materials and on any plant or equipment "contaminated" by supplied material were required. In addition, the guidelines stipulated not only that supplied nuclear materials and equipment must be subjected to IAEA safeguards and inspections, but also that no material or equipment could be transferred to a third country without the concurrence of the supplier nation and that supplied technology could not be used to duplicate facilities unless these too were subjected to IAEA safeguards.

How does the Third World view the guidelines adopted by the London suppliers club and the way they have been enforced? Although it has been said that the signatories of the NPT will not be affected by these guidelines, because their programs are fully safeguarded, most of the Third World countries view them with skepticism and concern. They see this step as yet another example of an organized and deliberate effort on the part of the advanced countries to deny technology to the developing countries and impose onerous restrictions on its transfer, which is tantamount to constricting the growth of their civil nuclear programs. Some have spoken of the club as a "cartel" whose political and commercial purposes have been given a moral facade of non-proliferation. This is strongly disputed by the suppliers. Others believe that the guidelines are directed primarily against nonsignatories of the NPT and are designed to convince them that any development of independent nuclear programs without joining the fold of the NPT would be both difficult and expensive. Yet, in practice, these guidelines also adversely affect signatories of the NPT. Many of the LDCs that have signed the NPT resent them and consider them contrary to the spirit of article IV of the treaty, wherein they were promised unrestricted access to and sharing of nuclear technology for their safeguarded nuclear programs.

In analyzing these guidelines, one should not ignore the fact that nuclear technology is not confined to one specific narrow discipline. It is essentially interdisciplinary and encompasses such diversified engineering fields as the electrical, mechanical, chemical, process control, metallurgical, civil, and mining and such basic sciences as physics, chemistry, and computing. The capability to produce a crude nuclear device or relevant materials, therefore, does not require sophisticated knowledge of purely nuclear subjects. Know-how in chemical and mechanical engineering and electronics could enable a sizable group of specialists to develop such a device. This means that, in order really to block the transfer of technology for potential military nuclear applications, one would have to restrict the flow of basic knowledge in numerous technological areas not necessarily confined to the nuclear field.

A country that has no nuclear weapons program, and that for political reasons cannot formally join the NPT, would face the denial of basic technological information needed for its normal industrial development through a strict application of these guidelines. For the members of the London suppliers club, it will become increasingly difficult to draw the line between what is exclusively related to nuclear technology and what is not. Commercial firms will tend to exploit the situation by indulging in dubious practices and charging a premium for supplying equipment that could have applications in both nuclear and nonnuclear fields.

The guidelines apply not only to the supplied facilities, equipment, and physical processes, but also to new facilities utilizing technology derived directly from transferred facilities and know-how. They also safeguard for a period of twenty years or more certain basic physical processes that are not entirely nuclear. Essential information on these processes has already been published, and it seems illogical to impose restrictions on its use.

One of the difficulties the supplier states could face in enforcing controls on basic nuclear technology is that it runs counter to a domestic tradition of wide public dissemination of scientific and technological knowledge. One of the key factors that has helped accelerate technological development in advanced countries such as the United States is the free and prompt dissemination of new know-how that has enabled industries to develop new systems, products, and equipment. It is not easy to foresee the implications of a scientific discovery ten to twenty years hence, nor is it possible to deprive domestic industry of the opportunity to exploit it for fear it might spread to other countries. Unless governments of supplier states impose severe restrictions on the exchange and release of basic scientific information and data, both internally and externally, the spread of new technologies that could contribute indirectly toward the advancement of nuclear technology cannot be contained.

An understanding has also been reached among the supplier states to strengthen IAEA safeguards by giving them necessary technical and political support. The guidelines stipulate that if there is a dispute over the violation of an international bilateral safeguards agreement with any recipient country, then all the supplier states will join to restrict supply of equipment and services to that country. This tends to support the view that the group acts like a cartel. It is, however, noteworthy that, while the suppliers have agreed to act against a violating recipient state, they are silent about how to deal with a defaulting supplier state that may unilaterally abrogate its supply agreement with a recipient country.

The effect of the club's guidelines has also spilled over into the IAEA's procedures. Although the agency is not a party to the NPT and had nothing to do with the evolution of the guidelines, it has been asked by the members of the club, who are also the most influential members of the IAEA, to establish revised procedures for granting technical assistance. They want the IAEA, contrary to its Statute, to discriminate in favor of signatories of the NPT and against nonsignatories in matters of technical assistance. Furthermore, the IAEA has been asked by certain prominent supplier states to accept the concept of "sensitive" technologies and not to grant technical assistance related to them without asking for safeguards. They even want to extend coverage to fellowships and scientific visits related to sensitive technologies. Insistence on these measures has generated considerable debate within the IAEA and has resulted in strong disagreement not only between the signatories and nonsignatories of the NPT, but also between supplier and recipient states.

One reason the guidelines are so restrictive and perhaps difficult to implement is that they were evolved by supplier nations without taking into account the points of view of recipient states. It has been reported that in the evolution of the guidelines different viewpoints were expressed among the suppliers. The suggestion that there should be a dialogue with the recipient states was considered but was thwarted by the hard-liners. If these guidelines had been evolved with the participation of at least some of the recipient states that are signatories to the NPT, they could have been more realistic and acceptable.

CASE HISTORIES IN THE EROSION OF CONFIDENCE

After mid-1974, a number of restrictive measures were taken by advanced countries vis-à-vis recipient states, including LDCs. Canada, incensed by what it felt was a breach of the understanding over the use of Canadian-supplied facilities and technology for nonexplosive use only, canceled its nuclear cooperation agreement with India. The latter strongly disagreed with the Canadian interpretation. India was already well advanced in nuclear technology, did not depend critically on Canadian supplies, and was, in any case, able to obtain certain supplies elsewhere. However, delay in the completion of its heavy water plants led India to seek supplies of heavy water from the Soviet Union. Being a member of the London suppliers club, the Soviet Union referred the matter to the IAEA, which applied its latest safeguards, with the result that both the RAPP-I and the RAPP-II reactors, in which the heavy water was to be used, were brought under full IAEA safeguards.

Canada continued its nuclear cooperation with Pakistan for some time, but in late 1974 it decided at the last minute to withhold the dispatch of equipment for a fuel fabrication plant for which a contract had been signed two years earlier. Pakistan was also asked to renegotiate the 1959 bilateral nuclear cooperation agreements and the 1965 Karachi nuclear power plant agreement. When the two sides could not concur, Canada unilaterally terminated these agreements in December 1976, canceling any further shipments of heavy water, nuclear fuel, and spare parts.

In 1975 France and South Korea signed an agreement for a fuel reprocessing facility. It was approved by the IAEA Board of Governors in September 1975. The agreement was unique in the sense that, for the first time, an IAEA member state was asked and agreed to accept the concept of safeguards on the use of a physical process per se, rather than just on a facility or materials. In line with the rapidly evolving guidelines of the London suppliers club, this agreement also provided for safeguards on specified technical information, unilaterally designated as such by the supplier state, irrespective of whether such information was already publicly available. In other words, the supplier state was given the right to demand control over the use of a known technical process as a precondition for the supply of a facility or plant. This caused considerable consternation among the recipient states because it set a precedent for the future. That, indeed, was proved when, a few months later, Brazil and Pakistan were asked to accept similar conditions, with the phrase "specified technical information" replaced by "relevant technological information not already published." Additional riders were attached that required that any other nuclear facility built within an agreed number of years (say twenty years) based on the same physical process (such as the solvent extraction process) should automatically attract safeguards, making absolute the restriction on the construction of a similar facility without safeguards.

In February 1976 the IAEA Board of Governors approved the West German–Brazilian and Franco–Pakistani agreements. However, the United States did not wholeheartedly accept them as adequately safeguardable from the point of view of non-proliferation. The United States continued to exert pressure on the suppliers concerned not to proceed with the implementation of these agreements and on the recipients to abandon the projects for building reprocessing or enrichment facilities. Either as a consequence of the United States attitude or on its own accord, France also decided to take another look at the Franco–Pakistani agreement and began bilateral discussions with Pakistan on this issue.

Earlier, South Korea had "voluntarily" agreed to cancel its agreement with France for building a pilot-scale reprocessing facility. The United States claimed that it was done by "persuasion" on its part, while France stated that the real decision had been taken in Paris. In any event, Brazil and Pakistan were also approached for similar "voluntary" cancellations. Many Third World countries have viewed the controversy surrounding the Pakistani,

Brazilian, and South Korean agreements with great misgiving, because three developing countries that had fully accepted strict IAEA safeguards were being pressed to give up the construction of plants or facilities that they regarded as necessary for their peaceful nuclear programs.

The United States and Yugoslavia signed an agreement to build a 600 MW reactor plant at Krsko. This project was approved by the IAEA Board of Governors in June 1974. In 1976, however, two years after construction began, the United States decided to withhold further supplies of major components and asked for renegotiation of the safeguards agreement. This created serious differences between the two countries. Yugoslavia felt that, since the project had been duly approved by the IAEA Board, and since it was also an NPT party, there was no need for it to accept any additional obligations, particularly those calling for prior United States clearance for reprocessing of the fuel, which was to be supplied through the IAEA. The matter was later resolved by negotiation, and the question of reprocessing was deferred.

The Soviet Union has always followed its own concept of nuclear assistance and non-proliferation. As a matter of policy, even before the NPT, it restricted the flow of sensitive nuclear technology to East European partners and discouraged them from developing their own nuclear fuel cycle programs. It insisted on the return of fuel irradiated in reactors it had supplied, which it agreed to replace with new fuel from the Soviet Union. Romania, on the other hand, has continued to seek a certain measure of independence in its nuclear policy. It has recently concluded an agreement on cooperation with Canada for the construction of CANDU power reactors. With the possible exceptions of India, Cuba, and North Korea, the Soviet Union has insisted on prior adherence to the NPT as a condition for concluding bilateral supply agreements concerning nuclear materials and equipment.

THE IMPACT OF FORCED RENEGOTIATION OF CONTRACTS

Several case histories have been cited in which the nuclear supplier countries have unilaterally broken their existing agreements on cooperation with LDCs. They have behaved similarly toward certain advanced recipient states. And pressure from supplier states has been directed not only against nonsignatories to the NPT, such as Pakistan and Brazil, but also against South Korea and Yugoslavia, which have signed the NPT. All this has badly shaken confidence in the credibility of any long-term supply agreements with supplier states in general.

In March 1978 the United States passed the Nuclear Non-Proliferation Act, which has contributed to the feeling of insecurity among recipient states, further affecting the image of the United States as a reliable supplier. This act stipulates a period of eighteen months during which any country seeking assistance from the United States must agree to begin negotiations on "full-scope" safeguards, whereby all its nuclear facilities would be brought under the purview of IAEA inspection. The act, together with the guidelines adopted by the London suppliers club, will make the situation far more difficult for the developing countries that are not parties to the NPT or are not willing to accept full-scope safeguards, even in cases where contractual obligations already exist. It has also created serious difficulties vis-à-vis the advanced recipient states, namely European countries and Japan. The passage of this act exemplifies how a leading nuclear country can seek through national legislation to modify its existing international commitments. If other countries followed the same approach, there would be no continuity of international cooperation, since any state could resort to the adoption of new domestic laws that would absolve it of all its existing international commitments. This tendency would be especially disturbing if recipient states used similar means to get out of their obligations.

NON-PROLIFERATION TREATY

The Non-Proliferation Treaty is generally recognized to be discriminatory. Although the formal text of the NPT remains unchanged, there have been many de facto modifications of it. The nuclear powers have tried to impose their own interpretations on certain articles and have deferred implementation of others. The nonimplementation of article IV of the NPT has caused disenchantment among signatories. During the 1975 NPT Review Conference, a number of countries voiced their concern about the nonfulfillment of the obligations undertaken by nuclear-weapon states in articles III, IV, and VI of the NPT. More may be heard on this issue at the 1980 meeting. It is no exaggeration to say that the NPT in its original form has undergone important de facto modifications. The obligations of non-nuclear-weapon states have been unilaterally increased while the undertakings by the nuclear-weapon states have been diluted. This has undoubtedly seriously undermined confidence and faith in the NPT among the non-nuclear-weapon states, and particularly among the LDCs.

PERCEPTIONS IN THE THIRD WORLD

In light of the foregoing discussion, let us see how the Third World views the interrelated problems of the energy crisis, nuclear power, transfer of technology, and proliferation.

Energy

The Third World represents a wide spectrum of countries at different stages of development and with varying resource bases. They could be grouped into two broad categories, OPEC and non-OPEC LDCs. The latter could be subdivided into relatively advanced LDCs that are nearing the take-off stage and the most seriously affected (MSA) LDCs. Their respective reactions to the energy crisis and attitudes toward the use of nuclear energy could be summarized as follows.

The least developed of the poorer countries, particularly those in Africa and Asia that have just started their economic development programs, generally have small electric power grids and will have limited power requirements in the immediate future. They do have a potential interest in nuclear energy, but it is over the long term. They wish to start with training technical manpower and setting up basic nuclear facilities for research, as first steps toward utilization of nuclear power. However, some of them are uranium and nuclear raw material producers and therefore have a role to play as increasingly important suppliers of materials. As such, they do not wish to be ignored in the setting up of norms for the exploitation of their materials.

The remaining non-OPEC LDCs, particularly the MSAs, are seriously concerned about the rising cost of energy and have genuine apprehensions about their ability to compete with affluent states in obtaining oil supplies in a progressively tightening and uncertain market. They are therefore most anxious to look for alternative energy-related technologies. They share the assessment of many industrialized countries, particularly in Western Europe, that the only practical and viable alternative available at present for meeting a good part of their electricity needs is nuclear power. This group of countries will therefore continue to show increasing interest in the acquisition of nuclear technology and nuclear power plants, largely for near- and long-term economic considerations.

The OPEC countries represent a special group of LDCs that do not appear to have immediate energy problems and are preoccupied with building their economic and industrial infrastructures and finding secure investments for their surplus earnings. The more developed among them (such as Iraq) are turning toward nuclear energy partly because they desire to conserve their oil and acquire alternative technologies while they can. However, these OPEC countries are beginning to feel concerned about certain issues, including the following:

- The buying power or real worth of their oil earnings is diminishing rapidly, and their savings are being eroded by inflation, currency fluctuations, and economic uncertainties.
- They are being pressed to produce oil at a rate commensurate with the requirements of the industrialized countries and not in accordance with their own economic needs and abilities to assimilate oil revenues.

- Their nonrenewable oil resources are rapidly being depleted. In two decades or so they are not likely to have alternative technologies, while their energy needs will have increased considerably and there is not likely to be much oil left for export and domestic use.
- They are not getting worthwhile basic technologies and know-how, including nuclear power technology, in return for oil.
- The papers presented and the discussion held at the meeting on energy held in March 1979 under the auspices of the Organization of Arab Petroleum Exporting Countries voiced similar views, and in fact they expressed grave concern about the rapidly disappearing oil resources of these countries. It is therefore not unlikely that they and other OPEC countries may press more strongly for rationalization of energy policies in the advanced countries and reduction in the increasing consumption of oil in the world. They could also ask for a linkage between assured oil supplies and transfer of technology.

As for the safety and environmental impact of nuclear power and its acceptance by the public, the LDCs do not demonstrate such deep anxiety as is evident in the advanced countries. This may be partly due to lack of adequate comprehension of the potential hazards and partly due to the limited size of their programs. So far nuclear power has met with much less resistance in the LDCs than in the developed countries. However, with the latest incident at the Three Mile Island plant in Harrisburg, the educated elite and policymakers in these countries have become much more aware of what can go wrong, and this is bound to influence their thinking.

The developing countries are watching with keen interest the results of research and development work on nonconventional energy resources—particularly solar, wind, and biomass. Some of them have initiated indigenous national programs for utilizing these resources in the rural and remote areas where they can find immediate applications. They also realize that these new technologies are not yet sufficiently well developed to provide solutions to their near-term energy problems, particularly for meeting large-scale electric power needs for industrial purposes. They are allergic to any attempted overselling of these "soft" technologies to them and regard such moves as yet another way of persuading them to give up their interest in nuclear power programs.

The LDCs face major economic and political constraints on the development of their programs. The cost of nuclear plants has risen by a factor of five during the past ten years. They do not have the capital resources to cover such high initial investments, and international financing has also become very hard to obtain. Conventional plants have also risen in cost, but nuclear units are still at least 50 percent more capital-intensive. As far as overall economics are concerned, however, a typical 600 MW plant suitable for LDCs is still more economical than an oil-fired plant that burns over $100 million worth of fuel oil per year.

The LDCs face serious political constraints, too, because of the presumption of certain supplier countries that a nuclear power plant, even under IAEA safeguards, could contribute toward proliferation. This impression persists despite the fact that no country that has exploded a nuclear device has done it through the use of a nuclear power reactor or by violating IAEA safeguards. Obviously, a country wishing to proliferate would not adopt the cumbersome civilian nuclear power route because of its economic and political cost.

Proliferation

Most of the Third World countries regard the existence of nuclear weapons as a very serious problem that has been created largely by a few advanced countries and about which developing countries are powerless. They feel that, unless the superpowers give a definite lead in this, not much can be done to eliminate or reduce the inherent dangers. They note with increasing dismay the continuing testing and development of nuclear weapons and further vertical proliferation. Lack of progress in limiting nuclear stockpiles and their destructiveness, as envisaged in the NPT, casts doubts on the intentions and seriousness of purpose of the nuclear powers. This affects their credibility and diminishes their influence in preventing horizontal proliferation. Their stand on non-proliferation is further compromised by lack of action on their part to enforce nuclear discipline on their client states, such as Israel and South Africa, and to bring them under the non-proliferation regime. As for further horizontal proliferation, the Third World countries do feel a genuine concern and wish to see this danger eliminated.

To put things in proper perspective, the developing countries do not possess the trained manpower and the economic and technological wherewithal to engage in any significant nuclear weapon development program. Such a program would preempt resources sorely needed for their development. Since they cannot attain a credible nuclear deterrent, possessing a few nuclear weapons will not add to their security and may in fact endanger it. Horizontal proliferation can cause serious security problems, particularly in a regional context, lead to nuclear competition between rival states in sensitive regions, and cause global repercussions. This has led to several proposals for the denuclearization of different regions and the establishment of nuclear-free zones in various parts of the world. The success of that approach requires the full backing of the non-nuclear-weapon states in the regions concerned and the active interest and support of the nuclear powers. So far, the latter have been reluctant to give such support.

The incentive toward horizontal proliferation could also be lessened if the non-nuclear-weapon states that forswear the acquisition of nuclear weapons, through signing the NPT or subscribing to equivalent regimes, could obtain security guarantees from the nuclear-weapon states against nuclear attack or blackmail. The nuclear-weapon states are not willing to give such guarantees except as a part of military alliances or pacts, which limits the nuclear umbrella to the members of opposing power blocs.

The LDCs, along with a number of advanced countries, believe that the proliferation of nuclear weapons is essentially a political problem that should be tackled primarily on the political plane rather than through the self-defeating policy of trying to apply "technical fixes." That is why they regard the London suppliers club with considerable suspicion. They do not accept that the right to make use of sensitive technologies should be the privilege of a few advanced countries considered to be politically safe by the suppliers group. They want to be free to exercise their right to develop nuclear fuel cycle programs, based upon their own needs and requirements, under IAEA safeguards. The overemphasis on restricting sensitive technologies has actually been counterproductive because it has generated a feeling of insecurity among the have-nots who do not wish to remain forever dependent on others for fuel cycle services. This feeling could, as mentioned in various international meetings, lead to the indigenous development of the fuel cycle in a number of recipient states, which would be much harder to safeguard.

Respect for Contracts and Agreements

The efforts made to force a number of recipient countries to renegotiate existing nuclear cooperation agreements have seriously undermined the sanctity of international contracts and agreements in the eyes of the Third World countries. They fear that, if they now subscribe to any international agreements, these too would be subjected to unilateral changes by supplier states through various mechanisms including domestic legislation. That is why they cannot rely upon such arrangements and are skeptical about the proposed international fuel bank that is purported to ensure a continued supply of fuel, materials, and services for their nuclear power programs.

The Third World realizes the urgency of establishing the rule of law for the development of nuclear energy, which should apply equally to advanced and developing countries, and to both suppliers and recipients. Instead of dividing the world into small interest groups, we should think together for our common survival. We must share both the potential benefits and the hazards of nuclear energy.

Technology Transfer

Third World signatories and nonsignatories to the NPT are equally concerned about the consequences of denial of nuclear technology. Third World signatories to the NPT have not received the fullest possible exchange of scientific and technological information related to the peaceful uses of nuclear energy that they had been guaranteed under article IV of the treaty.

Their access has been restricted to the so-called nonsensitive nuclear technologies, which has introduced yet another discrimination between the NPT parties. The refusal of nuclear powers to faithfully fulfill their promises under article IV does not encourage the nonsignatories of the NPT to join its fold. The nonsignatories have, no doubt, lost some of the benefits that have accrued to the signatories, but they have been compensated by being free to develop their own indigenous nuclear technology. All in all, there is a feeling among some of the LDCs that, while the signatories to the NPT have gained little, the nonsignatories have not lost much.

The policies of the London suppliers club and the restrictive measures adopted by certain advanced countries regarding transfer of technology have increased the existing feelings of mistrust between the advanced and the Third World countries. That is because this issue has somehow been linked with the larger question of technology transfer. The developing countries are not fully convinced that the denial of sensitive nuclear technologies and other restrictions are based entirely upon altruistic considerations of strengthening the non-proliferation regime. Instead, they view this development in the broader context of outstanding issues between the North and South.

Cooperation between Third World and Advanced Countries

A quid pro quo relationship can and should be established between the LDCs and the advanced countries on the basis of the exchange of technology for raw materials. The advanced countries should be at least as generous in sharing their renewable technology with the needy developing countries as the oil-producing LDCs have been in sharing their diminishing oil resources with the industrial world.

There should be a realization that the advanced and the developing countries are really interdependent in energy, raw materials, and technology, and that no single region can find a satisfactory solution for its long-term energy problems except in the context of a rational global energy policy. Instead of competing for limited oil, and thereby raising its price and accelerating its depletion, one should work toward a partnership between the OPEC and industrialized countries. The OPEC countries should provide the needed oil to enable the advanced countries to maintain their industrial growth and have the time and resources to develop alternative energy technologies that, when ready, should be freely shared by the two sides. This will give the industrialized countries the immediate energy security they desperately need and assure the OPEC countries of the availability of alternative technologies to meet their future needs when the oil runs out.

Nuclear technology should be shared between the advanced and the developing countries under appropriate safeguards and without discrimination.

Proliferation problems, which are extremely serious, should be handled through political means, not by a deliberate policy of denial of nuclear technology, which increases tension and distrust. Both supplier and recipient states should honor their undertakings – supply agreements and safeguards commitments – so that a climate of confidence is developed that reinforces the rule of law and develops a sense of security and interdependence.

REFERENCES

Deese, David A. 1981. The oil-importing developing countries. In *Energy and security,* ed. David A. Deese and Joseph S. Nye. Cambridge, Mass.: Ballinger.

Fallen-Bailey, D. G., and Byer, T. A. 1979. *Energy options and policy issues in developing countries.* World Bank Staff Working Paper no. 350. Washington, D.C.: World Bank.

Hughart, D. 1979. *Prospects for traditional and non-conventional energy sources in developing countries.* World Bank Staff Working Paper no. 356. Washington, D.C.: World Bank.

Independent Commission on International Development Issues. 1980. *North-South: A programme for survival.* Cambridge, Mass.: MIT Press.

Stockholm International Peace Research Institute. 1978. *World armaments and disarmament: SIPRI Yearbook 1978.* Stockholm: Stockholm International Peace Research Institute.

World Bank. 1978. *World Bank atlas 1977.* Washington, D.C.: World Bank.

World Energy Conference. 1978. *World energy resources 1985–2020: Executive summaries of reports on resources, conservation and demand to the conservation commission of the World Energy Conference.* Guildford/New York: IPC Science and Technology Press.

4

International Safeguards 1979

David Fischer

THE NATURE OF THE PROBLEM

The purpose of this paper is to examine the extent to which the risk of further "horizontal" proliferation of nuclear weapons is being contained by international agreements and by the application of International Atomic Energy Agency (IAEA) safeguards under these agreements. This will require an examination of the geographical scope of such safeguards, the gaps in their coverage, and their political and technical effectiveness. Before turning to these matters, however, we must look at the nuclear proliferation problem itself.

Political Aspects

The late Homi Bhabha, when chairman of the Indian Atomic Energy Commission, coined the terms "horizontal proliferation," to describe the spread of nuclear weapons to additional countries, and "vertical proliferation," to indicate the stockpiling of additional nuclear weapons by the existing nuclear-weapon states. Today the chief purpose of international safeguards is to help the international community deal with part of the problem of horizontal proliferation: in the words of article III of the Treaty on the Non-Proliferation of Nuclear Weapons (NPT), to verify that there is no "diversion of nuclear energy from peaceful uses to nuclear weapons or other nuclear explosive devices" and, in the words of the IAEA's model NPT safeguards agreement, to deter such diversion "by the risk of early detection." It is obvious that international safeguards can be an effective deterrent only in countries where all nuclear material is placed under safeguards, and that safeguards do not address what may now have become the most likely mode

Date of origin, September 1979.

of horizontal proliferation, the construction of unsafeguarded facilities capable of producing nuclear weapons material in countries that have not yet accepted the NPT or similar obligations.

The risks of horizontal proliferation arising from peaceful nuclear activities will be addressed in this paper, but we should not overlook the risks posed by existing nuclear armaments and by vertical proliferation. The vast stocks of increasingly ingenious and devastating nuclear weapons in the nuclear-weapon states represent, at least potentially, a far greater risk to the survival of man than do the imponderable consequences of the spread of nuclear explosive capacity to additional countries.* The vertical proliferation of these stocks and the testing that accompanies it may constitute only a minor addition to each scale in the "balance of terror," but some countries consider this form of proliferation as deplorable as the horizontal variety. Nevertheless, horizontal proliferation could have very serious consequences.

By further horizontal proliferation I mean the acquisition of a nuclear explosive device by countries other than the six that have already demonstrated that they have nuclear explosive capability. This approach is essentially in accordance with the concept of "nuclear-weapon states" as defined in article IX.3 of the NPT.† The spread of the capacity to make nuclear explosives, such as by the acquisition of reprocessing or enrichment facilities or by the acquisition of significant quantities of separated plutonium or highly enriched uranium, cannot be considered proliferation in the sense of the NPT. Such an interpretation would in any case render the discussion of horizontal proliferation largely devoid of meaning, because of the difficulty of ascertaining precisely which states have acquired such capacity and when they did so. In what category should one, for instance, put a country that has a "hot cell" or is carrying out experimental work with centrifuges or laser fusion?

Nevertheless, the possession of a significant *unsafeguarded* reprocessing or enrichment plant, or of significant quantities of *unsafeguarded* weapons-grade fissile material, must raise questions about the plans of the country concerned. In other words, acceptance of safeguards is an earnest expression of the intention not to acquire nuclear weapons.

Consequences of Further Proliferation

The impact of further horizontal proliferation would depend very much upon the country involved, the way it acquired its nuclear explosive capacity, and whether and how it had demonstrated that capacity. Acquisition of

* Estimates by the Stockholm International Peace Research Institute (SIPRI) indicate that well over ten thousand nuclear warheads stand ready to be fired by the nuclear-weapon states.

† It should be noted, however, that only five of the six countries concerned qualify strictly under article IX.3 of the NPT, which limits nuclear-weapon states to those that had manufactured and exploded a nuclear weapon or other nuclear explosive device before 1 January 1967, the definition that will be adopted in this text.

nuclear explosives by a major industrial country in a sensitive geostrategical or geopolitical situation might lead to a political crisis of the first order, threatening the fabric of regional arrangements, security agreements, and alliances that has been built up since the Second World War. In these circumstances it seems that whatever the country concerned might gain by proliferation would be totally out of proportion to what it would certainly lose. In the case of small or developing countries, proliferation might lead to a chain reaction of further proliferation within the region concerned, a development that has already been seen in Asia and may go further. Such proliferation might also make it difficult for some of the industrial non-nuclear-weapon states to maintain their policies of nuclear self-restraint.

It also seems likely that, in certain circumstances, further proliferation would have severe consequences for the international nuclear industry. One of the main arguments of opponents of the civil nuclear industry is that it inevitably leads to proliferation (cf. H. Alfvén's statement that "nuclear power and nuclear weapons are Siamese twins"). Again, the impact would depend upon which country acquired nuclear explosives and how. The perceived impact would also vary from country to country; thus it might have a severe impact upon the export policies of countries like Australia but little effect on the nuclear industries of Socialist countries.

Varying National Perceptions of the Consequences of Proliferation

It is obvious that perceptions of the likelihood and consequences of proliferation vary widely. The main preoccupations of the Soviet Union on this score are presumably being met by the ratification of the NPT by West European countries and Japan, and by the introduction in these countries of IAEA safeguards.* The concern of the OECD countries appears to be more global and less specific. From recent United States, Canadian, and Australian attempts to limit the spread of reprocessing and enrichment plants from those industrial countries that already have them to the developing countries that do not, one might infer that North American and Australian concern is now focusing chiefly on the Third World. This would be a misreading of their policies. Australia and Canada, for instance, insist on IAEA safeguards even in nuclear-weapon states, while the United States is making a particular effort to secure South African ratification of the NPT and played a leading role at an earlier stage in promoting such ratification by West Germany and Japan.

In Western Europe, Ireland took the lead at the United Nations by introducing the resolution that eventually led to the NPT (General Assembly

* It was a Soviet space satellite, however, that detected a reported nuclear explosive testing facility in South Africa in 1977, and the Soviet Union has taken a strong position in the Nuclear Suppliers Group in support of the principle that nuclear exports should be made only to countries that have accepted full-scope safeguards.

Resolution 1665 (XVI) of 5 December 1961). Ireland's concern was shared by Britain, the Scandinavian countries, and the Netherlands. Recently, West Germany and France have adopted increasingly strict non-proliferation policies. Finally, the common interest of leading West European countries (and Japan) in reconciling the production and use of separated plutonium with the requirements of non-proliferation and their differences of opinion with recent United States policy in this regard have helped further to unify their approach to the risks of proliferation.

In some developing countries, there are publicly expressed apprehensions about nuclear development in Israel and South Africa, but for many others the problem of proliferation must seem to be remote while that of energy supply is immediate and overwhelming.

Despite the wide variety of views, it can be said that concern about the consequences of proliferation has risen remarkably during the past two decades, particularly since the Indian explosion of 1974. In the 1950s, when the danger of nuclear war between the superpowers sometimes seemed close, the problem of further proliferation was hardly perceived. The acquisition of nuclear weapons by Britain, France, and China may have been a subject of regret or apprehension in Washington or Moscow, but the rest of the world tended to regard this proliferation as inevitable. The climate of opinion is very different today. The mere prospect of proliferation is enough to stimulate a sharp response, as recent developments in regard to South Africa and Pakistan have shown.

Technological Relationship of Proliferation to Nuclear Power

A nuclear power (or research) program provides access to materials, plant, and knowledge that may be of use in a weapons program. However, with the passage of time, nuclear weapons programs have diverged from nuclear power, and today there is no necessary link between them. It is very difficult to gain the public's understanding of this, although a few illustrations will show the error of statements such as that of Alfvén quoted earlier.

The most compelling arguments for looking at the two questions separately are those of relative cost and complexity. To make weapons-quality material, all that is needed is natural uranium fuel, a research reactor, a moderator (which might be industrial-grade graphite or heavy water), and a simple reprocessing plant. United States studies have estimated the total cost of the plant required at about $50 million. The infrastructure needed is that of a large nuclear research project.

In contrast, a nuclear power plant of today's standard size involves a capital investment of the order of $1–3 billion, twenty to sixty times as much as the minimum weapons material complex. It also requires the creation of a complex administrative and regulatory infrastructure and the training of professional and specialist cadres. Furthermore, for most countries other than the main industrial powers, the plant and its fuel would have to be imported and placed under international safeguards. The investment of funds and technological resources becomes even vaster if a commercially viable reprocessing or enrichment plant is included in the nuclear power program. Indeed, only one reprocessing plant for light water reactor fuel is now in operation anywhere in the world on a commercial scale, at Cap de la Hague in France.

It is, of course, obvious that a country with experience in the technology of producing and handling nuclear materials would have an advantage in time if it were to decide to acquire nuclear weapons and had the technology to do so. But a very wide range is covered by the idea of experience in nuclear technology: it stretches from the operation of small imported reactors or nuclear power plants, on the one hand, to the indigenous construction and operation of advanced fuel cycle facilities on the other. Neither necessarily includes nuclear weapons technology. Nevertheless, consideration of the potential military value of a particular technological choice may affect national decision making. As Bertrand Goldschmidt has pointed out, the French decision in the early days to adopt the gas-graphite fuel cycle rather than alternatives such as the light water enriched uranium cycle was intended to enable France to take the weapons route whenever it chose to do so (Goldschmidt 1967, p. 204).

It follows that the selection a country makes within the wide range of choices of nuclear technology that are open to it may be an indication of whether its intentions are purely commercial or whether it also has in mind a military option. At one end of the spectrum, the choice of a light water power reactor without other fuel cycle facilities, which is dependent on imported enriched uranium, could be made only on commercial grounds; at the other end, choice of a graphite or heavy water moderated natural uranium reactor, or investment in enrichment research and development that is not commensurate with national requirements, could reflect military interests.

These qualifications do not in any way modify the conclusion that a country wishing to embark on the manufacture of plutonium explosives would have every reason to take the far cheaper, quicker, simpler, and less conspicuous route described earlier rather than invest the comparatively vast expense and effort involved in a commercial nuclear power program. If enriched uranium rather than plutonium were the chosen nuclear explosive, the direct route of building a small pilot enrichment plant that would directly upgrade natural uranium into a nuclear explosive would also be far simpler and less costly than launching a commercial power program with associated enrichment capacity. In short, the sad truth today is that it is cheaper, easier, and less conspicuous to make nuclear-weapons-grade material than to produce nuclear power.

The ability to make weapons-grade material today is probably within the reach of some twenty to twenty-five countries. This list does not coincide with the list of the countries that have nuclear power programs. We have already referred to Israel and South Africa as countries that have been reported to have nuclear explosive capabilities. Yet they have no operating nuclear power plants. On the other hand, several countries like Canada, Sweden, and Switzerland have large nuclear power programs but have no present intention of engaging in reprocessing or enrichment and thus of acquiring the ability to make weapons-grade material. It should be obvious that no action by countries such as Canada, Sweden, or Switzerland to curtail or limit their own nuclear power programs would have any effect on the policies of other countries with regard to the manufacture of nuclear explosives.

It is equally clear that the time has passed when any restraints of secrecy or export policy could serve as more than a short-lived brake on the spread of nuclear technology. In the long term, in fact, embargoes are likely to increase the desire of the embargoed country to obtain the forbidden fruit. The technology of enrichment offers an example. This was strictly guarded as a military secret by each nuclear-weapon state that acquired it, though each in turn had learned it with little or no help from its forerunner in the field. Moreover, the policy of embargo has induced non-nuclear-weapon states to develop their own enrichment technologies, so that today at least ten countries have mastered the secret of enriching significant quantities of uranium.* The technology of reprocessing is, of course, more widely known. It was declassified in 1955 and was vigorously propagated among OECD countries during the thirteen years of operation of their joint reprocessing enterprise, Eurochemic, in Belgium.

In the long term, the only form of restraint that has any prospect of success is one that is voluntarily accepted (though if such restraint is embodied in a formal agreement it becomes relatively impervious to any subsequent change of mind). For restraint to be accepted voluntarily, say by adhesion to the NPT, the countries concerned must clearly perceive that their own national interests are served by it, because, for instance, they believe that their own security is thereby increased or, more mundanely, because they may receive compensatory benefits from other countries.

Prospects of Additional Proliferation

We must therefore conclude that the number of countries that are technologically able to make nuclear explosives today has already risen from two or three in the late 1940s to more than twenty in the late 1970s, and that the list of such countries will steadily grow longer as the century progresses. Efforts to erect barriers against the spread of nuclear technology, besides

* Papers presented at the IAEA Conference on Nuclear Power and its Fuel Cycle at Salzburg in May 1977 indicated that, besides the five nuclear-weapon states, enrichment technology had been mastered by West Germany, Italy, Japan, the Netherlands, South Africa, and probably Australia. Press reports suggest that Pakistan may be joining this group.

being inconsistent with article IV of the NPT and with the general premises of North-South cooperation and the industrial development of the Third World, will not be effective in the long term and are of doubtful value even today.

Just as the reasons that cause countries to embark on nuclear weapons programs are political, so today the only effective barriers against proliferation are political.* In the first place, there must be a political calculation by the government concerned that the acquisition of nuclear explosive capacity is or is not contrary to the national self-interest. The reasons for forgoing weapons development may vary widely. In some cases the perceived incentive to acquire nuclear weapons is very small, in some the potential penalty may be very severe. National security may be seen to be better served under the umbrella of an alliance, or by a policy of strict neutrality.

In many if not most cases, the problem may not be that of dissuading a country that is intent on acquiring nuclear explosives from following that course. It may rather be that of helping a government reach the conclusion that, by keeping open the option to acquire nuclear weapons in case of need, it is no longer serving the national interest, since by this very act it is arousing suspicions about its own eventual nuclear intentions. Such suspicions may trigger a competitive acquisition of nuclear explosive capacity, with its attendant costs, and thus damage regional security in a manner quite contrary to the initial wish of the countries concerned. In other words, out of fear that a potential competitor may acquire nuclear explosives, a country may refrain from accepting binding non-proliferation restraints or may take preemptive actions that have precisely the undesired effect by creating a situation in which both countries engage in competitive nuclear programs.

The vast majority of non-nuclear-weapon states have committed themselves to not acquiring nuclear explosive capacity. Today, therefore, any sudden reversal of the declining rate of proliferation — any sudden spurt — seems unlikely. Additional proliferation, if it does occur, seems more likely to take the form of isolated demonstrations of nuclear explosive capability (or more probably secret stockpiling of nuclear weapons material) in regions of political insecurity. Only a major and probably worldwide political crisis, involving the nuclear-weapon states themselves and leading to a widespread denunciation of the NPT, seems likely to precipitate rapid proliferation. In that case, no restraints that one can conceive of, technical or institutional, would save the day.

However, the present situation gives no cause for complacency. As the next section of this paper shows, there are twelve non-nuclear-weapon states that have not accepted full-scope safeguards of one form or another and that are currently operating nuclear plants. In seven of these states, all known

* It is no accident that the first three cases of proliferation were the direct outcome of the Second World War and the Cold War, or that the fourth was the outcome of the 1956 Israel-Arab War. It is also no accident that the most obvious candidates for further proliferation are in regions of political tension.

plants are under safeguards; in five, they are not. Moreover, there have been press reports that the number of states operating significant unsafeguarded nuclear plants capable of making nuclear weapons material may increase in the near future. If this should happen, it would reverse a trend, which began in 1970 with the entry into force of the NPT, toward placing all civilian nuclear activities throughout the world under IAEA safeguards.

The NPT Consensus and Reciprocal Obligations

In some cases the renunciation of nuclear weapons has been tacit and perhaps provisional, but most countries have explicitly demonstrated their renunciation by becoming parties to the NPT and by joining formally in the consensus the NPT represents, or at least represented in 1970.

Briefly, this consensus was that, in return for renouncing nuclear weapons and nuclear explosives and for accepting the application of nuclear safeguards across its entire national nuclear industry, the country concerned would benefit from the commitments in articles III.3 and IV of the NPT that its peaceful nuclear development would not be impeded and that it would have the right to take part in "the fullest possible exchange of equipment, materials and scientific and technological information for the peaceful uses of nuclear energy." Moreover, other parties to the treaty in a position to do so accepted a commitment to help their less advanced companions develop their peaceful nuclear energy programs. The NPT also placed an obligation on the three nuclear-weapon state parties to strive in good faith to end the nuclear arms race, and the preamble to the NPT recalls the determination of the parties to the Moscow Partial Test Ban Treaty of 1963 (which include the three NPT nuclear-weapon states) to seek to achieve "the discontinuance of all test explosions of nuclear weapons for all time."

It may be argued that, while the non-proliferation and safeguards provisions of the NPT are very clear-cut and explicit, as befits the explicit non-proliferation purpose of the treaty, the obligations to cooperate in peaceful nuclear development and to pursue nuclear disarmament are considerably more vague and general. Nevertheless, the NPT was perceived as a mutual exchange of obligations: as a bargain that entailed commitments by nuclear-weapon states and advanced industrial countries as well as by the developing world. In other words, it was interpreted as meaning that the renunciation of nuclear explosives and acceptance of full-scope safeguards should be balanced by an equal commitment to maintain nuclear supplies (in the form of technology as well as hardware) and thus to enable the "renouncing" country to develop its peaceful nuclear energy program without let or hindrance of a political character. The consensus further implied not only that safeguards were to be "full-scope" but also that they were to be generally accepted and were politically and technically effective. These questions will be examined in subsequent sections.

Proliferation and the Developing Countries

An unfortunate trend in recent years has been that of seeing proliferation as a temptation to which developing countries are especially likely to succumb. This was certainly not the assumption on which the NPT was founded. The first object of the NPT, in the eyes of many countries at least, was to ratify and make permanent the non-nuclear status of a number of major industrial countries. Opposition to the treaty was also most outspoken and persistent in those countries, where certain groups made much of the supposed risks of industrial espionage that would result from the application of safeguards, the loss of technological spin-off that would be entailed by eschewing the manufacture of nuclear explosives, and the consequences to national security of renouncing nuclear weapons. A major political effort was needed until 1975 to secure ratification of the treaty by the countries concerned. On the other hand, it was a group of developing countries that took the first international initiative, even before the NPT, to prevent further proliferation: namely, the countries of Latin America that concluded the Tlatelolco Treaty in 1967.

Historical developments have contributed to the illusion that proliferation is a developing-country problem. By now, almost all industrial countries either are themselves nuclear-weapon states or have ratified the NPT. Although some seventy-two developing countries have also done so, ten of the twelve significant non-nuclear-weapon states that have not yet ratified the NPT (or accepted full-scope safeguards under the Tlatelolco Treaty) are "developing," and these include some of the most populous and politically influential developing countries. One may speculate at length about the reasons the twelve countries concerned have so far chosen not to follow the path already taken by most other nations of the world. As implied elsewhere in this paper, it is probable that the paramount consideration is often political insecurity; unlike the majority of countries of Europe, North America, and the Far East, which enjoy, by treaty or otherwise, the protection of a "nuclear umbrella," many of the twelve live, voluntarily or not, outside the protective wall of any military alliance.

THE ROLE OF SAFEGUARDS IN A NON-PROLIFERATION REGIME

Those of us whose daily work is with safeguards and the negotiation of safeguards agreements are sometimes puzzled by the scene we see today. To use an analogy, it is like a house that has taken a long time to build and is not far from completion. Some important parts of the roof, however, are still missing, and, for no obvious reason, work has almost come to a standstill. Yet, apparently oblivious of the pressing need to finish the job in hand, several people are scurrying around with blueprints for all sorts of elaborate additional edifices, most of which will be built on top of the house or at least lean on it for their main support. The incomplete house is, of course, the international non-proliferation regime.

Safeguards are only one element in this non-proliferation regime. Even more important, in my view, are the formal commitments that governments make, when acceding to the NPT or similar agreements, to forgo the acquisition of nuclear explosives. Safeguards verify this potentially far-reaching political commitment.

Various supplementary non-proliferation measures have been developed and implemented. There is one case in which international safeguards are being supplemented by regional safeguards. It is also becoming a common practice of some governments to insist on provisions for "fall-back" bilateral safeguards in case international safeguards should cease to be regarded as adequate by the supplying party. Additional measures include the Nuclear Suppliers Group guidelines, the national controls that certain exporting states require in addition to those guidelines, and, presumably, those means of surveillance that go under the euphemism "national technical means of verification" in the language of the strategic arms limitation (SALT) agreements of 1972.

However, international safeguards are the only non-proliferation measures that have received the support of a broad international spectrum of supplier and recipient states. Moreover, most of the new schemes that are now being discussed depend, in the last resort, on the full application of efficient international safeguards. Projects such as an international fuel authority, a multinational fuel cycle center or an international plutonium management scheme, as well as the suppliers' guidelines and the United States Nuclear Non-Proliferation Act, would have little value from a non-proliferation point of view if effective international safeguards could not be applied to all the nuclear material involved. Even the so-called technical fixes such as the proposed Civex reprocessing technique or the once-through fuel cycle presuppose effective full-scope safeguards in all the countries concerned.

This does not, of course, imply that IAEA safeguards have achieved mature perfection. Far from it. The concept that an international body should have a continuous, accurate, and timely overview of the location and flow of all nuclear material in a complex industry is barely ten years old, and the basic ideas as to how this should be done (and, indeed, whether it can be done in certain cases) are under constant review, while the techniques for doing it are rapidly evolving. Moreover, applying safeguards is an exercise in politics. In the development of the safeguards systems, as well as in IAEA negotiations with member states, the interests of industrial efficiency, safe operation, and commercial competitiveness, as well as those of effective safeguards, must be taken into account. The very idea that a technically and politically sensitive industry, playing an increasingly vital role as a source of energy, should open its doors to international inspection is hardly two decades old. It is surprising that so much has been achieved so quickly.

Geographical Coverage of Safeguards

The positions of the nations of the world regarding the NPT and IAEA safeguards are outlined in tables 4.1 through 4.3. At the time of writing there are 110 parties to the NPT.* Three of the five nuclear-weapon states are parties. A fourth, France, has stated that, though it has not become a party, it will pursue policies consistent with the treaty, and it has done so. The fifth nuclear-weapon state, China, has expressed opposition to the treaty but does not appear to have taken any actions inconsistent with it.

The 107 non-nuclear-weapon states that are parties to the NPT include 78 developing and 29 developed countries.† The developed countries include all the major industrial nations except Spain. It is understood, however, that Spain will have to become a party to the treaty, or at least accept NPT-type safeguards, if and when it joins the European Community; otherwise the common market in nuclear materials that is one of the foundations of the Community could not fully embrace Spain, and this would be incompatible with the Treaty of Rome.

The geographical coverage of the NPT and of international safeguards is also of interest. All non-nuclear-weapon states in Europe, both East and West, except Albania and Spain, are now parties to the treaty. IAEA safeguards are in fact being applied in every nuclear plant in the non-nuclear-weapon states of Europe with one exception, the Vandellos power plant in Spain (Albania having no nuclear facilities). In the non-nuclear-weapon states of the European Community, the IAEA's safeguards are applied jointly with those of Euratom, which has had its own safeguards system since the early 1960s.

Two further countries are in the process of accession to the treaty, Bangladesh and Turkey. Turkey's accession would mean that all the northern tier of Near Eastern countries—Turkey, Syria, Jordan, Iran, Iraq, Afghanistan—were parties to the treaty. The remainder of the Near East presents a very mixed picture. Countries not party to the treaty include all the nations of the Arabian peninsula (except the People's Democratic Republic of Yemen), as well as Israel and Egypt. Only the last two have any nuclear plants at present.

* By the end of 1981 five additional states (Bangladesh, Barbados, Egypt, Saint Lucia, and Turkey) had become parties to the NPT. Meanwhile, however, the government of Vietnam had indicated that it did not regard itself as bound by the decision of the former administration in South Vietnam and should not be considered a party to the treaty. The number of parties to the NPT had thus risen to 114, including three nuclear-weapon states (ed.).

† By "developing countries" I mean those countries in Asia, Africa, the Middle East, and Latin America that are eligible for assistance from the United Nations Development Program.

TABLE 4.1. Non-Nuclear-Weapon States Party to the NPT That by 31 August 1979 Had Concluded the NPT Safeguards Agreement with the IAEA Required by Article III of the Treaty

Afghanistan	Lebanon
Australia	Lesotho
Austria	Luxembourg
Belgium	Madagascar
Bulgaria	Malaysia
Canada	Maldives
Cyprus	Mauritius
Czechoslovakia	Mexico
Denmark	Mongolia
Dominican Republic	Morocco
Ecuador	Nepal
El Salvador	Netherlands
Ethiopia	New Zealand
Fiji	Nicaragua
Finland	Norway
Gambia	Paraguay
Germany (East)	Peru
Germany (West)	Philippines
Ghana	Poland
Greece	Portugal
Holy See	Romania
Honduras	Samoa
Hungary	Singapore
Iceland	Sudan
Iran	Surinam
Iraq	Swaziland
Ireland	Sweden
Italy	Switzerland
Jamaica	Thailand
Japan	Uruguay
Jordan	Yugoslavia
Korea (South)	Zaire

Except for Sri Lanka, Southern Asia (Pakistan, India, Bangladesh, and Burma) generally remains outside the treaty, although Bangladesh is taking steps to accede to it. Nuclear plants exist in only two of these countries—India and Pakistan.

With Indonesia's accession, the entire band of countries stretching from Japan through South Korea, Taiwan, Vietnam, Laos, Kampuchea, Thailand, Malaysia, the Philippines, Singapore, Australia, and New Zealand are now parties to the treaty.

The pictures presented by Africa and Latin America are mixed. Non-parties in North Africa include Algeria, Mauritania, and Niger, none of which has any nuclear plant. In Africa south of the Sahara, Uganda, Tanzania, Mozambique, Malawi, Zambia, Zimbabwe, and Angola, as well as South Africa, remain outside the treaty. In all of Africa, only Egypt, Zaire, and South Africa have at present any nuclear plant, but there have been reports that Libya (which is a party to the NPT) will acquire a nuclear power

TABLE 4.2. Non-Nuclear-Weapon States Party to the NPT That by 31 August 1979 Had Not Concluded the NPT Safeguards Agreement with the IAEA Required by Article III of the Treaty

Country	Date of Ratifying or Acceding to NPT		Country	Date of Ratifying or Acceding to NPT	
Nigeria	27 September	1968	Chad	10 March	1971
Cameroon	8 January	1969	Burundi	19 March	1971
Botswana	28 April	1969	Tonga	7 July	1971
Syria	24 September	1969	Vietnam	10 September	1971
Taiwan	27 January	1970	Kampuchea	2 June	1972
Malta	6 February	1970	Benin	31 October	1972
Mali	10 February	1970	Ivory Coast	6 March	1973
Laos	20 February	1970	Bahamas	10 July	1973
Togo	26 February	1970	Gabon	19 February	1974
Tunisia	26 February	1970	Grenada	19 August	1974
Upper Volta	3 March	1970	Sierra Leone	26 February	1975
Costa Rica	3 March	1970	Rwanda	20 May	1975
Liberia	5 March	1970	Libya	26 May	1975
Somalia	5 March	1970	Venezuela[a]	26 September	1975
Bolivia	26 May	1970	Guinea Bissau	20 August	1975
Haiti	2 June	1970	Panama[a]	13 January	1977
Kenya	11 June	1970	Liechtenstein[ab]	20 April	1978
San Marino	10 August	1970	Congo[b]	23 October	1978
Guatemala	22 September	1970	Tuvalu[b]	19 January	1979
Central African Republic	25 October	1970	Sri Lanka[b]	5 March	1979
Senegal	17 December	1970	Yemen (South)[b]	1 June	1979
			Indonesia[b]	12 July	1979

[a] Agreement negotiated and approved by IAEA Board of Governors but not yet in force.
[b] Grace period of eighteen months for concluding the agreement, provided for in article III.4 of NPT, not yet expired.

plant from the Soviet Union. The plants in Egypt and Zaire are small research reactors; South Africa's nuclear plant is discussed below.

Mexico and all countries of Central America and the Caribbean (except Cuba), as well as Venezuela, Ecuador, Peru, Bolivia, Paraguay, and Uruguay, are parties to the treaty. Argentina, Brazil, Chile, and Colombia are not. Each of the latter countries has at least one research reactor. Argentina and Brazil have extensive nuclear power programs that in time will encompass the whole nuclear fuel cycle, and Cuba is acquiring a nuclear power plant from the Soviet Union.

The map may also be looked at in another way. At present, twelve non-nuclear-weapon states that have at least one significant nuclear facility (e.g., a research reactor) are not parties to the treaty. However, in seven of the twelve non-parties, all nuclear plant of which the IAEA is aware is at present under IAEA safeguards because of the application of non-NPT safeguards agreements. The seven are Colombia, Chile, Argentina, and Brazil (so that all significant nuclear plant in Latin America as far as we are aware is at present under safeguards), together with Turkey, Pakistan, and North Korea.

In each of the remaining five of the twelve non-parties, at least one significant nuclear facility is not under IAEA safeguards. In two of the five,

TABLE 4.3. Non-parties to the NPT on 31 August 1979

Nuclear-Weapon States	
China	France
Non-Nuclear-Weapon States without Safeguardable Nuclear Facilities or Materials	
Albania	Mauritania
Algeria	Monaco
Angola	Mozambique
Bahrain	Nauru
Bangladesh[a]	Niger
Barbados	Oman
Bhutan	Papua New Guinea
Burma	Qatar
Cap Verde	São Tomé and Principe
Comoros	Saudi Arabia
Cuba	Seychelles
Djibouti	Solomon Islands
Equatorial Guinea	Tanzania
Guinea	Trinidad and Tobago
Guyana	Uganda
Hong Kong	United Arab Emirates
Kuwait	Zambia
Malawi	
Non-Nuclear-Weapon States Where All Known Nuclear Activities are Covered by Safeguards	
Argentina	Korea (North)
Brazil	Pakistan (?)
Chile	Turkey[a]
Colombia	
Non-Nuclear-Weapon States Having Significant Unsafeguarded Nuclear Facilities	
Egypt[b]	South Africa
India	Spain[c]
Israel	

[a] In process of acceding to NPT.
[b] Unsafeguarded plant: small research reactor provided under Soviet safeguards.
[c] Unsafeguarded plant: power reactor operated jointly with France.

however, this probably has no significance from a proliferation point of view. The two are Egypt, which has a small unsafeguarded research reactor, and Spain, which has the unsafeguarded (Vandellos) power plant mentioned above. This leaves us with three states where there are significant agglomerations of unsafeguarded facilities, namely India, Israel, and South Africa.* Each of the three is capable of producing unsafeguarded nuclear-weapons-grade material: plutonium in the case of India and Israel and enriched uranium in the case of South Africa.

From this description, the earlier reference to an incomplete structure will be clear. While one might say today that the desirability of stopping any further horizontal proliferation of nuclear weapons is now widely recognized (and this was by no means true a decade or two ago), further progress will

* There have been numerous reports that Pakistan is acquiring a centrifuge enrichment plant and may also have some unsafeguarded pilot reprocessing capacity. If these reports are correct, Pakistan should join India, Israel, and South Africa.

have to be made before this principle becomes binding upon all those states that have at least one significant nuclear facility. As I indicated above, such progress will probably depend to a considerable extent on the establishment of a new consensus regarding supply (irrevocable safeguards being balanced by irrevocable supply arrangements) and possibly on some progress in nuclear disarmament or arms control itself, although the counterargument can be made that precipitate withdrawal of "nuclear umbrellas" could lead to horizontal proliferation.

At present it seems highly unlikely that more than one or two of the twelve significant non-parties will formally accede to the NPT. In Latin America, Brazil has stated that it regards the NPT not only as discriminatory but also as sanctioning the possession of nuclear weapons by the nuclear-weapon states; since Brazil is opposed to all nuclear weapons, it cannot, for reasons of principle, accede to the NPT. Principled objections of a similar character have been placed on record by Argentina and India, where the treaty is seen to be discriminatory, and India has also made it clear that its accession would depend upon significant action by the nuclear-weapon states to limit vertical proliferation of nuclear weapons.

Egypt has made accession to the treaty contingent upon similar action by Israel, which gives no indication of any intention to accede. On the contrary, Israel has apparently decided to do without nuclear power (for which the "price" would probably be accession to NPT) for the time being.

The only reactor at present in operation in South Africa is under IAEA safeguards. However, a pilot enrichment plant is not. South Africa has indicated its willingness to place a commercial enrichment plant under safeguards but has been silent about the pilot plant. Two power reactors are under construction near Cape Town and, at least initially, will require fuel from the United States, which is most unlikely to be forthcoming in the absence of South African accession to the NPT.

One further point requires comment. Cuba is acquiring a nuclear power plant from the Soviet Union and is in touch with the IAEA about the conclusion of a safeguards agreement. Apparently, for political reasons, Cuba will remain outside the NPT.

Conclusion of NPT Safeguards Agreements

The NPT requires each non-nuclear-weapon-state party to conclude a full-scope safeguards agreement. Sixty-four of the 107 states concerned have concluded such agreements. Of the remaining forty-three, only two, Libya and Venezuela, have or are reported to be acquiring a nuclear facility. Venezuela has a United States supplied research reactor operating under non-NPT safeguards, and its NPT safeguards agreement has been negotiated but not yet approved. Libya is reported to be purchasing a nuclear power plant from the Soviet Union as well as certain research facilities. Although the safeguards agreement should have been concluded by December 1975, negotiations have not yet been started by Libya.

The IAEA Secretariat is regularly in touch with NPT parties that have not yet concluded their safeguards agreements or have not yet brought them into force.

Geographical Summary

Summing up, the prospect is in sight when all leading industrial countries will have accepted full-scope safeguards, and when they will apply to all those non-nuclear-weapon states that have at least one *significant* nuclear facility in most of the main regions of the world (Western Europe, Eastern Europe, North America, the Far East, and Oceania). In Latin America this situation has been achieved de facto but not de jure. The chief areas where non-proliferation safeguards are not applied are part of the Near East, the South Asian subcontinent, and Southern Africa.

An alternative route to completing the non-proliferation structure in Latin America is offered by the Tlatelolco Treaty. Brazil has ratified the treaty. Argentina is contemplating doing so and has begun discussions with the IAEA about a full-scope safeguards agreement pursuant to the treaty. However, the treaty does not enter into force for the ratifying country unless it waives four requirements that have to be met to bring it into force throughout the region. Brazil and Chile have not waived these requirements.*

If the application of the Tlatelolco Treaty (or NPT) becomes universal in Latin America, a completely nuclear-weapon-free and nuclear-explosive-free status, ratified by international law, would have been achieved by that large region of the world. No such prospect is at present open in any other region.

THE EFFICACY OF SAFEGUARDS

The efficacy of safeguards in providing continuous assurance that no diversion is taking place depends upon both political and technical factors, of which the political are of course far more important.

The International Political Factor

Continuous efficacy obviously depends upon the willingness of the international community, represented in this case by the Board of Governors and

* Three of the four requirements for the full entry into force of the Tlatelolco Treaty are close to being met. The only ratifications still outstanding for this purpose are those of Argentina and Cuba. The nonregional states concerned have ratified the two protocols or are in the process of doing so. The fourth requirement is that all parties shall conclude safeguards agreements with the IAEA.

the General Conference of the IAEA, to continue to make available the financial resources and the qualified manpower needed by the expanding safeguards operation. The Board of Governors has a crucial role, too, in the safeguards operation itself since, under the IAEA Statute, it is the board that must perform the quasi-judicial task of determining, in cases of doubt, whether there has been any breach of a safeguards agreement. It is the board, moreover, that has the responsibility of reporting noncompliance directly to the Security Council and the General Assembly of the United Nations and of deciding which of the sanctions foreseen in the agency's Statute should be imposed in case the state concerned fails to remedy the noncompliance. It is significant that the General Conference is not involved in these procedures, and that the board itself decides on sanctions irrespective of whatever action might be taken by the Security Council or the General Assembly. How effective these sanctions would be is of course a separate question.

The unique role of the IAEA's Board of Governors explains, in part, both the desire of developing member states of the IAEA to increase its size, so that they may be more extensively represented on it, and the concern of many developed member states that a further increase in the board's present size (thirty-four members) could promote the introduction of extraneous political considerations into its discussions and detract from its capacity to exercise impartially the quasi-judicial function of determining compliance or noncompliance. Whatever solution is found to this matter, it is clearly of crucial importance that the solution should continue to provide adequate assurance to safeguarded states that the board will exercise this function impartially, irrespective of the group of states to which the state in question belongs and irrespective of its relations with other states or with the international community.

The National Political Factor

The effective application of safeguards depends on the full cooperation of the safeguarded state concerned in setting up and operating an effective system of accounting and control of nuclear material, in submitting accurate and timely reports, and, in particular, in permitting international inspectors to carry out their functions without undue impediment, indeed in helping them in their tasks.

The Technical Efficacy of Safeguards

The technical efficacy of safeguards can be assessed against the technical objective, as stated in the basic IAEA text on full-scope safeguards agreements: "the timely detection of diversion of significant quantities of nuclear material from peaceful nuclear activities to the manufacture of nuclear weapons or of other nuclear explosive devices or for purposes unknown, and deterrence of such diversion by the risk of early detection" (INFCIRC/153 1972, p. 9). The three points that stand out are that detection must be "timely," that it must be possible to detect the diversion of a "significant" quantity of nuclear material, and that it is not necessary to prove that the nuclear material is being used for making nuclear explosives; it is sufficient to establish that it has disappeared "for purposes unknown."

There has been considerable discussion of what is meant by "timely" and "significant quantities." Significant quantities have normally been regarded as the quantities of plutonium or enriched uranium needed to manufacture a single nuclear explosive. In some circumstances and in some states, the unexplained disappearance of such quantities of material might indeed be significant. In states having elaborate fuel cycles, the fact that a "significant" quantity could not be accounted for at a given moment might not in fact mean that the material had been diverted but might merely result from measurement uncertainties in a large bulk-handling facility. Moreover, for the state concerned, the diversion of a single "significant" quantity might make very little sense, politically speaking.

The difficulty here is that an international organization cannot discriminate between its member states, and so it must generally apply the same yardsticks in each of them. However, this problem should not be exaggerated. While the inspector's findings must be based, as far as possible, on standard objective criteria and, in particular, on so-called quantitative (i.e., quantifiable) criteria, these findings will always be subject to internal evaluation within the IAEA Secretariat and, in the final analysis, to evaluation by the Board of Governors, which must afford the state concerned "every reasonable opportunity to furnish the Board with any necessary reassurance" (INFCIRC/153 1972, p. 7).

With regard to timeliness, it is obviously desirable that safeguards should be able to detect a diversion before it is consummated by the explosion of the diverted material. (In practice, of course, a diverting state might stockpile weapons-grade material and not use it for an explosion; indeed this is the more probably scenario in most cases.) Now that safeguards are being increasingly applied to plants that handle nuclear explosive material (reprocessing plants, enrichment plants, and plants that fabricate or otherwise process highly enriched uranium or plutonium fuel), concern about the timeliness of detection has increased. It has been realized that the concept of materials accounting, on which emphasis was laid in the early 1970s, might entail delays of several months before a diversion was detected. The safeguards operation has therefore placed much more emphasis on measures such as continuous inspection (or continuous instrumental surveillance) and on physical or other forms of "containment" that would either physically prevent diversion or make it possible to detect it very promptly. In practice (i.e., in the safeguards arrangements agreed to for individual plants), the soundness of this new approach has generally been accepted, but there is still considerable difference of opinion about the precise figures that should be set as measurements of "timeliness" and about what can be achieved in this respect in large bulk-handling facilities.

It should also be borne in mind that these large bulk-handling facilities dealing with nuclear explosive material are still confined to a very small group of countries, most of which are nuclear-weapon states. Outside the nuclear-weapon states, such plants are operating, and are under IAEA safeguards, in West Germany, Italy, and Japan. A reprocessing plant operating in India comes under IAEA safeguards from time to time when it handles safeguarded irradiated fuel. Pilot-scale reprocessing plants are planned or under construction in Argentina and Brazil and will be placed under safeguards, but, because of their small size, these do not present the technical problems referred to in the previous paragraph.

It has been affirmed recently that the central political value of IAEA safeguards is to give time for diplomatic pressure to be brought to bear on a diverting country after the IAEA has detected the diversion but before the state concerned has been able to consummate its intentions by exploding a nuclear device. If the state is operating a reprocessing plant or has a store of plutonium or highly enriched uranium, the time needed to make the explosive may be a matter of weeks or even days – too short for diplomatic action to be effective. This has been one of the main arguments used in proposing a delay in the construction of reprocessing plants and the introduction of fast breeder reactors.

Most of us would agree that it is desirable to concentrate the production of enriched uranium and the separation of plutonium in as few locations as possible, and to undertake it in truly commercial large plants operated, if possible, under multinational auspices. In other words, nations should perceive that their interest lies in voluntary restraint on the number and dispersion of plants that produce weapons-grade material. The concept that the main purpose of safeguards is to serve as a burglar alarm that would give the diplomatic police time to rush in before the safe is blown open represents, however, a complete reversal of the basic concept of IAEA safeguards that has been accepted for the past twenty-four years. It also focuses attention on the risk of "abrupt" diversion in supposedly unstable Third World countries, ignoring the fact that most of the wars of the past two hundred years have been launched by the richest nations. Finally, it represents a rather improbable diversion scenario.

If clandestine diversion ever does take place, the more probable scenario is that of the "trickle" diversion, under which relatively small quantities of weapons-grade material are quietly salted away for use in case of desperate need. Abrupt diversion (or abrupt seizure of a stock of plutonium or a reprocessing plant), followed by an early explosion, could hardly be perceived as serving the national interest of any country. Indeed, once a country was caught in the act, and even if it were prevented from finishing it, it would be likely to forfeit international credibility for decades, and any safeguards applied thereafter in that country would be likely to have lost their principal political value. In other words, the cardinal object of safeguards is to give assurance of the *absence* of diversion: to give assurance that the safeguarded state is not steadily diverting material for use in an emergency. If a diversion occurs, this assurance is lost.

CONCLUSIONS

From this analysis, we may identify a set of priorities. The first and foremost is to universalize the non-proliferation regime. It is of particular importance to ensure that full-scope safeguards are applied in those countries that have significant unsafeguarded plant. In three cases, as we have seen, an unsafeguarded plant is producing or is capable of producing weapons-grade material. In some of these and other cases, it may not be possible to secure accession to the NPT, at least until there is further progress in constraining vertical proliferation, and the more limited object of de facto application of full-scope safeguards may be all that can be attained at present. De facto application of full-scope safeguards signifies that, because of the existence of a complete mosaic of non-NPT safeguards agreements, all nuclear materials and all significant nuclear plants in the country concerned are, at a given time, under IAEA safeguards. This is certainly better than having significant unsafeguarded plants, but from a non-proliferation point of view it is much less satisfactory, for legal as well as practical reasons, than de jure application of full-scope safeguards. De facto application does not, for instance, preclude the future development of an unsafeguarded domestic nuclear fuel cycle. Moreover, the individual agreements constituting the mosaic are likely to be different in many important respects, such as the precise undertaking by the country concerned, duration of the agreement, technical procedures followed under its terms, and coverage of subsequent generations of nuclear material or of replicated plant.

Second, every effort must be made to secure de jure acceptance of the non-proliferation regime, by ratification of the NPT itself (or of the Tlatelolco Treaty, with waiver of the four conditions), by more limited agreements between neighboring countries coupled with IAEA verification (for instance, regional or subregional nuclear-free zones), or by unilateral acceptance of full-scope safeguards agreements. It is, of course, also necessary to verify that states that have ratified the NPT conclude without delay the safeguards agreements and more detailed safeguarding arrangements ("subsidiary arrangements and facility attachments") that are needed to implement these agreements. These arrangements are also required (and are sometimes difficult to negotiate) under non-NPT agreements.

Third, since the actions that would have to be taken to achieve both the first and the second priorities are political, each government concerned must itself perceive that its best interests are served by accepting a nonproliferation commitment, and that the political factors at work in each country will be quite different. In other words, neither general exhortation nor negotiation by legislation is likely to have much effect; patient and persuasive diplomacy may have a better chance.

Fourth, it is essential to ensure that the political base of IAEA safeguards remains secure. This means that the governing organs of the IAEA must continue to perceive that it is in the common and individual interest of their members to provide adequate funds and resources to maintain an effective operation. It is particularly important to ensure that safeguarded countries will continue to perceive the Board of Governors as a body that will impartially discharge the quasi-judicial functions that devolve on it in approving and interpreting safeguards agreements and in dealing, if it has to do so, with questions of noncompliance.

Fifth, the safeguards operation must be conducted in such a manner as to maintain and, if possible, increase the confidence of states that, in opening their nuclear programs to international safeguards and inspection, they are not exposing themselves unduly to the risks that valuable industrial or commercial information will be purloined, that plant operations will be impeded, that they will be penalized in any other way in relation to their competitors, or that they will suffer any other form of prejudice to their national interests. Responsibility for these matters falls chiefly upon the Secretariat of the IAEA, which must continue to be aware of and responsive to the concerns and sensitivities of safeguarded states and of the nuclear industry.

Finally, the technical effectiveness of safeguards must be maintained and enhanced. This is desirable not only to fortify the assurance given by safeguards that no diversion is taking place, but also to permit economies in the use of manpower by the introduction of instrumental surveillance and containment. Improved safeguards will permit greater reliance on quantifiable and objective criteria in reaching more precise conclusions with less delay than at present, although it is very doubtful that it will ever be possible entirely to eliminate an element of human judgment in reaching such conclusions. It will also be necessary to develop new safeguards approaches for the large bulk-handling plants (reprocessing, enrichment, etc.) that are now being planned or are coming under safeguards in nuclear-weapon states. Responsibility for the technical improvement of safeguards rests partly on the Secretariat of the IAEA but also on governments, whose support is indispensable in research and development and in providing expertise to the agency. Yet, while technical improvement of safeguards is highly desirable, it will achieve its goals only if the overriding political priorities can be met.

In time it will also be necessary to consider new verification approaches, to take account of the increasing differences in the levels of sophistication of the nuclear industries of safeguarded states. Safeguards approaches that may be quite reasonable for verifying the absence of diversion in a country that is operating only research or power reactors may no longer be appropriate in countries with highly developed nuclear industries embracing all or large parts of the nuclear fuel cycle. The evolutionary conceptualization of such new approaches will emerge in the IAEA's advisory bodies, but it must eventually be endorsed by its governing organs.

In conclusion, it should be recalled that IAEA safeguards are the cutting edge of almost every non-proliferation measure that has so far been applied or put forward. Safeguards would also play a part in any international scheme for limiting vertical proliferation. If the cutting edge of safeguards is blunted or if, for one reason or another, safeguards cannot be or are not being applied, the non-proliferation regime will suffer commensurate damage.

REFERENCES

Goldschmidt, Bertrand. 1967. *Les rivalités atomiques 1939–1966.* Paris: Fayard.

INFCIRC/153. 1972. *The structure and content of agreements between the agency and states required in connection with the Treaty on the Non-Proliferation of Nuclear Weapons* (INFCIRC/153 [corrected]). Vienna: International Atomic Energy Agency.

International Atomic Energy Agency. 1980. *The Annual Report for 1979.* Vienna: International Atomic Energy Agency.

Szasz, Paul C. 1970. *The law and practices of the International Atomic Energy Agency.* Legal Series no. 7. Vienna: International Atomic Energy Agency.

5

Nuclear Non-Proliferation: Failures and Prospects

Ryukichi Imai and Robert Press

INTRODUCTION

For the purpose of this paper, the term nuclear non-proliferation implies a cessation of growth in the number of sovereign states proceeding to develop or otherwise acquire nuclear weapons—that is, a cessation of "horizontal" proliferation, as distinct from "vertical" proliferation, which refers to a qualitative or quantitative increase in existing nuclear weapon arsenals. It does not imply any cessation in the development of nuclear technology for peaceful application, and particularly as a contribution to the civil energy needs of any sovereign state. From various political, technical, and economic points of view, so much has been written and said on this subject that it is difficult to say anything new, even though the subject has rather acquired the nature of a moving target. The objective of this paper is to examine the evolution of combined political and technical attempts to achieve worldwide acceptance of a commitment to non-proliferation, to note failures to date, and to identify essential factors to be satisfied if the greater success necessary is to be achieved in the immediate future. To do this, it is important to separate realism from the unrealism so often involved in discussing the concept of non-proliferation, as defined above, particularly if treated as a moral principle rather than as part of a general security issue reflecting shifts in regional and global stability.

The political nature of the non-proliferation problem is underlined by the fact that, whereas five nuclear-weapon states are currently accepted, any threatened increase in that number is discouraged by every possible peaceful means. This fact combines political acceptance of an existing international situation with a belief that any addition to the present number must lead to international instability. Success in preventing additions may be more readily

Date of origin, March 1980.

achieved through political understanding, and perhaps some compromises in particular cases, than through seeking a universal solution to a generalized problem. For example, in a conflict between the application of nuclear science and technology for beneficial economic purposes, on the one hand, and measures to prevent a military application of that same science and technology on the other, all countries may not have precisely the same perception of what the optimum balance of their national interests requires. But the more they perceive this to lie in developing a full fuel cycle program for civil power, the more they may be perceived to be acquiring a military potential. Provided their specific economic and security interests are seen to be given full consideration, some countries that nurture a belief that the acquisition of nuclear weapons could serve their national interests may well be open to political persuasion that the price and security cost of such a course would outweigh the perceived gain. At the same time, it has to be recognized that some countries may persist in their belief in a political if not military need for nuclear weapons, regardless of the price to be paid.

EVOLUTION OF POLICIES FOR ACHIEVING NON-PROLIFERATION

Ever since nuclear energy literally burst upon the world with explosive force in 1945, nations have sought acceptable means of controlling the development and dissemination of nuclear technology as a source of energy for peaceful industrial use, while at the same time preventing, or at least minimizing, encouragement for the proliferation of nuclear weapons. The first attempt was based on a United States report, known as the Acheson-Lilienthal report, prepared in 1946 under the then United States Under Secretary of State (U.S. Department of State 1946). This report emphasized the positive aspect of control, namely cooperative international development, rather than mere policing operations. At the same time it insisted on the need to reserve certain "dangerous" activities in nuclear energy development (e.g., the separation of fissile material for bombs) as matters for international ownership and control; it called for the creation of an international atomic energy authority that would own and operate all nuclear facilities, ranging from uranium mines to research and production plants, throughout the world.

In 1946 the United States presented to the United Nations Atomic Energy Commission a control plan based on these proposals. The plan (known as the Baruch Plan) failed to gain acceptance by the Soviet Union, which feared a United States monopoly of this new technology, and an impasse quickly developed. The difficulty was political, and the Soviet Union was not alone in being apprehensive on political grounds. Thereafter, nuclear weapon development proceeded apace in the Soviet Union and in the United States, with the Soviet Union subsequently breaking the United States monopoly by exploding its first nuclear device in 1949. At the time of the Baruch Plan, the United States was the only known possessor of the technology for fabricating a nuclear explosive weapon. It had already refused to share the "A-bomb secrets" with its main wartime associates (Britain and Canada) in the Manhattan Project responsible for the development of the bomb. By its Atomic Energy Act of 1946, it even cut off the flow of information on civil nuclear power development anticipated under options on technical cooperation provided for in wartime agreements. This action merely increased the determination of others to "go it alone." Britain and, later, France embarked on full-scale military and civil programs, and Canada began its civil program.

Britain and France separately and independently joined the so-called nuclear club (those with a demonstrated ability to develop and explode a nuclear device) in 1952 and 1960 respectively. China followed in 1964. The limitations and counterproductive effects of the principle of denial of technology as a main element of policy had thus been established, and horizontal proliferation was already a reality. At that stage, however, it was the threat of continuing pollution of the atmosphere by increasing weapon testing that was primarily becoming alarming. Concern about the public health hazards of radioactive materials being deposited in the atmosphere as a result of exploding nuclear weapons there, rather than a direct campaign in non-proliferation terms, was the genesis of pressure for a ban on nuclear weapon testing. This pressure ultimately led in 1963 to an international treaty banning nuclear weapon tests in the atmosphere, underwater, and in outer space, but not underground. This Partial Test Ban Treaty (PTBT) was signed and ratified not only by three of the four nuclear-weapon states then conducting such weapon tests, but also by 107 states neither possessing nor developing nuclear weapons.

Following the failure to gain international acceptance of United States proposals for the ownership of designated nuclear facilities, and about ten years in advance of the PTBT, President Eisenhower launched the Atoms for Peace program. His declared purpose was to encourage all nations, particularly those in the undeveloped world, to plan for the use of nuclear energy as a great new resource that would hasten their industrial progress and solve many of their social problems. This program was internationally acclaimed and was subsequently developed through the International Atomic Energy Agency (IAEA) set up in 1957.

The IAEA was expected to help all member states acquire safeguarded nuclear technology and have "due consideration for the needs of the underdeveloped areas of the world" (IAEA Statute, article III.A.2). In fact, in the early days, nearly all the assistance in the form of research reactors and training, together with associated safeguards, was arranged bilaterally without involving the IAEA. Notwithstanding a perception of possible misuse of nuclear energy, the Atoms for Peace program necessarily led to the "nuclear" training of an important number of scientists and technologists from many developing countries, and scores of research reactors were built. Similarly, the Soviet Union initiated close cooperation with China, trained many Chinese scientists, helped in developing uranium resources, and supplied China with a research reactor. Given time and resources, it had to be expected that these various nuclear-trained people would develop skills leading not only to the development and operation of civil nuclear power plants but also, at least potentially, to explosive devices.

These facts, combined with ongoing nuclear proliferation in Britain, France, and China, generated fears of the further spread of nuclear weapons to an increasing number of additional countries. By 1968, fears of so-called horizontal proliferation, or the "*n*th-country problem," led to an international Non-Proliferation Treaty (NPT). This treaty was regarded not only as a supplement to the PTBT but more widely and hopefully also as a forerunner of effective measures to halt the nuclear arms race by the superpowers, to prevent other industrialized countries from acquiring nuclear weapons, and, ultimately, to achieve nuclear disarmament.

The NPT has now been signed and ratified by more than a hundred states, and all non-nuclear-weapon states that are parties to the treaty are obliged to open all their nuclear facilities to IAEA safeguards. However, France and China, with a considerable group of "near-nuclear" countries (i.e., potentially near to a capability to explode a nuclear device), are not parties to the NPT. Thus, India, Pakistan, Israel, and South Africa have all refused to sign the treaty. In Latin America, certain states with major nuclear programs (Argentina, Brazil) are also not parties to the NPT. They have, however, indicated their willingness to undertake comparable obligations under the Treaty of Tlatelolco that will, when fully implemented, create a nuclear-weapon-free zone in the whole of Latin America. Argentina, Brazil, and Chile have not yet taken all the necessary steps to bring the latter treaty into force on their territory, and Cuba has not yet signed it. For the purpose of the Tlatelolco Treaty, as for the NPT, acceptance of safeguards on all nuclear facilities would be regarded as a clear indication of intention not to acquire nuclear weapons.

The NPT, at the time of negotiation, not only recognized the urgency and importance of preventing the spread of nuclear weapons, as a measure to reduce the danger of nuclear war, but also accepted the importance of intensifying international cooperation in the development of peaceful applications of nuclear energy. In return for cooperation in promoting the application of IAEA safeguards, the treaty recognized that all signatories had an "inalienable" right to engage in research, production, and use of nuclear energy for peaceful purposes and would be able to acquire source and special fissionable materials, as well as equipment, for the processing, use, and production of nuclear material for peaceful purposes.

Taken together, the NPT and IAEA safeguards worked satisfactorily for a period of time, but in the mid-1970s the prospect of an increasing spread of nuclear power, the separation of plutonium from spent nuclear fuel, the advent of more economical methods of enriching uranium, the projected acquisition of sensitive facilities by several countries with limited nuclear power programs, and the potential introduction of fast reactors all led to a growing concern about the adequacy of the current NPT/IAEA safeguards regime. Moreover, parties whose non-weapon intentions were already enshrined in their ratification of the NPT appeared to be incurring disadvantages compared with non-parties to the treaty. The emergence of this concern, coupled with the Indian nuclear explosion in 1974, led in turn to the creation of a Nuclear Suppliers Group that sought to ensure that suppliers of nuclear equipment, materials, technology, and services would not compete for export business by offering commercial contracts with differing constraints in respect to safeguards and related factors. Not surprisingly, these increasing measures of control over supplies caused concern and suspicion in some developing countries. They interpreted them as an attempt by the industrialized world to use general fears about weapon proliferation to retard technological progress in undeveloped countries by denying them a transfer of nuclear technology and imposing upon them undue inspection. Measures to encourage non-proliferation generally were thus unfortunately beginning to be seen in a North-South perspective.

By this time, prospects for the development of nuclear power appeared very different to the non-nuclear-weapon states, and to the undeveloped countries in particular, compared with those presented to the world at large in the mid-1950s. In earlier years there was a great deal of optimism regarding both the future of the technology and the economics of nuclear power, while acceptance of the IAEA as a medium for the acquisition of nuclear technology and materials, with associated safeguards, was considered sufficient condition for full participation in the nuclear future. Nuclear non-proliferation was not then a subject of major international concern. On the contrary, the evolution of all non-proliferation policies had recognized a need to encourage national acquisition and development of full-scale nuclear technology for the development of nuclear power. Alternative reactor concepts were studied and advanced without much regard to the proliferation potential of the back end of the fuel cycle, involving reprocessing and the management of plutonium.

When negotiating the NPT, the terms "proliferation" and "non-proliferation" were respectively applied specifically to the actual spreading and the prevention of nuclear weapons development. Since the early 1970s, and particularly since the Indian nuclear explosion in 1974, there has been a marked tendency to regard all activity in the field of nuclear energy as contributing to proliferation. In becoming the sixth country to demonstrate a nuclear explosive capability, India claimed her achievement was a development for peaceful purposes. Whatever justification India may have perceived for this development, she employed material and assistance explicitly provided for non-weapons purposes. This, coupled with increasing efforts by several countries to acquire "sensitive" nuclear technologies, triggered a process of major reappraisal throughout the world concerning general proliferation risks and the adequacy of existing non-proliferation measures. Canada saw the Indian explosion as a breach of a bilateral understanding on the use of an unsafeguarded reactor and technology supplied by Canada for peaceful purposes only, and it consequently canceled its nuclear cooperation agreement with India. In 1976 President Ford declared that avoidance of proliferation must take precedence over economic interest, and that commercial reprocessing should be deferred until there was sound reason to conclude that the world community could effectively overcome risks of proliferation. This was followed in 1977 by President Carter's call for a new examination of the risk of proliferating a nuclear weapons capability through the growth of nuclear power. Specifically, he called for indefinite deferment of the reprocessing of spent fuel elements, with its consequent widespread availability of separated plutonium.

In aiming at the exclusion of separated plutonium from the nuclear fuel cycle, as the main cause of concern in a proliferation context, the Carter proposals included an ending (at least for the time being) of United States commercial reprocessing for civil power purposes, cancellation of the demonstration fast breeder project at Clinch River, and a call to all other nations to adopt a similar policy. Moreover, and perhaps with some prejudice to its purpose, the United States implied an enforced compliance by other countries, based on controls that the United States, as an earlier monopoly supplier of enriched uranium, could, under most of its agreements, exercise over the treatment of spent fuel resulting from use of that uranium. President Carter saw his proposals as a responsible course between forgoing the energy benefits of nuclear power and a commitment to the commercialized use of plutonium before knowing how to deal safely with its risks. On the other hand, his proposals caused worldwide resentment among developed and undeveloped countries alike, all of whom have widely differing energy needs and differing resources to satisfy them. In Europe and Japan, the anti-plutonium proposals were seen as prejudicing efforts to enhance security of energy supplies by developing nuclear technologies that would diminish requirements for uranium, of which they had none while the United States had plenty. In less developed countries, the Carter proposals were seen both as an effort to deny them access to technology that could help their economic development and as a sign of United States suspicion of their underlying intentions. This latest phase in the evolution of non-proliferation policies was a significant factor in a collapse of confidence between suppliers and supplied in the nuclear field. The Eisenhower initiative of 1953 had opened the way to national reprocessing of spent fuel as a legitimate and necessary commercial activity, to be undertaken as the need arose, with the inevitable utilization of plutonium. Indeed, the idea of plutonium fueled breeder reactors was then pushed enthusiastically. By recommending policies that would defer these activities, the Carter initiative created political issues that are incredibly complicated by major considerations of energy requirements, economics, technology, ecology, state security, and, in the end, sovereignty. The ultimate end of non-proliferation — that is, increased national and international security — was being endangered not only by the means for its achievement but also in that the means were becoming ends in themselves. It remains to be seen whether and, if so, to what extent the Carter proposals, together with the stringent control provisions of the subsequent United States Nuclear Non-Proliferation Act of 1978, may be modified by the report of the International Nuclear Fuel Cycle Evaluation (INFCE).

RESULTS OF NON-PROLIFERATION POLICIES

Thirty-five years after the first demonstration of the feasibility and military use of nuclear weapons, international concern about their possible proliferation is more urgent and widespread than ever: more urgent because the requisite technology and materials are no longer the closely guarded possessions of only one state or even five states; more widespread because the inevitable dissemination of requisite technology and materials has caused, on the one hand, supplier nations to fear the proliferation potential of their actions and, on the other, a growth of nuclear suspicions between nations where dangerous political tensions already exist. Reasons for continuing international concern about the possible proliferation of nuclear weapons, despite the negotiation and coming into force of both the PTBT and the NPT within the second half of this thirty-five-year period, are discussed in the following paragraphs.

It can be said that the PTBT was not primarily a non-proliferation measure, even though that may have been seen as a possibly associated advantage. As already noted, it was negotiated under public pressure primarily to end polluting the atmosphere with damaging radioactive products. By failing to put an end to underground nuclear testing, it did not legally close the doors to any signatory state wishing to develop nuclear weapons or other nuclear explosive devices. For example, India, as a non-nuclear-weapon state, signed and ratified the PTBT but not the NPT (which did prohibit non-nuclear-weapon states from acquiring nuclear weapons or other nuclear explosive devices) and proceeded to develop and to test underground a nuclear explosive device. In doing so it claimed to be seeking the potential benefits from the peaceful application of nuclear explosions allowable to non-nuclear-weapon signatories of the NPT. France, on the other hand, which did not sign the PTBT but subsequently and voluntarily accepted its constraints on atmospheric testing, continued its nuclear weapons development through underground testing. The superpowers, the United States and the Soviet Union, as founder signatories of the PTBT, have relentlessly pursued their weapons development through extensive underground testing. On a very much smaller scale, Britain has also engaged in underground testing. Thus the underground route to the actual testing of a nuclear weapon or an explosive device is legally open to any state wishing to do so, even though that state may have signed and ratified the PTBT.

The NPT has given rise to much debate concerning its implementation, even though its basic obligation has been observed. This in turn has been reflected in the non-proliferation difficulties generated in recent years. In essence, the NPT of 1968 sought to limit the number of nuclear-weapon states to the then-existing five—the United States, the Soviet Union, Britain, France, and China. Although it explicitly recognized the possible connection between civil and military uses of nuclear energy, it did not prohibit any nuclear energy activity whatever, except that defined in articles I, II, and III. So-called peaceful nuclear explosions by non-nuclear-weapon states were precluded by the treaty, but article V provided that the potential benefits from any peaceful application of nuclear explosions should be made available to them on a nondiscriminatory basis. Otherwise, in return for forgoing the acquisition of nuclear weapons and accepting safeguards to verify their commitment, the non-nuclear-weapon states believed they would be freer than they had recently found themselves to engage in any form of peaceful nuclear activity. They were also to have rights of access to "the fullest possible exchange of equipment, materials and scientific and technological information for the peaceful uses of nuclear energy" under articles III.3 and IV of the NPT. In addition, under article VI, they looked forward to more rapid progress than there has been toward a major reduction in stockpiles of nuclear weapons and a cessation of the nuclear arms race.

On that last point, and notwithstanding the current SALT II agreement, the non-nuclear-weapon states have charged the nuclear-weapon states with encouraging vertical proliferation rather than reversing it, thereby undermining the implied quid pro quo for adherence to the NPT. But the non-nuclear-weapon states have been even more concerned about what they have alleged to be a switch in the concept of proliferation from meaning actual acquisition of nuclear weapons to meaning engagement in particular nuclear activities that, though represented by them to be essential for civil nuclear power programs, might be suspected of facilitating the manufacture of material for such weapons. Largely as a result of this, the original differentiation between nuclear-weapon and non-nuclear-weapon states has been overlaid by a differentiation between suppliers and recipients of nuclear materials, equipment, and technology. For the non-nuclear-weapon states, and particularly for the less developed countries, the NPT bargain has been eroded. Restoration of confidence between suppliers and supplied needs now to have a high priority if the NPT is to remain the effective cornerstone of international non-proliferation arrangements.

Whether the meaning of proliferation is to remain as originally accepted in the NPT, and as still maintained by nuclear recipient states, or is to be extended to require special restraint on a range of civil nuclear activities that they consider to be required for their energy programs (e.g., enrichment, reprocessing, and the use of separated plutonium), there can be no doubt that there is now a clear and pressing need to underpin the Non-Proliferation Treaty. Before discussing how that might be achieved, it is necessary to discuss some issues that have either been too readily ignored in evolving non-proliferation policies to date or have emerged as a result of changes in the political, economic, or military background during the past decade. At the same time, it is worth noting that five questions touched on in the original Acheson-Lilienthal report thirty-four years ago remain significant issues for any international non-proliferation regime. They are:

- Who are the potential proliferators?
- To what extent can capabilities and motivations for proliferation be constrained even if such constraint is perceived as transgressing the sovereign rights of a nation?
- What is to be the mechanism for international control?
- What is to be the required effectiveness of international safeguards?
- To what extent can monopoly/discrimination fears be met, as between suppliers and recipients?

INESCAPABLE ISSUES FOR A VIABLE NON-PROLIFERATION POLICY

National Sovereignty

Nuclear proliferation or non-proliferation is fundamentally a matter of national security and strategic concern, where security involves political, military, and economic assessments. The military factor is probably the paramount consideration for those who have traditionally been outside a central or regional security system and do not now have the protection of a credible military nuclear "umbrella." For them, non-proliferation is therefore ultimately dependent on national political perceptions rather than on any "technical fixes," however sophisticated or institutionalized these may be. This has been demonstrated throughout the evolution of non-proliferation policies, from 1946 up to and including the Carter initiative in 1977. In this context it is worth bearing in mind that the "technical fix" of denatured plutonium, which was a major point in the Baruch Plan, proved to be ineffective, thus illustrating the possibly very temporary validity of any "technical fix" that may be claimed. National sovereignty may be infringed in respect not only of a nation's own judgment whether to develop nuclear weapons, but also of its right to engage nationally in particular nuclear activities, such as enrichment of nuclear material or reprocessing of spent fuel, both of which may admittedly lead to national stocks of weapons-usable uranium or plutonium. It is counterproductive not to accept that such activities are perfectly legitimate political and technical entitlements when they are in fact for declared peaceful purposes and when obligatory safeguards are imposed. Whether they are prudent and economic may in some circumstances be a matter for political judgment, but that is an entirely separate issue from entitlement. They cannot be considered to be proliferation in the sense of the NPT. Prudence may be negotiable. Sovereign rights are not negotiable, and, whatever illustrations there may be of apparently accepted violations of sovereignty, an appeal to sovereign rights remains a powerful political factor, particularly in the developing countries.

In accepting the NPT, and thereby forgoing the option to acquire nuclear weapons, a number of nations cherished suggestions of possible intervention by the United Nations Security Council in the event of their being menaced by nuclear attack. The erosion of this possibility with the passing of time has enhanced ultimate concern for the sovereign right to avoid any possible threats to national security interests. Such threats may be perceived in forms varying according to regional political circumstances. For example, immediately after the announcement of an intended withdrawal of United States ground combat troops from South Korea, the latter's president publicly hinted that his country might have no alternative but to acquire nuclear weapons of its own as a means of nuclear deterrence. This is not the place to discuss the wisdom of such a decision in this particular case, but the incident illustrates the interplay of nuclear proliferation and quite separate developments in regional political or military situations. An example of a regional "domino" effect in proliferation is provided by India and Pakistan. It now matters little whether the Indian explosion of a nuclear device in 1974 was associated with earlier Chinese development of nuclear weapons; recently reported Pakistani nuclear activities can be seen as a sequel to that explosion. These are examples of how the most highly powered international non-proliferation campaign can prove insufficient to prevent actual or apparently imminent regional proliferation when issues of national security are at stake. They are in addition to reports, for similar reasons, concerning an Israeli bomb, and apprehension about the nuclear weapon intentions of some other states that have not signed the NPT, notably in Latin America and South Africa. There are a number of forms the resolution of these concerns might take, for example acceptance of the NPT, the Tlatelolco Treaty, or nuclear-weapon-free zones, but it must be recognized that they will in fact be resolved only if national and regional security concerns are taken fully into account. Otherwise, fear that a potential competitor may acquire nuclear explosives may lead a country to refrain from accepting binding non-proliferation restraints or to take preemptive action that creates a situation in which both engage in competitive nuclear programs.

Important as is the need to assure regional stability, the possibility of proliferation in a still wider context cannot be ignored. Japan may be cited as a case in point, being heavily dependent on the security of sea communication routes for oil supplies from the Persian Gulf across the Indian Ocean, South-East Asian seas, or Pacific Ocean.

In the wider strategic respect, it may in general be argued that reasonable worldwide nuclear stability, through a "balance of terror," has been achieved between the nuclear superpowers and within their subsidiary NATO and Warsaw Pact theaters. These theaters at present include nearly all the major industrial nations of the world, and horizontal proliferation among them is very unlikely; any major disturbances in the existing world order would be contrary to their national interests. Nevertheless, vertical proliferation continues, through quantitative and qualitative increases in the nuclear arsenals of nations already in possession of nuclear weapons. Thus, even among the nuclear-weapon states, the facts of vertical proliferation and continuing failure to achieve a comprehensive nuclear test ban may be said to underline an ultimate sovereign determination to protect national security.

Energy Requirements

As noted earlier, the issue of sovereignty may arise not only in a direct military sense but also in respect to nonmilitary matters. Some countries may consider the most important current example of these to be the development of nuclear power as a source of energy for peaceful purposes. For developed and undeveloped countries alike, adequate and secure supplies of energy at tolerable prices are an essential mainspring of economic life. This needs no underlining with the politics of oil so vividly demonstrating the precarious nature of oil as a source of energy. Any survey of the future foresees a need within the next few decades to replace oil by alternative sources of energy at acceptable costs; but the long lead time required to bring a radically new source to commercial use means that, during the transition from fluid fuels (oil and gas), reliance must be placed on greatly increased use of already proved sources of energy. In this respect conservation can be helpful, but only to a limited extent. It is not readily accepted in developed countries, and there is little or no scope for it in many developing countries struggling to raise their standards of living.

Whatever the attractions and feasibilities of such alternative options as wave and tidal power, solar energy, or even thermonuclear fusion, the fact remains that the only two substantial and proved sources of energy, other than the oil and gas soon to be replaced, are coal and nuclear fission. Neither of these is without considerable problems in practice. Setting aside limitations on the resource availability of coal and the increasing social problems of manpower, health, and safety in mining it, a very large expansion in burning coal foreshadows major environmental problems. These range from despoiling the immediate countryside to increasing local pollution of the atmosphere and, even more important, to predicted global climatic effects, resulting from significant increases in the emission of carbon dioxide to the atmosphere. As regards nuclear fission, disquiet has been expressed about a "plutonium economy"—a risk for the proliferation of nuclear weapons and a radiation hazard for the future of mankind. Nevertheless, objective environmental comparison between nuclear energy and other energy sources has hitherto consistently resulted in showing the advantage to lie with nuclear energy rather than coal. It is not a simple source, but, whether one likes it or not, it is the only new one capable of contributing significantly to energy supplies, while also being the least costly solution for electricity generation in economic terms and in most locations. Governments may be justifiably reluctant to deprive themselves of the nuclear option for the period when the transition from oil and gas is likely to occur, but if they seek to maintain the nuclear option they must demonstrate to their publics that they can deal effectively with requirements for the safe operation and siting of nuclear plants and for long-term waste disposal or management, and at the same time satisfy non-proliferation requirements.

For developing countries, conscious of their relatively low energy consumption as compared with industrialized countries, nuclear power is seen as an opportunity for the industrial development encouraged by the United States Atoms for Peace program of more than twenty-five years ago. Hitherto, they have not demonstrated such deep anxiety as in some advanced countries concerning related safety and environmental factors. There is as yet little evidence that even the recent Three Mile Island accident in the United States has led to any new influence on their thinking in these respects. For them, it is natural to aspire to long-range programs for the indigenous manufacture of components and execution of the various activities in the fuel cycle of nuclear power plants, which can be expected to have a useful life of thirty years or more. Failure to pursue national self-sufficiency in, for example, enrichment or reprocessing may be seen by some of them to entail an unacceptable vulnerability to the political or economic whim of a supplier government. Fear of an implied stigma of inferior status may also be a factor. Yet, contrary to how these countries interpret the NPT, recent international concern that some of these activities, particularly the urge to procure independent national control of the fuel cycle, might serve to facilitate weapons programs has led to denial policies on technology, materials, and equipment. The effect of this has been to generate in undeveloped countries a hostility based not only on deliberate discrimination against them but once again on a perceived sense of the infringement of national sovereignty.

Bearing in mind that, historically, a main cause of international conflict has been a need or desire to acquire resources from possessors unwilling to supply them, it is no exaggeration to say that energy has become a critical theme for future world peace. Discrepancies in energy availability between rich and poor must increase social pressures as well as discontent among the less developed countries. They believe a transfer of nonrecoverable energy sources from themselves to industrialized countries should oblige the latter to give a measure of reciprocal help, in the form of appropriate new technologies, as oil and gas resources run out. Energy policy, whether national or international, is an art of "uncertainty management" rather than of optimization based on forecasts of demand and costs of supply or crisis management at a particular time, and it should not be allowed to induce a crisis for world peace. As time horizons have lengthened, policy emphasis has shifted to retaining options as an insurance against undesirable outcomes. The key decisions are long-term and should not be dominated by short-term considerations. The validity of the current international pressure apparently aimed at denying national aspirations to engage in particular nuclear activities is therefore open to serious doubt.

Restrictions on Transfers of Technology

Following the Carter initiative in 1977, wide public debate concerning proliferation and the acceptability of nuclear power centered on the need for reprocessing spent fuel from civil nuclear power reactors and the so-called plutonium economy, including the use of plutonium in fast breeder reactors. In proliferation terms, and important though the possible abuse of the civil power fuel cycle remains, the initiative appeared to give undue weight to the production of weapon-usable materials in the nuclear power fuel cycle, as compared with the attraction, to a potential proliferator, of alternative routes outside the nuclear power industry. Some of these could be both cheaper and faster: for instance, the development of small unsafeguarded enrichment facilities under national control, or the production of plutonium in small dedicated or research reactors. This omission has since been recognized, and alternative routes to the acquisition of materials for nuclear weapons were fully recognized in INFCE. INFCE was launched in 1977 as a technical and analytical study, but without a political mandate. Its task was essentially to set down technical and institutional options concerning the nuclear fuel cycle. It was not to negotiate a new political framework to govern nuclear trade and non-proliferation. Sixty-six countries and five international organizations participated in this work, but the participants were not committed to INFCE's results.

The arguments for imposing severe restrictions, or even a complete ban, on national reprocessing and on the transfer of associated technology and equipment have rested on:

- the possibility of national diversion of plutonium produced in civil power reactors for use in fabricating nuclear weapons;
- concern about the possibility of plutonium being seized by terrorist groups;

- suggestions that measures necessary for the protection of plutonium would infringe individual civil liberties.

These considerations have all been debated at length in various forums. While the potential for national diversion from the civil power reactor fuel cycle should not be underestimated as a future route to proliferation of weapons, it can be argued that a determined government anywhere in the world could, in technical terms and if it so desired, equip itself with nuclear weapons whether or not it had civil nuclear power stations and whether or not it possessed plutonium. Moreover, China in 1964 and India in 1974 are examples of developing countries taking a political decision to invest technical and material resources in developing a nuclear explosive notwithstanding major domestic and social problems with which they were and are beset.

No country now possessing nuclear weapons has chosen to develop them by diverting fissile material from its civil nuclear power facilities, and fears that Israel and South Africa may be capable of building nuclear weapons are sustained though neither yet has a commercial power reactor in service. The absence of a necessary link can be demonstrated by the following facts from a recent article (Greenhalgh 1979). Between 1945 and 1954, three countries (the United States, the Soviet Union, and Britain) developed nuclear explosives before the entry into service of their first prototype commercial power reactors. Between 1955 and 1964 two more countries (France and China) did likewise, and between 1965 and 1974 only one country (India) followed suit. Yet, over the same time span, the world nuclear generating capacity grew from 5 MW in 1955 in one country to 54,000 MW in 1974 in nineteen countries, and it is currently 110,000 MW in twenty-two countries. Although the civil fuel cycle would be a disproportionately expensive way of obtaining weapon quantities of fissile materials for military purposes, it should be recognized that a country with experience in the technology of producing and handling nuclear materials would have a material advantage in time if it were to decide to acquire nuclear weapons and had the technology to do so. Over and above that consideration, however, a system of international safeguards can be readily applied to monitor whether civil facilities are being used for weapons purposes.

A meaningful level of nuclear armament, as distinct from a "political" demonstration of an exploded nuclear device to serve a possible regional politicomilitary purpose, would require many large-scale, complex, and simultaneous developments, according to the degree of credible deterrence to be achieved. These would create problems in successful concealment. Even a limited offensive level would require a considerable number of nuclear weapons engineered to dimensions critical enough to be carried in modern delivery vehicles, indigenously produced or purchased elsewhere with associated command and control systems for reliable operation. If defensive measures were simultaneously taken, then still more complex military activities would be required and would be likely to become apparent long before suspected nuclear intentions could be implemented by a proliferator. Meanwhile, the latter would be exposed to the risk of preemptive action by others, with severe jeopardy to its whole nuclear energy program. Given these various considerations, one may ask whether legitimate nuclear power development has not been unduly frustrated by the fear of proliferation; they certainly should be taken into account in any measures in aid of a non-proliferation policy.

Plutonium, about which so much proliferation concern has been expressed, is produced in uranium fueled thermal reactors, which are likely to be the main base of the world's nuclear power capacity for the remainder of this century. This plutonium is completely inaccessible while inside a reactor. When the fuel elements containing it are removed from the reactor, they are so radioactive as to leave the contained plutonium effectively inaccessible for a considerable time. Even after radioactive decay, it remains difficult to perform the handling and chemical steps required to extract plutonium from the spent fuel elements. Its subsequent storage as a mixture of extracted plutonium and uranium oxide would render the material difficult to use for weapon purposes by terrorist groups but would constitute a relatively minor impediment to such use by nations. Hence an assessment of the proliferation aspects of producing plutonium as a constituent material for a nuclear weapon is significantly dependent on the nature of the risk being considered (i.e., whether national or subnational), on whether the plutonium is in extracted form, and on the level of radioactive and chemical protection given to it. In these respects it is the machinery for the regulation of custody and the use of extracted plutonium that is important.

Present-day plans for the use of plutonium produced in thermal reactors call either for its use as fuel for fast breeder reactors or for recycling it in thermal reactors. It is the former that has given rise to fears about the large-scale handling, processing, and transportation of extracted plutonium, but it has been argued that use as fuel in the breeder reactor has an attraction from a non-proliferation point of view in that it puts extracted plutonium back inside a reactor where it again becomes inaccessible. It is then "burned" under operating conditions that allow either slightly more or slightly less new plutonium to be produced than is "burned" in producing that new plutonium. On the other hand, thermal reactors operated on a once-through fuel cycle steadily produce more extractable plutonium as a by-product of nuclear power, and therefore add more quickly to the growth of plutonium in the world, than would a comparable number of breeder reactors. This is a non-proliferation disadvantage, whether seen in terms of storage within spent fuel or in extracted form; but, like the non-proliferation advantage of fast reactors in limiting the world total of plutonium, it tends to be lost in disputes concerning the need for fast reactor development. Nevertheless, both fuel cycles do involve the production of extractable plutonium, and this has been a major factor in the proliferation arguments of those who resist a "plutonium economy."

These points have been argued in contending that proliferation dangers in the use of plutonium as a fuel are considerably less than the proliferation dangers of not using it, and that the mere availability of plutonium is not necessarily the most serious proliferation problem facing the world. If that is so, one may ask whether the option of developing reprocessing technology, for its fuel resource advantages, has not been unduly frustrated by fears of proliferation. In any event, as in the more direct case of national security already discussed, neither technical nor institutional "fixes" will necessarily survive a perceived erosion of sovereign political judgment on the nuclear power requirements essential to assure economic welfare. This will be all the more so in countries that fear that currently developed uranium-based reactors may in the longer term be subject to restraints similar to those now being experienced with oil, and that consequently perceive their long-term independence in energy resources to be more assured through breeding plutonium. According to the relative merits of incentives and disincentives, including the economics involved, countries will certainly reserve the right to decide whether to have national reprocessing facilities, or to purchase reprocessing services or engage in multinational arrangements for the production and/or handling and storage of separated or extractable plutonium. This is a subject on which the findings of the INFCE study may prove helpful.

The foregoing paragraphs on sovereignty, energy, and restrictions on the transfer of technology are not meant to downplay the importance of efforts to achieve non-proliferation. They are intended to illustrate the complex interactions between national security, energy requirements (for materials and technology), non-proliferation, and international politics. Non-proliferation cannot successfully be treated as though it were an independent and abstract subject or moral issue. Achieving it means reaching an acceptable balance of interests, risks, and security, rather than merely seeking to impose some detailed technical procedures on a narrow sector of activity. An acceptable balance clearly encompasses political, economic, military, and even prestige assessments, all of which inevitably carry uncertainties of a nature few sovereign countries will readily place at the mercy of "technical fixes" or fuel supply monopolies.

PROSPECTS

A lesson of history is that technology will advance whether one likes it or not. One's aim should therefore be to master it rather than suppress it. In the case of nuclear energy, more than thirty years' experience of scientific and technological achievement has already been amassed; a worldwide nuclear power industry exists, and a framework of international control, or at least safeguards, has been established. Neither the world's energy needs nor the issue of non-proliferation would be well served if the nuclear option were to be suppressed by major developed countries. The former would be seriously aggravated by more aggressive competition for alternative energy resources, leading in turn to increased pressure for nuclear expansion and indigenous development in those very countries on whose cooperation the success of non-proliferation policies must depend. It is a fairly safe assumption that plutonium recovery through reprocessing will occur in a number of states, including some non-nuclear-weapon states, and any practical and acceptable regime for non-proliferation should accept that reality.

Prudent international policies of collaboration in the development of nuclear energy, in accordance with article IV.2 of the NPT, may well succeed in promoting non-proliferation where policies of technological denial or imposed conditions have been seen to fail and are likely to continue to fail. Notwithstanding the less developed countries' strong economic compulsion toward nuclear energy as a practical and viable source to meet long-term needs, a clearly collaborative climate could be strongly reinforced by the undeveloped countries' growing recognition of the difficulty of the technology involved: the need for highly trained, and thus scarce, manpower resources; the high capital cost of nuclear power plant; the fact that both the technology and the economics of nuclear power have so far pointed toward the deployment of larger and larger reactor units, of a size appropriate only to countries able to use thirty to fifty central electricity generating stations, each of the order of 1,000 MW. This is not the place to discuss optimum patterns for developing countries' industrialization and energy supply, but, given these factors, the number of countries in a position to deploy nuclear power could remain strictly limited by considerations of size, industrial distribution, level of industrial development, and industrial infrastructure.

The detailed future international deployment pattern of nuclear industries and the nuclear fuel cycle is, however, likely to be a compromise between the "interdependence" logic of economics and the "independence" arguments of national status and security of supply. One important result of the INFCE study may be to enable countries to identify their own optimal course of action for the introduction of reactors, having regard to their own energy demand and supply and to economic and other implications of different fuel cycles. But, as noted earlier, INFCE had no political mandate, nor were its participants committed to the results of the study. Its results will be transmitted to governments for their consideration in developing their national nuclear energy policies and in international discussions concerning cooperation in nuclear energy development, related controls, and safeguards. It will, we hope, provide a basis on which to develop a new international consensus and political framework to govern nuclear trade and non-proliferation. The time is past when restraints of secrecy or export policy alone can serve as more than a short-lived brake on the spread of nuclear technology. Embargoes generate increased desire for self-sufficiency; for instance, originally heavily guarded enrichment technology is now available in at least ten countries, and the technology of reprocessing is widely known. However, a clear distinction must be drawn between agreed evaluations in INFCE, at a technical working level, and the subsequent development of national and international policies. The latter will be the acid test of whether the obvious pressing need to underpin the present NPT regime can be approached constructively, collaboratively, and successfully in seeking to reconcile the role for nuclear energy in meeting world energy needs with the non-proliferation objective.

From the lessons of recent history and from general trends in the INFCE discussions, some essential principles for a new approach seem already clear. First, there must be a restoration of confidence in non-proliferation objectives and policies between suppliers and customers, based on the bargain implicit in articles III and IV of the NPT. This is a major priority, and it will require greater political trust in assisting those whose established economic need is for an assured national energy supply. The approach should be evolutionary, leading to greater certainty about the terms for nuclear trade and the transfer of technology, particularly nonsensitive technology, and more effective measures on nuclear arms control. Not only must suppliers be prepared to recognize the prime importance less developed countries attach to nuclear power and its associated technology to meet their energy needs, but the recipients of necessary supplies and technology, whether sensitive or nonsensitive, must also be ready and willing to support universal acceptance of effective full-scope safeguards as one extremely important mechanism in satisfying non-proliferation requirements. They must be no less ready to accept such technical improvements in safeguards techniques as may occasionally be possible and desirable. Failure to restore confidence in the credibility of long-term supply agreements can be expected to encourage developing countries to establish their own nuclear facilities free from external controls. On the other hand, guaranteed supplies of materials and enrichment or reprocessing services might well postpone pressure to transfer these technologies, at least until they are shown to be economically necessary.

Second, since both suppliers and recipients have a vested interest in ensuring that there is no diversion of nuclear supplies to military purposes, a stable and credible international non-proliferation regime will require an agreed basis on which not only the technology of different phases of the nuclear fuel cycle can be transferred, but also fissile materials usable in weapons can be produced, stored, and used in national civil power programs. Discussions leading to such an agreed basis, preferably under the auspices of the IAEA, should be open to all customers and suppliers alike. Sensitive technologies may require different safeguards arrangements than those applicable to nuclear materials and equipment in general. In respect to the storage and use of fissile material, the outcome of current IAEA studies on the principle, practicality, and efficacy of possible multinational institutional arrangements for storing both excess separated fissile material and spent fuel will be of the highest importance. The practical managerial difficulties and financial implications of establishing and operating a multinational venture for this purpose are not to be underestimated, but there is little doubt that the principle has strong attractions for public acceptance and international confidence in non-proliferation policies. It would be particularly attractive if measures to achieve the perceived minimization of proliferation risks could be coupled with an assured supply of nuclear materials and thereby provide both material supply and financial disincentives to embarking on multiple national facilities.

Third, it should be accepted that, as in general arms control agreements, there can be no assurance of the absolute effectiveness of any agreed verification measures. A progressive series of practical confidence-building and cost-effective steps, based on mutual understanding and reasonable compromise, could be more profitable than insistence on an immediate and comprehensive requirement that in itself may imply an unacceptable measure of distrust. It may also fail to provide the flexibility required to deal with the fact that nuclear power development will be on differing scales and at different rates of growth in different countries.

If a new framework to govern the development of world nuclear energy needs is to be based on a renewal of the bargain in articles III and IV of the NPT, what should be done about any countries that, for one reason or another, refuse to adhere to the NPT? The answer might depend on how many and which countries remained outside an NPT network. In certain circumstances, there could be dangers in ostracism. But such countries should not be enabled, through supplementary arrangements, to obtain less onerous trading terms than those that accept the non-proliferation regime of the day. Agreement that, whether the potential customer was inside or outside the NPT, IAEA safeguards would be applied to all its civil nuclear facilities would be an important step in this direction.

Stable arrangements for preventing the proliferation of nuclear weapons should:

- recognize the primarily political rather than technical nature of the problem, since a decision by a government to develop nuclear weapons is a political decision;
- be based on political commitments voluntarily undertaken by sovereign governments in arriving at a cost-effective framework within which the security, energy, and political elements of a non-proliferation policy can be mutually reinforcing;
- avoid undue discrimination between industrial and developing countries, but accept that some discrimination is unavoidable merely through the prior existence of nuclear-weapon states;

- recognize that the main deterrent to overt proliferation lies in the potential political reaction of the international community;
- recognize that there are no technical means by which any one fuel cycle can be made proof against abuse;
- recognize that the problem will not go away merely by banning the construction and operation of nuclear power stations or by relying on policies of material and technological denial.

REFERENCES

Greenhalgh, Geoffrey. 1979. Positive steps towards non-proliferation. *New Scientist,* 30 August.

United States Department of State. 1946. *A report on the international control of atomic energy.* Publication no. 2498. Washington, D.C.: Government Printing Office.

6

International Custody of Plutonium Stocks: A First Step toward an International Regime for Sensitive Nuclear Energy Activities

Russell W. Fox and Mason Willrich

INTRODUCTION

It is widely acknowledged that perceived and real risks of the further proliferation of nuclear weapons and capacities for nuclear weapons production constitute the greatest obstacle to the proper development and expansion of international trade and assistance related to the civil uses of nuclear energy. Accordingly, measures to limit to acceptable levels the weapons proliferation risks arising from the nuclear industry must be developed, agreed upon, and widely implemented as an integral part of the future development of civil uses of nuclear energy.

A number of nations are already depending to a large extent on nuclear energy to meet their growing requirements for electric power. More will begin using nuclear power in the near future, when plants now under construction begin operation. Additional countries want to ensure, at this stage, that they will be able to acquire the raw materials, services, equipment, and technology essential to developing nuclear energy programs in the future.

Concerns about the security of energy supplies have raised apprehensions within nations and tended to destabilize international relations. Consequently it is important to recognize the need to restore confidence in international nuclear energy relations among exporters and importers and among developed and developing nations. The requisite confidence and stability are lacking in the present international political environment.

The risks of nuclear weapons proliferation arising from nuclear power and the development and use of related technology cannot be reduced to zero. However, adherence to the Treaty on the Non-Proliferation of Nuclear Weapons (NPT), the administration of safeguards on civil nuclear activities by the International Atomic Energy Agency (IAEA), the implementation by

Date of origin, November 1978.

the Nuclear Suppliers Group of agreed guidelines on exports of nuclear materials, equipment, and technology, and other measures are being widely employed to limit the risks. The Euratom Treaty applicable to the members of the European Community and the Treaty on the Prohibition of Nuclear Weapons in Latin America (Treaty of Tlatelolco) also contribute importantly to this end.

This paper rests on the following assumptions:

1. The world energy situation will be more difficult to manage, nationally and internationally, without nuclear power than if nuclear power is used wherever and whenever it can be reasonably justified on economic and other relevant grounds.

2. Nuclear power is being developed along a variety of technological paths, including once-through fuel cycles based on light water and heavy water reactors, and fuel cycles involving reprocessing to recover plutonium from spent fuel for possible future use in breeder reactors and, depending on the timing of breeder reactor commercialization, in light water reactors. No common path is likely to be followed by all countries. Instead, nuclear power will continue to be developed in a growing number of countries along diverse lines according to different schedules.

In view of the present lack of confidence in the established non-proliferation regime, and the urgency of creating an international consensus on an acceptable regime so that necessary development of nuclear power may proceed, we suggest that a useful first step would be to reach the broadest possible agreement on arrangements for international custody of civil plutonium. Such an agreement could, we hope, accelerate further steps toward an effective and acceptable international regime for the conduct of sensitive nuclear energy activities involving the production or processing of weapons-usable materials.

THE PROBLEM OF PROLIFERATION

A full exposition of the subject of nuclear weapons proliferation and the resulting dangers is beyond the scope of this paper. It is necessary at this point, however, to indicate what we mean by "proliferation." "Proliferation," as used here, refers to an increase in the number of governments having national control of nuclear explosives. This is often referred to as "horizontal" proliferation. The term "proliferation" is also used in relation to countries that are already known to possess nuclear weapons: that is, to be nuclear-weapon powers. Any increase in the quantity or quality of their weapons is often referred to as "vertical" proliferation. The paper does not deal with this related and important subject.*

* Furthermore, the paper does not deal primarily with the risks of seizure of weapons-usable materials by subnational or terrorist groups. That problem must be prevented primarily

In considering nuclear proliferation risks, we are necessarily concerned with the capacity to make nuclear explosives. The capacity of a national government to make only a few nuclear weapons is a major concern, even though sophisticated means of delivery may be lacking. The technical knowledge required to make a nuclear explosive device is widely available. The capacity for weapons manufacture also requires possession of or access to the facilities required to produce fissionable materials suitable for use in nuclear explosives, or possession of or access to such fissionable materials—namely, separated plutonium or high-enriched uranium. Such a capacity may be achieved through construction and operation of a small plant designed specifically to produce materials for use in explosives or, more indirectly, by diversion from a civil nuclear power program.

The proliferation risks associated with nuclear power development are mainly represented by unsafeguarded production of plutonium or its diversion from a commercial reprocessing plant or civil plutonium stock and by unsafeguarded production of high-enriched uranium or its diversion from civil facilities.

In focusing attention on the technical requirements for a nuclear weapons capacity, we must not lose sight of two major qualifications. The first is that intent is important. Capacity does not constitute an actual risk of proliferation at any particular time unless it is coupled with some intention, or at least some option intentionally retained. But the second qualification, which largely overrides the first, is that what usually matters most is the perception in other countries of the capacity, possible intentions, and possible options of the particular country concerned.

We regard the matter of perception of risk—looking in from the outside—as requiring particular emphasis. Weapons proliferation risks may well be perceived where they may not in fact exist. This circumstance is in itself apt to produce international tension, if not worse. If, therefore, a nation has nuclear material or equipment in use for civil purposes only, it is desirable for that nation to take steps to engender in other nations a confidence that nuclear weapons capacity, or intention, or both, are absent. The burden of satisfying the international community by taking such steps should be upon the nation concerned. The possession of weapons-usable materials such as plutonium or high-enriched uranium, or the possession of sensitive facilities such as reprocessing or enrichment plants that are capable of being used or readily converted to produce weapons-usable materials, is likely to be regarded by other countries as providing a basis for suspicion. It is not neces-

by national authorities, though international cooperation is highly desirable and in the common interest of all established governments. International custody of civil plutonium stocks may further reduce the risks of diversion by subnational groups, but such custody is no substitute for effective physical security measures applied by the national government on whose territory the stock is located.

sary to have a nuclear industry in order to raise such concerns. Some experts think that the greater danger lies in the use of small-scale or "experimental" reprocessing or enrichment facilities.

Many governments that now have the capacity to produce nuclear explosives have not done so. Two observations are important in this regard. One is that, even with a nuclear weapons capacity ready at hand and even with well-developed military establishments, a government may not find it necessary or desirable to acquire nuclear weapons. This is obvious enough from current experience. A government's perception of the security of its nation is of prime relevance. Moreover, to become a recognized nuclear-weapon power is a very serious matter internationally, with major consequences of its own. Thus the governments of most nations without nuclear weapons have adhered to international treaties, such as the NPT or Treaty of Tlatelolco, and thus undertaken an obligation not to acquire such weapons in the future. The second observation is that a comprehensive non-proliferation regime must have as an objective the removal of incentives for nuclear weapons, the creation of deterrents, the containment of capacity, and the early detection of any clandestine activities directed toward the production or acquisition of fissionable materials for nuclear weapons.

There is substantial concern that existing international treaties and institutions may not provide an adequate basis for dealing with the nuclear weapons proliferation problem. The IAEA Statute, the Euratom Treaty, the NPT, and the Tlatelolco Treaty are all important and valuable in their respective spheres. But adherence to them is not regarded, by a significant number of nations, as providing adequate international control of proliferation risks. The Nuclear Suppliers Group has come into being with a limited and selected membership, and the guidelines the group's members have adopted provide for certain restrictions, in addition to IAEA/NPT safeguards, on transfers of nuclear materials, equipment, and technology in an attempt to cope with the problem. Furthermore, certain suppliers are inserting into their bilateral agreements for nuclear cooperation even more stringent provisions than the Nuclear Suppliers Group guidelines require.

Especially difficult problems are raised by efforts of some suppliers to retain the right to approve any disposition or treatment of spent fuel discharged from power reactors in recipient countries, even though the supplier is under no obligation to accept the return of such spent fuel if no other disposition is mutually acceptable. It is noteworthy in this regard that several nations are both suppliers to other countries of technology and equipment and recipients from others of uranium or enrichment services or both.

There is no universal or even general agreement on desirable restrictions, in addition to current IAEA safeguards, or, indeed, on whether such restrictions should be required at all. The most that the nuclear suppliers as a group have been able to achieve is agreement to exercise "restraint" in the export of sensitive nuclear materials or equipment. In any event, some of the additional restrictions are in the form of further promises, such as not to replicate sensitive facilities using transferred technology, when promises and assurances, even by treaty, have sometimes proved inadequate in the past. The additional restrictions some suppliers wish to impose are likely to be resisted by some recipients of nuclear materials and equipment. In particular, some governments take the position that the only conditions required for the full operation of article IV of the NPT and similar provisions in other treaties, providing for the fullest possible exchange among countries of nuclear materials and equipment for civil purposes, are those conditions expressly stated in the particular treaties.

Since nuclear weapons proliferation diminishes the security of the entire world community, the supplier nations, whether acting individually or as a group, cannot and should not be regarded as the sole guardians against proliferation risks. Cooperation is required among suppliers and recipients of nuclear materials, equipment, and technology. The need for assured supplies and stable markets is likewise an international problem that can be resolved only through substantial international cooperation.

Economic and technological factors also tend to work in favor of international solutions to the proliferation risks related to nuclear power development. The vast sums of money involved, and the scientific and engineering complexity of many operations, make it highly desirable in many cases to have at least some international cooperation. Examples of such cooperation are the various international consortiums for enrichment and for reprocessing that have been formed.

At this point, we conclude:

- The nuclear weapons proliferation problem relates to the security of every nation.
- The critical proliferation risk is the diversion, actual or feared, of weapons-usable materials.
- Assurances of supply of nuclear materials, services, equipment, and technology for civil nuclear purposes are dependent on the establishment of non-proliferation measures recognized as adequate.
- A number of nations do not regard existing treaties and international arrangements, embodied in the IAEA/NPT system and certain regional arrangements, as adequate to deal with the proliferation risks arising from the widespread use of civil nuclear power.
- Additional restraints in bilateral agreements are at best only a partial solution. Some suppliers insist on conditions that are more onerous than those required by others. The absence of uniformity not only tends to distort international trade patterns in nuclear materials, services, and equipment, but also may subject nuclear enterprises in recipient countries to conflicting conditions when several sources of supply are involved. Restrictions are most likely to be accepted where they are least needed.

- Considerations of finance and technology make international cooperation desirable in many, if not most, cases of nuclear development.

INTERNATIONAL PLUTONIUM CONTROL

Based on the preceding argument, we believe it is desirable to seek to establish an international regime to limit the proliferation risks arising from civil nuclear power that:

- is more universal and acknowledges the common responsibility to deal with the problem, thereby lessening the chance of discrimination against any country;
- strikes more directly and positively at the critical matter of diversion;
- places less reliance on restraints in bilateral agreements;
- acknowledges that existing treaty provisions and assurances are not always regarded as adequate;
- gives confidence to nations generally that diversion of weapons-usable material is not taking place;
- promotes greater international or multinational participation in sensitive nuclear fuel cycle activities;
- provides a basis for assuring to recipients the supply as needed of nuclear materials, services, equipment, and technology for the needed exploitation of nuclear energy for civil power purposes.

We believe that, in moving toward such an international regime, applicable to sensitive nuclear energy activities, priority should be given to the negotiation of measures providing for effective international control of civil plutonium stocks. Such control could occur either through a scheme implemented by several interested countries acting together on a multinational basis in accordance with prescribed international standards, or through more direct controls applied by an international agency such as the IAEA. Several reasons support this suggestion:

- Stocks of plutonium beyond immediate needs for specific civil power and research purposes are likely to exist in an increasing number of countries in which reprocessing plants are already operating or are being constructed. Such excess plutonium can be placed under international control, thereby reducing proliferation risks, without hampering the use of nuclear power or civilian nuclear research, development, and demonstration projects.
- Stocks of civil plutonium exist in some countries whose governments have already acquired nuclear weapons and in some whose governments have not. It is also noteworthy that some of these countries are parties to the NPT and others are not. Therefore a scheme for international control of civil plutonium stocks can be implemented on a non-

discriminatory basis in nuclear-weapon and non-nuclear-weapon states, and in parties and non-parties to the NPT.

- The need for special control applicable to plutonium beyond immediate needs for power and research has long been recognized in the Statute of the IAEA (article XII.A.5). Therefore, in consenting to special international controls over plutonium, national governments need not alter their existing positions regarding the need or lack of need for controls (in addition to verification safeguards) over other sensitive nuclear fuel cycle operations.
- International control of civil plutonium stocks is a logical step toward the possible creation of a broader regime providing for internationalization of other sensitive nuclear energy activities in certain circumstances. In comparison with internationalization of various other nuclear fuel cycle operations, international controls over separated plutonium stocks would be relatively simple and cheap to implement. On the one hand, if the negotiation of such international control measures is unsuccessful, it is unlikely that the negotiation of broader, more complex measures applicable to spent fuel or reprocessing or both will succeed. On the other hand, successful negotiations regarding international control of civil plutonium stocks may enhance the chances for success in further, more complex, negotiations regarding multinational control of reprocessing or spent fuel.

Despite the benefits to be gained from international control of civil plutonium stocks, we nevertheless foresee a number of objections that may be raised to such a step. The principal objections and our replies are outlined below:

- Those in favor of once-through fuel cycles and opposed to reprocessing may object that any agreement on international control over plutonium stocks would legitimize, by implication if not expressly, the conduct of reprocessing operations. We reply that, while some governments favor the once-through fuel cycle without reprocessing, many governments with important interests in nuclear power development do not. Accordingly, civilian reprocessing is in fact occurring nationally in an increasing number of countries. It may be added that the recovered plutonium may not be in a pure form that is readily usable in explosives. Measures such as coprocessing may have been adapted for reprocessing, thus requiring further complicated treatment before the plutonium could be used in nuclear explosives.
- It may be objected that any reduction of proliferation risks resulting from international control over civil plutonium stocks will be minimal, since the vast bulk of such stocks will be in nuclear-weapon states or in non-nuclear-weapon states that are not considered proliferation risks. We reply that it is important to place under international control even

small amounts of excess separated plutonium in various countries, and the likelihood of achieving such a result will be enhanced if such controls are implemented in a nondiscriminatory way.

- Owners of separated civil plutonium may object that the benefits for the world community as a whole to be gained from international control do not outweigh the cost to them of loss of national control of plutonium that will be valuable for later use in breeder reactor research and development activities and as fuel in power reactors. We reply that future availability of plutonium for specific research or power uses would not be jeopardized by international control of stocks if appropriate assurances regarding release can be agreed. Moreover, the owner of spent fuel may be under an obligation to the supplier of uranium, enrichment services, or reactor technology to obtain prior consent to any arrangements for reprocessing. International control of separated plutonium should facilitate obtaining necessary supplier consent to transfers of spent fuel and to reprocessing.

Many possible schemes for international control of civil plutonium stocks can be devised that may be deemed useful and acceptable. The characteristics of specific schemes may vary according to the identity of the controlling agency or participants in the international scheme, the location of the plutonium stock, and the international control measures applied. Other matters to be settled include release criteria, allocation of the costs incurred in the international control arrangement, and the quid pro quo (if any) for relinquishment of national control.

Controlling agencies may be established case by case, through negotiations among groups of countries, including one or more reprocessors and/or suppliers of spent fuel, uranium, enrichment, or nuclear equipment. Alternatively, pursuant to various statutory authorities, the IAEA may serve as the controlling agency, either as a safeguards requirement under article XII.A.5 of the Statute or as a service offered under article III. We believe that arrangements negotiated case by case, separately among various groups of interested countries, will probably prove more practicable. However, such arrangements should meet certain standards, specified if possible in advance by the IAEA, that are universally applicable and nondiscriminatory. In developing such standards, we believe two criteria should be used: (1) international control should enhance the effectiveness and credibility of other assurances against diversion and thereby significantly reduce the risk of weapons proliferation; (2) international control should not unduly hamper or restrain the development and use of nuclear energy for civilian power purposes.

Plutonium stocks may be located in nuclear-weapon states, in non-nuclear-weapon states with reprocessing facilities, or in non-nuclear-weapon states without reprocessing facilities. We believe that political discrimination between states should be minimized. Moreover, transportation of separated plutonium should be minimized. The preferred solution in most cases would be to locate international plutonium storage facilities on the same sites as commercial reprocessing plants. Nevertheless, in some cases the non-proliferation benefits of transport of separated plutonium from small reprocessing facilities to stocks under international control in countries with large reprocessing facilities may outweigh the risks of transportation.

The measures to be applied to civil plutonium stocks should necessarily include safeguards, administered by the IAEA, and physical protection, administered by the host state. The essence of the proposal, however, is that there be actual physical custody, direct and complete, by the agreed multinational or international agency. This will go beyond the degree of surveillance common to ordinary safeguards. Such international control may be extended to the provision of any fuel fabrication or transport services subsequent to release of plutonium for use as fuel in power or research applications. Ownership would normally remain with the individual owners, but other arrangements could be made.

An issue of critical importance in arrangements providing for international custody of civil plutonium stocks will be the criteria governing releases from a stock of previously deposited material. It is likely to be necessary to agree upon release criteria that are quite precise so that, on the one hand, owners can have high confidence that plutonium will be released promptly in appropriate circumstances and, on the other hand, other participating countries as well as interested countries outside the particular arrangement can be assured that plutonium will be released only when there is to be immediate use for a defined civil purpose. We believe that release criteria should specify: the use (in an identified reactor or research facility); the time (for immediate fabrication into fuel for use within a particular time period in a reactor that is operating or planned for operation or for immediate use in defined experiments); and the continuing safeguardability of the activities involving use of released plutonium.

With regard to assurances of release of their plutonium from international custody, it is noteworthy that, in the near future, large-scale commercial reprocessing will occur in only a few countries. These reprocessing states will offer services to others. Accordingly, the logical location for an international stock of civil plutonium will, in most cases, be on the same site as a reprocessing plant. Fuel fabrication facilities may also be located on the site. In such circumstances, without some form of international custody, excess plutonium would either remain in the custody of the reprocessing state, be returned immediately to the owner, or be sold and shipped to a third party. It seems that original owners might have as much confidence, if not more, in the release mechanism of an international or multinational agency, which no one state could alter on its own, as in the supply policy of the reprocessing state, which could be changed quickly and unilaterally.

As part of an international custody arrangement it may be useful to provide specially for resolution of any dispute concerning the authorization of releases of plutonium. It should be possible for such a dispute settlement mechanism to be negotiated by the participants, but the assistance of an international agency may be helpful.

The costs of arrangements for international custody of civil plutonium stocks would be relatively small compared with the costs of reprocessing, fuel fabrication, or waste disposal. Costs of a particular stock would include capital costs of the storage facility, operating costs of the facility, and any additional costs of international administration incurred by the IAEA, a special international body established by the participants, or both. Such costs may be apportioned among participants on the basis of user charges computed on a normal commercial basis. Alternatively, international custody of civil plutonium stocks may be viewed as an additional safeguard measure, contemplated in article XII.A.5 of the IAEA Statute, and therefore some or all of the costs of international custody may be allocated among participants (or among a larger group of countries) according to a specially negotiated formula. We believe most of the costs of civil plutonium storage should be allocated among the participants according to user charges, since, in view of the likely excess of separated plutonium over requirements, storage somewhere would be necessary in any event. However, we believe that, as an inducement to agreement by owners to deposit plutonium under international custody, a formula might be devised whereby that part of the costs attributable to internationalization could be allocated in a manner that provided owners of plutonium an appropriate "non-proliferation subsidy."

As we noted above, if excess plutonium would be kept in the reprocessing state until utilization in any event, then owners may have greater assurance of return from an international custodial facility in which they share authority with several other participating countries than from a national custodial facility operated by the reprocessing state. Furthermore, the costs of a relatively few international stocks of civil plutonium, colocated with large reprocessing plants, are likely to be substantially less than the costs of widely dispersed stocks under national control, assuming that international safeguards and national physical security measures afford acceptable levels of protection against proliferation risks. These benefits, together with the perceived non-proliferation benefits to the world community as a whole, may be deemed sufficient inducement to owners of civil plutonium for their agreement to international custody of their excess material.

Countries that rely heavily on foreign sources for supplies of uranium, enrichment services, reactor equipment, or technology may, however, seek more in return for their agreement to international custody of civil plutonium stocks. In particular, such countries may bargain for nuclear fuel assurances from supplier states and for the removal of restrictions imposed by their suppliers on reprocessing of spent fuel. As we noted previously, we believe agreement on international custody of civil plutonium is a positive first step that both recipients and suppliers of nuclear materials, services, equipment, and technology can and should be able to agree to now. Such agreement should pave the way toward further steps and compromises among suppliers and recipients, which will be necessary in order to create a stable international non-proliferation regime for the future. Such a regime should include credible supply assurances and also, where mutually agreed, internationalization of other sensitive civil nuclear energy activities.

Regarding implementation of an international custody scheme, some further suggestions are:

- The Board of Governors of the IAEA, as the most appropriate international body, should adopt a resolution giving approval and encouragement to international custody of civil plutonium stocks and establishing standards that such arrangements should meet. Among other advantages, this would help to dispel suspicion of discrimination against non-nuclear-weapon states.
- The resources of the IAEA should be used to assist with discussions between countries that are, or may be, interested. The countries participating in any particular custody arrangement would, of course, have to be so composed that improper collusion among them could not be reasonably suspected.
- There should be constantly on the site of an international plutonium stock a representative or representatives of each of the participating countries, or of the international controlling body. Close cooperation will be necessary between the security forces of the host country, the authorities having custody of the plutonium, and the international safeguards agency. The civil plutonium stock and the related reprocessing and fuel fabrication facilities, if any, should be subject to IAEA safeguards and, within the European Community, also to Euratom safeguards.
- Insofar as is practicable and acceptable to parties, arrangements for the international custody of civil plutonium stocks should be designed and implemented with a view to continuing the application of the arrangements after plutonium is released from the stock and during subsequent fabrication and transportation to facilities where it is to be used.
- The adequacy of arrangements for international custody of civil plutonium stocks should be certified by the IAEA. In some cases the periodic satisfaction of a supplier might be made a condition of the nuclear supply agreement.

Finally, it is conceivable that the host country where a civil plutonium stock is located may, in breach of its treaties, seize the stock and exclude the other countries or the international agency, as the case may be. This is a risk. Whether it should be accepted depends upon an examination of all the relevant circumstances. In any case, it is likely that such a seizure would be regarded as tantamount to a public declaration that the country was making nuclear weapons, and this would be likely to have major international repercussions.

CONCLUSION

In conclusion, we believe an international regime should be developed that will enable international nuclear commerce to proceed as freely as possible, without arbitrary or unpredictable inhibitions, provided non-proliferation requirements are satisfied. It is desirable that policies involved be non-discriminatory, so far as this is possible. This does not negate the application of sanctions in appropriate cases or the possibility of refusing supply to a country that fails to satisfy internationally accepted non-proliferation requirements or recommendations.

It is desirable that an international consensus be reached on acceptable non-proliferation requirements as soon as possible, so that the planning of national energy programs can proceed. The nuclear industry is in its infancy. Although uncertainties and conflicts in government policies have considerably weakened the industry, it is still capable of rapid growth. An appropriate balance must be struck between the expansion of energy supply from nuclear power, so far as that is desirable, and controls to reduce associated risks of weapons proliferation.

REFERENCES

Chayes, Abram, and Lewis, W. Bennett, eds. 1977. *International arrangements for nuclear fuel reprocessing*. Cambridge, Mass.: Ballinger.

Meller, Eberhard, ed. 1980. *Internationalization: An alternative to proliferation?* Cambridge, Mass.: Oelgeschlager, Gunn and Hain.

Smart, Ian. 1980. *Multinational arrangements for the nuclear fuel cycle*. Energy Paper no. 43. London: Her Majesty's Stationery Office.

Stockholm International Peace Research Institute. 1980. *Internationalization to prevent the spread of nuclear weapons*. London: Taylor and Francis.

Williams, Frederick C., and Deese, David A., eds. 1979. *Nuclear nonproliferation: The spent fuel problem*. New York/Oxford: Pergamon.

7

Nuclear Power Plant Lead Times

Richard K. Lester

INTRODUCTION AND SUMMARY

In several countries over the past few years, concern has been growing over the increasing length and unpredictability of lead times associated with the planning, design, and construction of nuclear facilities, including, most notably, nuclear power plants. These trends are seen to threaten the future economic viability of the nuclear power industry, at a time when many governments are acting to lessen the impact of anticipated fossil fuel shortages in the 1980s by increasing the proportion of electricity generated from nuclear energy.

The purpose of this paper is twofold: to assemble nuclear power plant lead-time information developed in several countries (some for the first time), and to conduct an international comparison of the trends in these data and the factors underlying these trends. I believe such an approach may shed additional light on the results of previous studies of the lead-time issue.

Much of the information reported in this paper has been obtained from the unpublished results of four preliminary analyses of nuclear power plant lead times in Japan, West Germany, Canada, and the United States, prepared earlier for the International Consultative Group on Nuclear Energy.* The paper also draws on several other relevant studies that have been published recently.

Date of origin, November 1978.

* Information from Japan and Canada was collated and provided by ICGNE members, assisted in the Canadian case by M. Joyce. Reviews of the situations in West Germany and the United States were carried out respectively by W. Jaek and W. Lenhardt of Kernforschungsanlage Jülich GmbH and by K. W. Sieving and E. A. Saltarelli of the NUS Corporation. Unless otherwise cited, it may be assumed that any lead-time data for those four countries appearing here have been drawn from those analyses, to which this paper owes a great deal. In addition, I am especially grateful for the invaluable encouragement, advice, and insight provided by Myron Kratzer and William Walker, whose contributions deserve far more than this brief mention.

The following section contains a brief illustrative discussion of the effect of lead times on the viability and competitiveness of the nuclear power industry. The quantitative evidence for national and international trends in lead times is reviewed, and the factors contributing to these trends are identified and analyzed in the two subsequent sections. The importance attached to the various factors must depend largely on qualitative judgment, since the detailed studies that would be required to analyze the frequently highly complicated interactions between them remain to be carried out. The paper's conclusions are contained in a final section. Among the principal conclusions set out there are the following:

- Nuclear power plant lead times have increased in the United States, West Germany, Japan, and Canada.
- The trend has been most pronounced in the United States and least severe in Canada.
- Increasing unpredictability in individual plant lead times has been identified as a particular problem in the United States and West Germany, and to some extent also in Japan.
- Lead times in many cases are approaching utility planning horizons and sometimes are exceeding them.
- The main reasons for longer and more unpredictable lead times include: the technical problems encountered when increasing the sizes of power plants; the lack of design standardization; the variable economic climate of the past decade; the increasing number and stringency of nuclear safety and environmental regulations and the continuing fluidity of the general regulatory environment; and, more generally, an inadequate level of political consensus on what constitutes an appropriate role for nuclear energy.

BACKGROUND

Increases in the length of power plant lead times adversely affect the economic attractiveness of nuclear energy in four major ways. First, delays require the purchase of replacement power by the utility, if it is available. Second, delays increase the length of time for which the utility is obtaining no return on its investment, thus increasing the contribution of interest charges to the overall cost of electricity generated by the power plant. Third, delays cause expenditures to be made later than originally planned, thereby increasing the risk of unanticipated inflationary effects.

The monetary costs of these effects depend on when the delay takes place and the need for the power that was to have been produced by the delayed plant. It is difficult to arrive at precise estimates of these costs, but, clearly, as plant construction proceeds and the total investment in the project increases, the impact of delays on interest charges grows. For example, for a 1,100 MW plant, one large United States utility has estimated a low cost of $50 million for a year's delay when the delay occurs early and no replacement power is required, and a high cost of $200 million per year of delay when the delay is at the end of the construction period and replacement power must be purchased (Murphy et al. 1978). At a fixed charge rate of 15 percent per year and a nuclear plant capacity factor of 65 percent, a $200 million increase in capital cost would add almost 5 mills/kWh to the cost of nuclear power. Thus, for a plant originally estimated to cost $1,000/kW, for which fuel costs were 7 mills/kWh, $200 million worth of delay would increase nuclear electric power costs by about 15 percent. (These numbers are meant only to indicate the magnitude of the effects of delay.)

The fourth major effect of lead-time increases is experienced during the utility planning process and is less easily measured in monetary terms. Nevertheless, if nuclear power plant lead times continue to grow, they will ultimately exceed the utility planning horizon beyond which reasonably reliable forecasts of energy demand cannot be made. In such a situation, confidence that proposed generating capacity will in fact come close to matching future energy demand and bring a satisfactory return on investment is inevitably eroded. The consequences will vary, depending on the nature of the utility and its relationship to the rest of the electric power sector. However, where a mandate prevails to provide power reliably and as economically as possible, utility planners faced with these prospects will prefer to turn to other energy sources with shorter and more certain lead-times, if these are available.

When a general trend toward lengthening lead times is combined with increasing uncertainty over the lead times to be expected for specific nuclear power plant projects, the problems just described are compounded. Growing lead-time unpredictability magnifies uncertainties over the ultimate cost and commercial viability of individual ventures. Furthermore, even if nuclear plant lead times remain within utility planning horizons on the average, the *possibility* that major delays will be encountered on individual projects may be sufficient to cause naturally risk-averse utility planners to select alternative energy supply sources with shorter and more certain lead times, even if longer-term economic penalties might be associated with such a choice.

There are, of course, several other uncertainties facing utilities investing in nuclear plants: changing social and political attitudes toward nuclear power, doubts about methods of disposing of spent fuels and nuclear wastes, and the price and availability of nuclear fuels, to name but a few. These uncertainties will not be discussed in this paper except where they have a direct bearing on reactor lead times, though it should be noted that at least some of them may themselves be exacerbated by the unpredictability of those lead times.

It should also be noted that this paper, which focuses exclusively on lead times for the planning, design, and construction of the existing generation of nuclear power plants, addresses only one aspect of a much broader subject. For instance, there is no discussion of the lead times associated with the deployment of the current generation of nuclear fuel cycle facilities and the development and introduction of new power plant and fuel cycle technologies. Such lead times would also be expected to exert a profound influence on the future viability and configuration of the nuclear power industry. Neither is there any discussion of the lead times associated with the installation of alternative energy supply and conservation systems; the important task of placing lead-time trends and problems in the nuclear industry within a broader energy policy framework is not attempted in this paper.

The scope is limited in another important respect. Most of the discussion is based on lead-time experience in four advanced industrialized countries—Japan, the United States, West Germany, and Canada—each of which maintains a major domestic nuclear power plant manufacturing and construction industry. The paper does not include an analysis of the lead-time issue in the context of countries, particularly in the developing world, without a large-scale nuclear industrial infrastructure, in which power plant projects are frequently characterized by a high degree of foreign participation.

One further qualification should be made. To the extent that lead-time trends are symptoms of the state of the nuclear power industry in a given country, analysis of the factors responsible for the trends fulfills a useful diagnostic function. But lead times are not necessarily the only, or even the most important, indicator of this state. In some countries, where lead times are not an important issue, analysis of other indicators and the factors responsible for them may be more illuminating.

Finally, although increases in lead-time length and unpredictability can serve as rough proxies for the economic costs that may be incurred by a utility, lead-time behavior does not in itself provide more than a partial indication of the broader social costs or benefits of the eventual outcome of nuclear projects. Estimates of the latter will be influenced by a multitude of additional factors, whose assessment will vary according to the views of the various estimators.

LEAD-TIME TRENDS

Definitions

Figure 7.1 shows in simplified, schematic form the principal stages required to bring a nuclear power plant into operation. These stages frequently overlap. For instance, site selection may be initiated before a decision to build a plant has been taken, and engineering typically might be half-completed by the time construction starts.

The interaction of the regulatory process with this sequence of activities varies between countries, and often between regions of the same country. Some of these differences are evident in figure 7.2, which illustrates

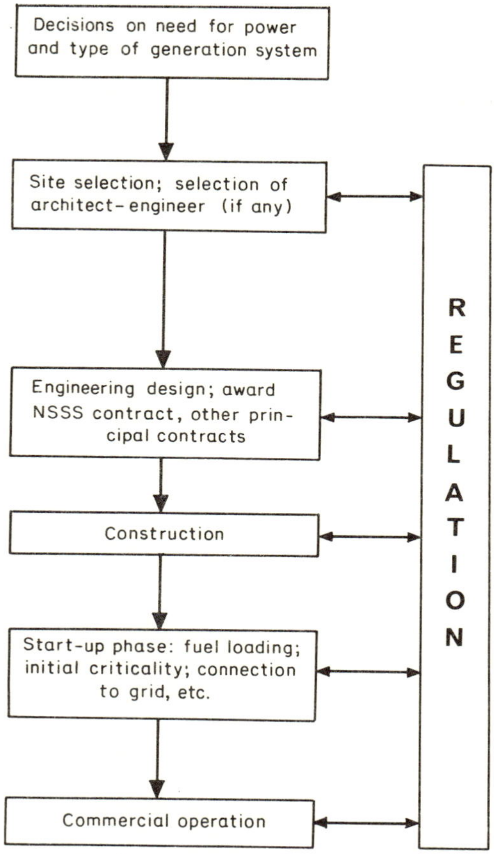

FIGURE 7.1. Reactor lead-time components.

regulatory milestones for power plant projects in Japan, West Germany, the United States, and Canada. For example, only one construction permit is issued for projects in the United States and Canada, whereas in both West Germany and Japan the construction permit is issued in several installments throughout the construction phase. Other differences can be seen at the operating license stage.

To avoid ambiguity in the interpretation of the available statistics, three measures of lead times will be used in the following discussion. The *overall lead time* will be taken to refer to the period from the utility's application for construction permits to build a nuclear power plant to the start of full commercial operation. This overall lead time will be divided into two component parts: the *preconstruction lead time,* which extends from the application for construction permits to the issue of the first (or, as in Canada and the United States, the only) construction permit, and the *construction lead time,* which

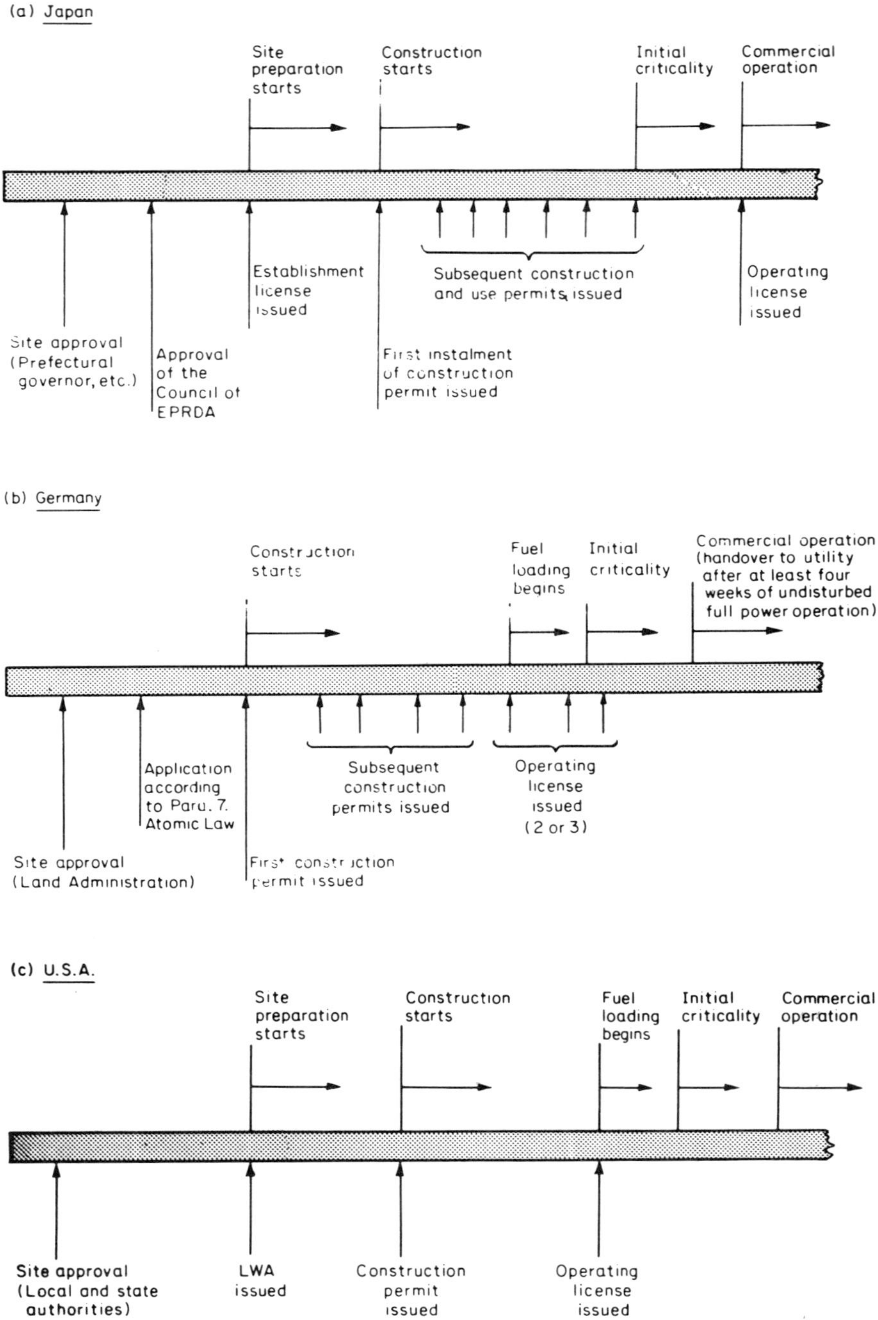
(a) Japan
Site preparation starts
Construction starts
Initial criticality
Commercial operation
Establishment license issued
Subsequent construction and use permits issued
Operating license issued
Site approval (Prefectural governor, etc.)
Approval of the Council of EPRDA
First instalment of construction permit issued
(b) Germany
Construction starts
Fuel loading begins
Initial criticality
Commercial operation (handover to utility after at least four weeks of undisturbed full power operation)
Application according to Para. 7. Atomic Law
Subsequent construction permits issued
Operating license issued (2 or 3)
Site approval (Land Administration)
First construction permit issued
(c) U.S.A.
Site preparation starts
Construction starts
Fuel loading begins
Initial criticality
Commercial operation
Site approval (Local and state authorities)
LWA issued
Construction permit issued
Operating license issued

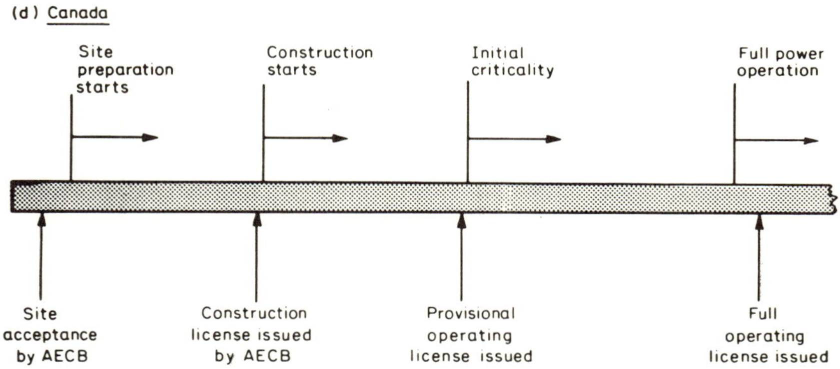

FIGURE 7.2. Major regulatory milestones: (*a*) Japan; (*b*) West Germany; (*c*) United States; (*d*) Canada.

extends from the issue of the first construction permit to the beginning of full commercial operation.

It should be noted that the definition of overall lead time does not include the period between the initial declaration of intent to build a nuclear power plant and the application for construction permits. This period includes part or all of the site selection and approval process, preliminary design work, and possibly other preparatory steps. Although delays experienced at this stage have relatively little impact on overall project costs, from the perspective of the utility planner the entire length of time from the formal decision to build a plant to the point of commercial operation is of vital importance, since it is over this period that energy demand must be projected with a high degree of confidence. Nevertheless, the phase preceding the construction permit application has been omitted from the formal definitions, since the declaration of intent may come in a variety of forms and is seldom easy to pinpoint in time; it therefore fails to provide an unambiguous reference point upon which to base a definition. However, the site selection and approval process that constitutes probably the most important element of this initial phase (although part of it may precede the formal decision to build a plant and other parts may take place in parallel with the construction permit application procedure) is addressed separately, and a rough estimation of *"total planning" lead times*, including the initial phase, is attempted at the end of the section.

The measures selected here are not ideal, but they appear to be the best available in the circumstances. A finer breakdown of lead times would allow more reliable identification of the stages at which delays are encountered, but the data at hand are not of sufficient quality to justify international comparisons on such a basis. Even with these broad divisions, considerable difficulties have been experienced in translating national statistics into a form appropriate for international comparison.

Site Selection and Approval Times

Before moving on to a discussion of overall lead times as defined, it is worth drawing attention to the evidence of increasing delays encountered in site selection and approval—certainly in Japan and the United States (data were not available for West Germany). In some extreme cases in Japan, it is reported that as many as three years elapsed after site selection before prefectural governments gave their approval and the proposed nuclear installation could be referred to the national Council for Electric Power Resources Development and Adjustment (EPRDA). The trend toward increased siting times in Japan is likely to be reinforced if the National Diet passes the Environmental Assessment Act, whereby additional agreement will be required from local inhabitants, mayors, village heads, and the minister for environmental protection. In the United States, the Atomic Industrial Forum has estimated that the time required for site selection and approval under current (1978) regulatory conditions (state and federal) ranges from 2.5 to 4 years, a substantial increase over the time required in earlier years (AIF 1978).

It is generally believed that multiunit siting would provide some check on these increases. In Canada, where multiple-unit stations are the rule rather than the exception, siting times in general have been shorter than in the United States.

Preconstruction Lead Times

As shown in figure 7.3, preconstruction lead times have become longer in recent years in Japan, West Germany, and the United States. The data points represent annual average lead times (i.e., for all reactors entering commercial operation in a given year). These lead times are identified as the intervals between the following events (the correspondence between which is only approximate):

	From	*To*
Japan	Approval of Council of EPRDA	Issue of first construction permit
West Germany	Application for permit according to paragraph 7 of the Atomic Law	Issue of first construction permit
United States	Application for construction permit	Issue of construction permit

Major increases in preconstruction lead times have been experienced in the United States, from an average of less than a year for plants receiving construction permits in the mid-1960s to over two and a half years more recently. Lead times for certain projects have ranged as high as five years, and, since there are a number of permit applications for the early 1970s that have still to

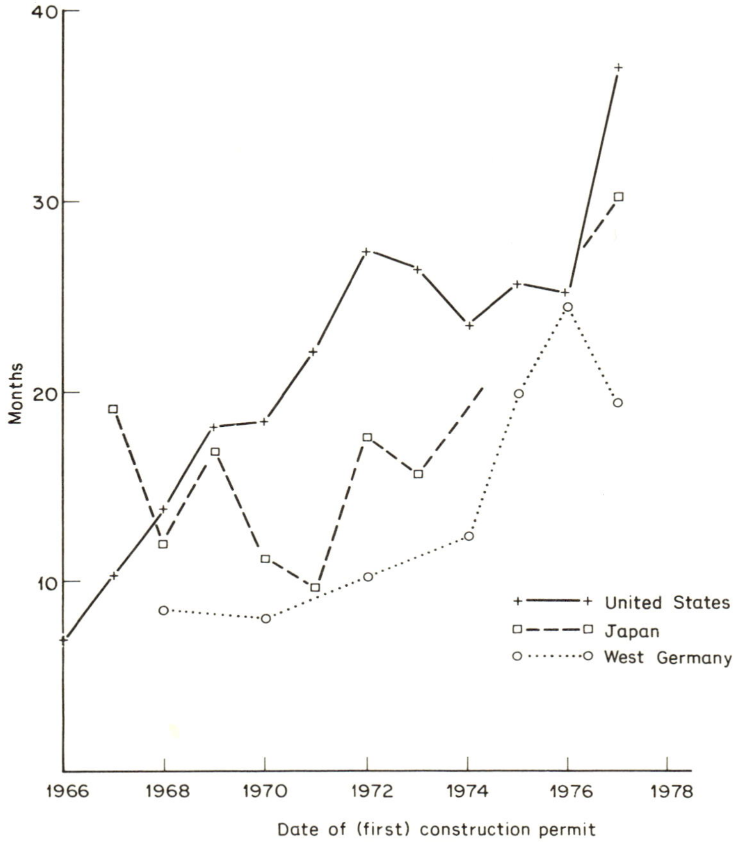

FIGURE 7.3. Average preconstruction lead times for nuclear power plants (calculated for all plants receiving [first] construction permit during year).

be granted approval (there are twenty-six outstanding from 1973/74 alone), it seems likely that this upper limit, if not the annual average, will be exceeded in the coming years.

In West Germany, preconstruction lead times have risen on average by over a year since 1973. While the average lead time for all West German plants with first construction permits is 1.4 years, lead times for the last five plants to receive permits have all exceeded 2 years.

Significant preconstruction lead-time increases have been experienced in Japan since 1974. Previously, these had never exceeded two years, with one year being fairly typical. Since 1973, they have never been less than two years, and the lead times for the permit applications in the pipeline may be considerably higher.

In Canada, the formal licensing process begins with the applicant's submis-

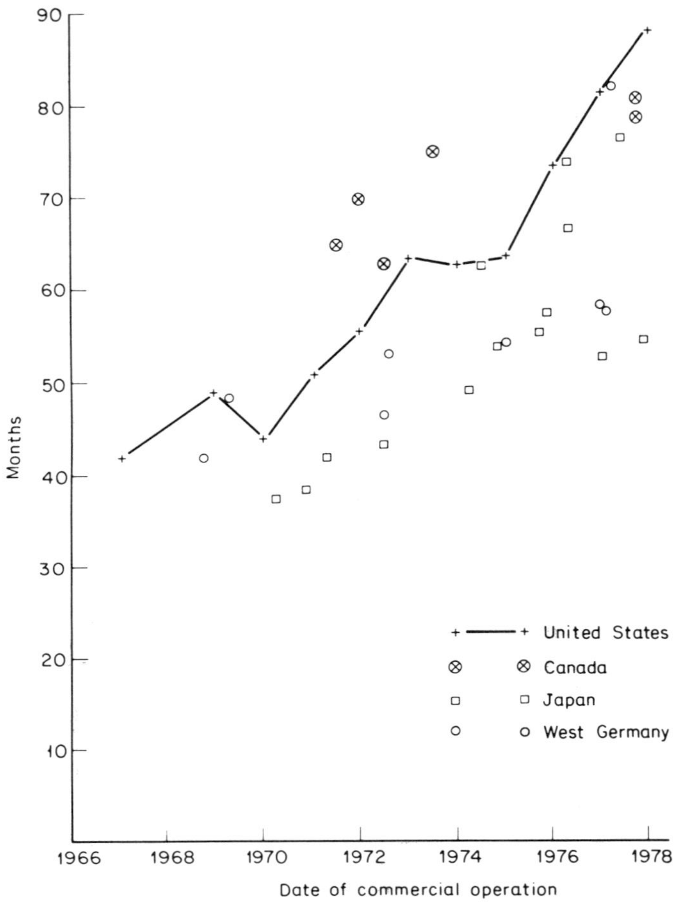

FIGURE 7.4. Construction lead times for nuclear power plants (data points represent annual averages for the United States and individual plant lead times for others).

sion of a letter of intent to the Atomic Energy Control Board (AECB). Following the submission, the procedures necessary to obtain site acceptance by the AECB are conducted in parallel with the first part of the construction licensing phase. Unfortunately, a formal application for a construction permit, which marks the beginning of the preconstruction lead time as defined previously, is not included among the procedural milestones for which Canadian data have been assembled, and it is not clear that the submission of the letter of intent corresponds well to the construction permit application milestone used for the other national cases reviewed here. Nevertheless, in the absence of more suitable data, it is convenient to redefine the preconstruction lead time in this case as beginning with the receipt of the letter of intent by the AECB. With this revised definition, the available information shows that preconstruction lead times in Canada have remained relatively stable, at an average of slightly under twenty-one months.

Construction Lead Times

Figure 7.4 shows how construction lead times for West Germany, Japan, the United States, and Canada have changed with time. The United States curve again represents annual averages, but the sample sizes for the others were too small to justify either curve-fitting or averaging, so that in these latter cases the data points refer to individual reactors. The data indicate that each country has experienced an upward trend in construction times (roughly from four to seven years in the case of the United States over the past eight years). It is interesting to note that construction lead times have been highest in the past in Canada. However, most plants in Canada are located at multiple-unit stations, and optimum project scheduling at such sites demands staggered completion of the individual units. For a four-unit station, for instance, two years might ideally separate completion of the first and last plants. This factor is partially responsible for the relatively high construction times experienced in this case.

Figure 7.4 gives no indication of the lead times being encountered on plants currently under construction. In the United States, there are thirty plants incomplete for which construction permits were issued before 1974; there are also substantial backlogs of incomplete plants in West Germany and Japan. The completion dates of these plants remain uncertain, but it seems unlikely that their construction lead times will show any marked improvement. For those plants still at the planning stage, the situation is more uncertain. It has been estimated, however, that the construction lead time for each West German plant currently awaiting its first construction permit is likely to exceed six years—substantially above previous experience. On the other hand, Canadian utility planning schedules indicate that construction lead times for plants currently under construction or still at the planning stage are not expected to increase significantly above recent experience.

Overall Lead Times

It is clear from the discussion above that, with the possible exception of Canada, where the trend seems to have been less pronounced, overall lead times (as earlier defined) have grown substantially over the previous decade in the countries we have been considering. In the United States, for instance, the Nuclear Regulatory Commission reports that overall lead times have increased from 51 months for plants completed in 1970 to 102 months for plants com-

TABLE 7.1. Estimated "Total Planning" Lead Times in Current Conditions for Nuclear Power Plant Projects

Country	Lead Time	Comments
United States	12–15 years	12–14 years (Murphy et al. 1978); 12.3 years median for plants coming on line in 1976 (NRC 1978); 13–15 years from decision to select site to commercial operation (AIF 1978).
Japan	9–12 years	From firm project definition to commercial operation, for projects receiving EPRDA approval in 1978.
West Germany	9.7+ years	Estimated average for nine LWR plants currently under construction, assuming at least one year for site selection and approval before application under paragraph 7 of the Atomic Law. (Construction at three of the nine sites halted by legal action; figures may be optimistic.)
Canada	7.3 years	Average from utility planning schedules for plants currently under construction, based on initial units entering operation at Pickering B, Bruce B, and Darlington A and individual plants at Gentilly and Point Lepreau. (Including subsequent plants at multiple-unit sites increases average lead time to 8.8 years.)

pleted in 1976 (NRC 1978). Estimates for overall lead times now typically lie between eight and twelve years in all four countries.*

Although it is not clear from the available data whether the underlying lengthening trend will continue beyond the mid-1980s, it seems almost certain that, for those reactors at the planning stage and under construction, lead times will remain protracted and may in fact show further increases.

"Total Planning" Lead Times

I mentioned earlier that the "total planning" lead-time period — from the declaration of intention to build a plant to its commercial operation — was of considerable importance to utility planners. Notwithstanding the difficulty of

* Preliminary data received from France (which was not included in the original international survey) indicate that projected French lead times lie at or even below the lower end of this range. Electricité de France (EdF) estimates that total lead times, including the time required to receive site authorization, for plants currently under construction or at the planning stage will range from 7.25 to 8.5 years, depending on how large the plant is and whether it is located at a multiunit site. It is noteworthy that projections of construction lead times are quite low relative to those in the other countries reviewed here, ranging from 56 to 66 months for the period from the first pouring of concrete to full commercial operation. (In France, the start of construction proper typically follows the issue of the *déclaration d'utilité publique* (DUP) and the construction license by one year, since preliminary site work — typically about a year in length — is not started until these are obtained.) According to these figures, EdF is thus expecting an improvement over the construction lead times achieved for the four French light water reactors (LWRs) already in service, which ranged from 75 to 86 months. Nevertheless, the utility is currently anticipating about a year's delay during construction for the first of the new plants to enter service (in 1979/80), although the lead times for subsequent plants built at the same sites are expected to be progressively reduced.

arriving at a consistent definition of the announcement of intent, it is interesting to compare estimates of total lead times for power-plant projects undertaken under current conditions, as is done in table 7.1. (A crude attempt has been made to account for the peculiarities of project scheduling at multiunit sites in the case of Canada.)

FACTORS RESPONSIBLE FOR INCREASING LEAD TIMES

Table 7.2 lists some of the principal factors that have been identified in the literature as exerting influence on power plant lead times. Although the list is not necessarily comprehensive, its character is nevertheless that of a "lowest common denominator": not all the factors are present in every country, but all of those mentioned in the national studies are included.

Two points concerning the taxonomy in table 7.2 need to be stressed. First, there are interdependencies both within each category of factors and between factors in different categories. For instance, the dynamic nature of the regulatory environment may influence standardization efforts in the industry, which may in turn influence labor productivity; there are also links between utility size, plant size, and multiunit siting policies, and so on. These interdependencies add to the difficulties of estimating the relative importance of the factors involved. Even for individual projects, it is not always easy to identify the particular cause of a delay or the full impact of a particular event on the overall schedule. But the task becomes much harder when data from different countries are being compared and national industrial experience is generalized.

Second, some of these factors may be responses to deeper and less tangible socioeconomic pressures. For instance, it is often argued that the growth of regulation and disruptive court action, which is by no means limited to the nuclear field, stems from social changes accompanying the rise of prosperity, the advance of education and communications, and increasing demands on the environment. Although such underlying issues are not addressed directly in this paper, their relevance to an analysis of the proximate factors affecting lead times should be kept in mind.

One further point should be emphasized at this stage. In weighing the contributions of different factors to lengthening lead times, all qualitative judgments must be treated with caution. In the past, for instance, those in the nuclear industry have tended to regard the regulators and that section of the public that opposes their objectives as the principal contributors to delay and uncertainty. Other sources have emphasized industrial deficiencies, financial constraints, and a reluctance on the part of the industry to adjust to generally diminished energy-demand growth rates. Thus, one of the purposes of this analysis is to identify areas of reasonable doubt as to causes, to the extent that this is permitted by the original sources of data.

TABLE 7.2. Factors Affecting Nuclear Plant Lead Times

Economic environment
Matching of supply and demand
Financial state of utilities
Availability of factors of production (industrial capacity, engineering resources, craft labor, etc.)
Technical and industrial performance
Utility size
Plant size
Plant type
Standardization
Labor productivity
Labor-management relations, labor disputes
Electric power industry organization: relationships between utility, reactor vendor, component suppliers, and architect-engineer (if any)
Project management experience, efficiency, and competence of utilities, manufacturers, and architect-engineers
Errors in engineering design and plant construction
Resolution of technical deficiencies identified in operating reactors
Errors and/or delays in the manufacture of long-lead-time components
Siting and regulation
Distribution of regulatory jurisdiction between national, regional, and local authorities: coordination and overlap
Mechanisms for public participation in regulatory process
Role of judiciary in regulatory process
Provisions for treatment of standardization in regulatory process
Accommodation of various local interests during siting stage
Provisions for multiunit siting, early site determination, etc.
Requirements for backfitting and regulatory ratcheting
Requirements for quality assurance and quality control
Volume of regulations
Volume of licensing applications
Efficiency and competence of regulatory authorities
Degree of separation of regulatory and promotional functions
Direct action at site by interest groups opposed to plant construction
Factors external to power plant siting, design, construction and regulation
Coupling of plant licensing to waste management, reprocessing, or other unresolved issues in the nuclear fuel cycle
Status of plant fuel supply coverage

Note: No attempt has been made to rank these factors in order of importance.

The organization of factors shown in table 7.2 is continued in the following discussion.

Economic Environment

A sharp change in the condition of the international economy has occurred over the past decade or so, from rapid and accelerating growth in the late 1960s and early 1970s to relative stagnation after 1973/74. Power plant lead times have been directly affected by these events in at least three respects: the rush of nuclear (and fossil fuel) plant ordering during the period of economic expansion in some cases placed serious strains on the resources available to utilities, equipment suppliers, construction firms, and regulatory authorities; heavy investment in present and future generating capacity subsequently led several utilities into financial crises; and the decrease in electric power demand growth rates associated with the economic recession has sometimes led to actual or probable grid overcapacity.

Evidence for resource shortages stretching construction lead times before 1975 has come mainly from the United States. Shortages occurred in a number of forms. Labor shortages caused disruption on several construction sites; a survey conducted in 1974 suggested that labor problems (mainly shortages) accounted for the slippages in as many as a third of the forty nuclear plants covered. Concurrently, there was a shortage of industrial capacity that reached a peak in 1974 with delays in the supply of many components. Perhaps even more important, serious shortages of professional and technical staff emerged between 1969 and 1974, with experienced skilled manpower being spread dangerously thin. These problems were particularly evident in architect-engineering firms, at a time when the number of man-hours required in the design of nuclear plant was growing rapidly. Conditions in the labor market were made even more difficult by the rapid expansion of regulatory bodies, which sought to overcome their inexperience and understaffing by recruiting skilled labor, especially at the managerial level, from utilities and plant suppliers. The resource shortages experienced by the regulators in turn added to the delays experienced in planning and construction.

Such shortages appear to have been less of a problem in Japan and West Germany in the past. In Canada, where construction lead times have been consistently higher, resource limitations, perhaps arising from the relatively small overall industrial capacity of the country, may have played a more important role.

In none of the industrialized countries with major nuclear infrastructures do resource shortages currently appear to be causing difficulties, however. If anything, the current problem is one of underutilization. But for some other countries, particularly in the Third World, that are developing nuclear power on a large scale without an extensive industrial infrastructure, resource availability, particularly of skilled manpower, may constitute a much more serious problem.

A related, more general question concerning the possibility of resource shortages is whether the current recession in the major nuclear industries of the world will erode the industrial infrastructure to the point where the prospects for meeting a resurgence in the demand for nuclear power, should one occur, may be adversely affected.

The impact on nuclear plant lead times of financial problems faced by electric utilities during the recession has varied from country to country and from utility to utility, according to the extent of overcapacity, relationships

TABLE 7.3. Average Construction Time (months)

Reactor Type	Calendar Year When Plant Started Commercial Operation						
	1970	1971	1972	1973	1974	1975	1976
PWR<600 MW	49(4)	47(1)	48(3)	61(5)	64(6)	61(4)	73(1)
PWR>600 MW	–	47(1)	57(5)	69(5)	74(6)	69(9)	74(5)
BWR<600 MW	46(1)	51(3)	64(3)	–	56(2)	55(2)	60(1)
BWR>600 MW	–	61(3)	60(3)	–	80(5)	84(5)	74(2)

Source: IAEA 1978.
Note: Figures in parentheses indicate number of units.

with financiers, the size of the utility, and so on. In West Germany, utilities have not yet encountered serious financial difficulties, possibly because they have been able to adjust their rates to cover debts and rising costs, and such difficulties are not anticipated in the future.

In the United States, however, it has been claimed that utilities' financial problems have had a significant influence on both construction and preconstruction lead times. It is difficult to gauge the severity and pervasiveness of this influence, partly because the utilities themselves have tended to understate its importance for political and economic reasons. Nevertheless, the analysis prepared for the ICGNE asserts that "beyond a reasonable doubt" many plant cancellations and postponements were made, and delays tolerated, because of the poor state of utility finances after the 1973 oil crisis. On the other hand, the United States Department of Energy has estimated in a recent report that only a very few utilities have had to extend construction schedules owing to such problems, though it acknowledges that several others are threatened (USDOE 1978).

Both sources are essentially in agreement on the proximate causes of these problems: the high costs of equity and debt capital coupled with rate structures that prevented adequate rate relief during periods of heavy construction expenditures. While general inflationary pressures were partially responsible for difficulties experienced in capital markets, both sources also blame an increasingly rigorous nuclear regulatory climate for the problems faced by utilities.

The effect of reduced electric power demand growth rates since 1973/74 on power plant lead times has also varied from country to country. In West Germany, growth rate reductions have been considerable, but the analysis prepared for the ICGNE indicates that the effect on plant lead times has been outweighed by licensing and other problems and has not been a controlling factor until now. On the other hand, the analysis of the United States situation shows that the construction schedules of a significant number of United States power plant projects in the mid-1970s have been deliberately stretched out because of grid overcapacity. There are reports that some delays in Japan between initial criticality and commercial operation (the most expensive period) have also been caused by this factor, but it has been pointed out, in this case, that utilities sometimes favor delaying the official start of commer-

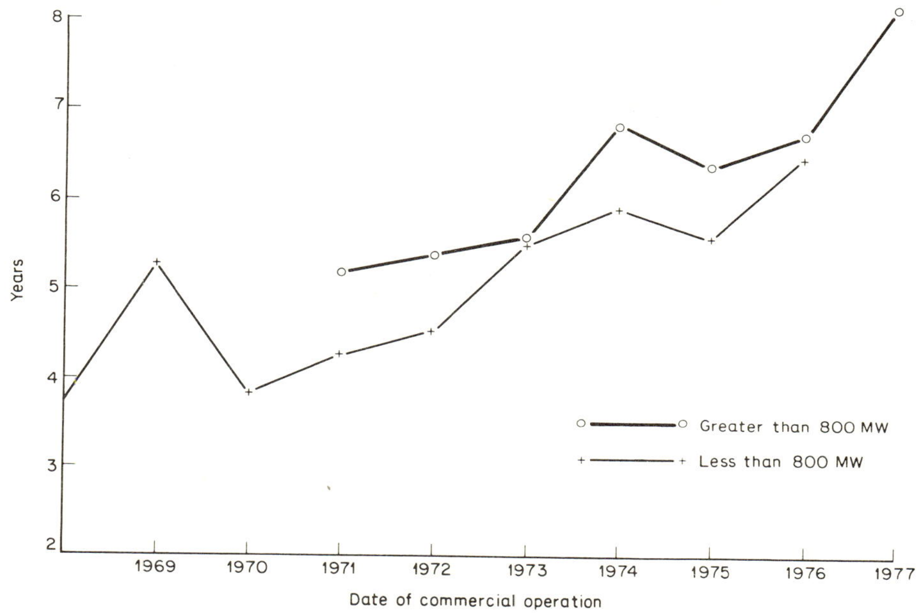

FIGURE 7.5. Average construction times for United States nuclear power plants under and over 800 MW (averages calculated for all plants entering commercial operation during year). From IAEA 1978.

cial operation for tax purposes. To date, reductions in electric power demand growth rates have not had a major impact on Canadian power plant lead times, but there is a possibility that this factor will play a more important role at the four-unit station at Darlington, currently scheduled to enter service during the second half of the 1980s.

Technical and Industrial Performance

Efforts to capture scale economies in the nuclear industry have resulted in a general trend toward increased unit sizes for power plants. Such a trend has complemented the generally increasing size of markets served by utilities. Fossil fueled generating stations have followed a somewhat similar trend. Larger plants take longer to build, however; evidence for this is shown in table 7.3, derived from IAEA data for all member states, and figure 7.5, for the United States in particular. Statistical limitations in the data available for Japan, West Germany, and Canada result in there being no clear evidence of this effect for these three cases.*

That plant size increases should result in increased lead times is not in itself

*In France, however, it has been estimated that construction lead times for 1,300 MW LWR units will be from eight to ten months longer than for 900 MW units.

TABLE 7.4. Materials and Labor Required for Nuclear Power Plant Construction in the United States per Unit of Generating Capacity

	Early 1960s	Late 1960s	Early 1970s	Mid-1970s
Plant Size (MW)	450	800	1,100	1,200
Concrete (10^3 cu yards/kW)	0.07	0.11	0.11	0.14
Structural steel (tons/MW)	4.3	3.7	8.5	9.1
Piping (10^3 feet/kW)	0.20	0.19	0.18	0.21
Wire (10^3 feet/kW)	3.0	2.9	2.7	3.7
Labor (man-hours/kW)	5.3	5.1	10.4	12.4

unexpected. Some studies have suggested, however, that extra delays have stemmed from a counterproductive positive feedback effect. As plants have been scaled up, designs have inevitably increased in complexity; the latter has led to an increase in regulatory complexity, which in turn has complicated the tasks of design and construction still further.

While the effect of plant size increases on the regulatory environment is difficult to determine, table 7.4 shows that, at least in the United States, the anticipated economies of scale do not appear to have materialized. Indeed, after slight falls in the 1960s, the quantities of materials and labor required per kilowatt installed increased substantially in the 1970s, in parallel with the growth of unit size. Increasing regulation has undoubtedly contributed to this situation, and it might be argued that significant scale economies actually have been achieved, only to be obscured by other, regulation-related factors that have tended to raise power plant costs generally. On the other hand, as implied above, plant size increases may themselves have contributed significantly to the more stringent regulatory climate. In any case, it also seems clear that the technical and managerial problems of scaling up were underestimated.

The plant size issue goes beyond questions of lead time and cost. There is some evidence that the performance of larger plants has been characterized by generally reduced reliabilities. While lower reliabilities are frequently experienced initially when any new technology is introduced, and while many of the larger new nuclear plants have not yet been in service long enough for the anticipated increase in reliability with plant age to become fully apparent, recent data show that the bigger plants have been less reliable than were their smaller predecessors at the same age and at a roughly equivalent stage of technological development (IAEA 1978). One conclusion that has been drawn from these various experiences is that increases in unit size have been implemented prematurely and may have contributed much to the problems faced by the nuclear industry.

It is interesting that the trend toward increasing plant size has been slower in Canada than elsewhere. The largest plants currently under construction are 750 MW units, and of the units still at the planning stage the biggest is 850 MW.

Related to the issues of the optimum *rate* of increase in unit size is the question of the *ultimately* optimal size. Indeed, in a dynamic industry the two are difficult to separate. Along with lead time, cost, and plant reliability criteria, the optimum unit size is also strongly influenced by the characteristics of the power grid within which the plant is to operate. In fact, the overall power system reliability and economic performance are of primary importance to the utility planner. Thus, in those countries with small or fragmented grids, the optimum unit size may be much smaller than the 1,200–1,300 MW average of most plant currently manufactured. It has also been suggested, however, that when all relevant factors are taken into account there may be a more general mismatch between the optimum unit size range and those sizes that are currently available commercially. Others argue that the evidence on which such conclusions are based is incomplete and necessarily biased, in part because not enough time has elapsed to permit a comprehensive assessment of the performance of the most recent generation of plants.

The question of plant size is not straightforward, and, as information on the subject continues to accumulate, a more comprehensive international assessment of the issue may be appropriate.

Returning once more to the discussion of the factors affecting lead times, the goal of moving plant construction down the learning curve by standardizing designs, thereby also speeding up licensing and improving quality control, has proved elusive. In the United States, some progress has been made, but there is general agreement that much more needs to be achieved. In Canada, no reactors have yet been true duplicates of previously built units, although some progress toward standardization is apparently anticipated. In West Germany, where boiling water reactors (BWRs) have consistently taken longer to build than pressurized water reactors (PWRs), the difference has been attributed to the greater susceptibility of the former to design changes and backfitting requirements during construction, and to the generally lower level of standardization than that achieved for PWRs. Japanese experience in this area is not known.

In all respects it is anticipated that standardization will bring savings in lead times, but, to date, the obstacles – in particular, the constant striving for technical improvement, geographical, and other local constraints that require individual tailoring, and the evolving demands of regulatory bodies – have proved stronger than the will to overcome them.

The contribution of more general industrial inefficiencies to rising lead times is difficult to ascertain. The experience and competence of the utility, nuclear supplier, and architect-engineer engaged in the project and the institutional relationships between them have all been identified in this context. One factor contributing to increased lead times in West Germany in the past has apparently been the emergence of new plant manufacturers. In the United States, the Tennessee Valley Authority's inexperience is thought to have contributed to delays at the first plant at which it served as its own architect-engineer (Keeny et al. 1977, p. 113). With subsequent projects, it has elected to do an increasing fraction of the engineering work before beginning construction — a decision that will probably result in a reduced need for time-consuming retrofitting. A related factor is the size of the utility engaged in the project. For instance, smaller utilities may be less willing to order large plants; on the other hand, small utilities may also be less likely to engage in multiunit siting, a measure that seems likely to result in reduced lead times. The countries selected for the study prevented an assessment of the possibility of significant differences between projects implemented by state-owned and privately owned utilities.

In the United States in the early 1970s, inefficiencies naturally arose from the shortage of appropriate managerial and technical skills, and the sheer pressure of work diminished the scope for organizational learning. In the United States and elsewhere, deficiencies in long-lead-time manufactured components, particularly those that already lie on the critical path, have been identified as significant contributors to delay.

The general question of industrial efficiency has received relatively little attention in lead-time studies, and further analysis may be worthwhile.

Siting and Regulation

With the exception of the Canadian analysis for the ICGNE, in which it is reported that regulatory delays have not generally appeared on project critical paths, all the national lead-time studies under review conclude that factors related to the regulatory process have been a major cause — perhaps the most important — of the recent growth in lead times. In several countries, both government and nongovernment constituencies have intervened increasingly to raise the level and scope of external regulation and control over the development of the nuclear power industry. While such trends have also affected other areas of technological investment, it is apparent that the effects have been unusually strong in the case of nuclear energy. In West Germany, Japan, and the United States, the volume of federal and regional regulation has increased substantially in the 1970s — spectacularly so in the United States, where it is reported that the number of codes and standards applicable to nuclear power plants rose from little more than 100 in the mid-1960s to about 1,700 in 1976. At the same time, the regulations have become stricter in their demands, adding to the time and economic cost of conforming to them.

At least as important as the volume and stringency of the body of regulation in its effect on lead times is the fluidity of the regulatory climate. Particularly contentious have been the practices of regulatory ratcheting during the design engineering phase and backfitting arising from regulatory change during, or even after, the construction phase. In fluid regulatory environments, the cycle of design and regulatory review will inevitably be iterative, although experienced designers will anticipate many regulatory changes. The iterations will be especially time-consuming if the ratcheting affects components on the critical path that must be ordered before the start of construction. Regulatory changes enforced during the construction period may cause even more serious delays. Upgraded seismic requirements have been cited as the source of particular backfitting difficulties. The dilemma is a familiar one: To what extent can the body of regulation be frozen in order to maintain a reasonably predictable environment during design and construction without destroying confidence in the responsiveness of the regulatory system to new technological developments and socioeconomic demands?

A more fundamental source of unpredictability in the regulatory climate in some countries has been confusion over the allocation of authority for nuclear power decision making. In the field of regulation, this has manifested itself in the conflict and lack of coordination between the various agencies of national, regional, and local governments. A related trend has been the devolution of authority to a local level, perhaps reflecting more general trends in this direction in a number of industrialized societies. In the United States and Japan, where siting problems have been identified as major contributors to lengthening lead times, the growing involvement of local and regional authorities is reported to be an important factor in this regard. Similarly, it is possible that "front-end" lead-time increases will be experienced in Canada as the participation of provincial jurisdictions in the decision-making process grows.

In addition to these jurisdictional trends, interest groups, dissatisfied with or otherwise reluctant to abide by established decision-making processes, have been resorting increasingly to courts of law to challenge the authority of the executive arm of government. The effect has been to cast the judiciary in the role of arbiter, a trend that has frequently had a direct bearing on planning and construction lead times and that has generally led to increased uncertainty within the industry. In West Germany, legal actions have been identified as the major contributor to lead-time unpredictability and have interrupted construction work at Wyhl, Brokdorf, and Grohnde. In the United States, several construction programs (e.g., Seabrook) have been affected by interest groups using the courts to bring about delays or cancellations. A recent Supreme Court decision may prove to be an important check on the role of the courts in the United States, however. On the other hand, by direct action at the plant site or elsewhere, interest groups have on occasion shown their willingness to outflank the constitutional system altogether (as at Wyhl).

Underlying all these problems is the thorny question of political consensus on the role of nuclear power. On the one hand, the increase in external regulatory control and jurisdictional claims has been prompted in large part by a sense that those enterprises with a vested interest in nuclear expansion have previously paid insufficient attention to the social and environmental impacts of their industry. On the other hand, it is argued that the current difficulties faced by the nuclear industry stem primarily from the activities of small minorities opposed to nuclear power who are unfairly exploiting legal and regulatory procedures to block progress.

Absolute consensus on any issue in any society is impossible. As a practical matter, however, a fundamental question concerns the degree of consensus required for the development of nuclear power. Requirements vary from country to country. In some, consensus is barely an issue. In others, the lack of it may be threatening the viability of the industry. The apparent inability of the governments of such countries to resolve the crisis of political authority and trust that has arisen lies at the root of much of the frustration experienced by the nuclear industry and its opponents, and arguably provides the key barrier to the successful reduction of nuclear power plant lead times.

External Factors

Unresolved issues elsewhere in the fuel cycle are having a major impact on power plant lead times in certain parts of the world. Most important, there is a growing tendency to link the nuclear power plant licensing process to the resolution of waste management and/or spent fuel reprocessing issues (as in Sweden, West Germany, and the United States). These trends reflect the deeply and widely held conviction that the institutional and technological status of waste management should be further advanced than it is at present if the nuclear power industry is to be allowed to develop.

Whatever the merits of these arguments, the fact remains that it will take several decades to resolve the waste issue: that is, to accumulate sufficient empirical evidence from a demonstration waste repository to show with reasonable confidence that it is performing and will continue to perform as intended. As a result, a major source of uncertainty with respect to the future of nuclear power will remain for at least the next twenty years. How nations choose to deal with this uncertainty—in particular, whether and how the technological uncertainty is transformed into institutional uncertainty associated with the nuclear plant licensing process—will be of crucial importance to the viability of the nuclear power industry in many countries.

Minimum Achievable Lead Times

An initial objective of the national studies prepared for the International Consultative Group on Nuclear Energy was to estimate the lead times that might be achievable in the absence of delays due to regulation and other external pressures of the types discussed in the preceding paragraphs. It was thought that such estimates might be useful for policy purposes, besides helping to attach weight to the various influences on lead times. The difficulties of disentangling the impacts of these factors notwithstanding, the review conducted for the ICGNE has estimated that minimum achievable lead times in the United States are currently of the order of 8.5 years. Another United States study concluded that the absolute minimum for a custom-built plant would be 8 years, although this could be reduced to 6.5 years if it were a standardized plant on a preselected site (Murphy et al. 1978). Finally, the Atomic Industrial Forum has suggested a time of about 9 years from a decision to acquire a site to the start of commercial operation (AIF 1978).

At first sight these estimates may seem high. However, preconstruction and construction periods in the United States were of the order of 1 and 4 years respectively during the late 1960s (see figs. 7.3 and 7.4), when regulatory delays were not generally regarded as having much impact. Adding to the construction time 6 months for the time between fuel loading and operation and an extra year to account for the larger plants that are now being built (see fig. 7.5), and adding a further 1 to 1.5 years for site selection, acquisition, and preliminary design work, the above estimates of minimum achievable lead times do not seem unreasonable. As actual lead times are now estimated to range from 12 to 15 years, this suggests that, *if siting and regulatory problems have been the controlling factors at all times,* delays from these sources have added from 3.5 to 7 years to the time from project definition to commercial operation in the United States.

No explicit estimates for minimum achievable lead times were presented for Japan, West Germany, or Canada.

CONCLUSIONS

This paper has been limited in scope. It has not considered lead times elsewhere in the nuclear fuel cycle or in the development and introduction of new technologies. The discussion has also been largely confined to the experiences of four countries: West Germany, Japan, Canada, and the United States. Further insight into the problem would undoubtedly be obtained by extending the analysis to other countries.

At the same time, the significance of lead times as indicators of the state of the nuclear industry varies from country to country. These differences would be clarified by a comparative analysis of the impact of lead times and lead-time modifications that has yet to be carried out. Moreover, factors whose major impact is on lead times in one country may assume a quite different (though equally important) role in another.

Despite these qualifications, a number of general conclusions can be drawn from the preceding discussion.

1. Over the past ten years, average nuclear power plant lead times have

increased in the United States, West Germany, Japan, and Canada. The trend has been most pronounced in the United States, where all phases of planning and construction have been affected, and least severe in Canada, where, despite the traditionally lengthy construction phase, overall lead times have remained relatively stable. An accompanying development, which has been identified as a particular problem in the United States and West Germany, and to some extent also in Japan, is the trend toward increasing unpredictability in individual plant lead times.

2. In the United States, the period between a decision made today (in 1978) to construct a nuclear plant and the ultimate date of entry of that plant into commercial service is likely to be twelve years or more under current conditions. Corresponding lead times elsewhere are somewhat lower, but they have been estimated to be as high as ten years in Japan and West Germany. Lead times of this magnitude approach and, in some cases significantly exceed, the planning horizons beyond which utilities no longer have confidence in their ability to forecast electric power demand accurately. Furthermore, lengthening lead times are liable to increase the capital cost of nuclear power plants considerably, particularly if delays are experienced late in the construction process.

3. The general economic climate of the past decade has exerted a significant influence on power plant lead times in many cases. In the United States, the rush of ordering in the years before 1974 placed undue pressure on productive resources, causing some delays. The subsequent economic recession undermined the financial health of some United States utilities thereafter, and it led more generally to reduced electric power demand growth and grid overcapacity, with consequent delays in some plant project schedules.

4. Prominent among the technical factors contributing to longer and more unpredictable lead times have been the problems encountered as a result of rapid increases in plant size over the past decade and the continuing lack of plant design standardization. The question of plant size transcends the lead-time issue and additionally involves considerations of capital cost and of plant and overall power-system reliability. As more empirical data on the performance of the larger new plants are accumulated, an international assessment of the plant-size question may be appropriate.

5. The increasing number and stringency of safety and environmental regulations, the linkage of power plant licensing to the resolution of the nuclear waste disposal issue, and the continuing fluidity of the general regulatory environment have been directly or indirectly responsible for many of the increases in lead-time length and unpredictability. Of particular importance have been the practices of regulatory ratcheting during plant design and backfitting arising from regulatory change during construction.

6. Difficulties encountered during siting have also been identified as major contributors to delay in the United States and Japan. In these and other countries, the growing involvement of local and regional authorities in the

decision-making process, particularly at the siting stage, may prove to be an increasingly important factor in the evolution of lead-time behavior.

7. Finally, the lead-time issue is inextricably linked to the more general question of the degree of consensus within societies over the proper role for nuclear power in the energy sector. This relationship differs from country to country and depends on the nature of national regulatory systems and the characteristics of underlying social, cultural, and political structures. Nevertheless, in countries where lead-time trends are threatening the economic viability of the nuclear industry, the success of efforts to halt or reverse these trends will arguably be determined primarily by the degree of national consensus on the general acceptability of nuclear power.

REFERENCES

Atomic Industrial Forum, Review Group on Design and Construction Project Lead-Times. 1978. *Licensing, design and construction problems: Priorities for solution.* Washington, D.C.: Atomic Industrial Forum.

International Atomic Energy Agency. 1978. *Operating experience with nuclear power stations in member states: Performance analysis report 1976.* Vienna: International Atomic Energy Agency.

Keeny, Spurgeon M., Jr., et al. 1977. *Nuclear power issues and choices: Report of the Nuclear Energy Policy Study Group.* Cambridge, Mass.: Ballinger.

Murphy, Arthur W.; La Pierre, D. Bruce; and Orloff, Neil. 1978. *The licensing of power plants in the United States.* Mount Kisco, N.Y.: Seven Springs Center.

Nuclear Regulatory Commission. 1978. *Construction status report: Nuclear power plant data for decisions (July 31, 1977).* NUREG-0030-77/8. Washington, D.C.: Nuclear Regulatory Commission.

United States Department of Energy, Office of Energy Technology. 1978. *A study of factors inhibiting effective use of domestic nuclear power.* Washington, D.C.: U.S. Department of Energy.

8

The Viability of the Civil Nuclear Industry

Måns Lönnroth and William Walker

1. INTRODUCTION

From the time its potential was first recognized in the early 1950s, nuclear energy was widely regarded as the successor to hydropower and fossil fuels in the generation of bulk electricity. The large-scale introduction of thermal reactors in the 1970s and 1980s, of breeder reactors in the 1990s and in the early decades of the twenty-first century, and ultimately of fusion reactors, would ensure that cheap, safe, and secure supplies of electricity and process heat were available into the distant future.

Significant industrial benefits were also considered to follow from the expansion of nuclear power. A competitive indigenous nuclear industry would bring international prestige, encourage the structural shift to more skill-intensive forms of production that many countries were seeking to achieve, and create substantial opportunities for foreign earnings. High-technology exports would also bring political influence over importing countries. Resources on an immense scale have been devoted to achieving those ends.

In the 1960s and early 1970s, expectations of a coming nuclear age seemed justified. Nuclear technologies appeared viable and economical, electric utilities were beginning to invest heavily in nuclear power, and nuclear fuel cycles were being established in several advanced countries. The rising price and insecurity of oil supplies in 1973–74 added further to the expansionist mood, with nuclear power plant ordering reaching new heights.

In 1979, five years later, there are some countries — notably France and Comecon countries — in which nuclear power continues to be promoted confidently and with determination. Elsewhere, perceptions of the future role of

Date of origin, December 1979.

nuclear power have changed dramatically. The almost universal faith and optimism of earlier years have been superseded by concern, pessimism, and, not infrequently, hostility. In the United States, West Germany, Sweden, and a number of other industrial and semi-industrial countries, nuclear programs and forecasts have been severely cut back since 1974, and it is unlikely that these countries will see a substantial resumption of nuclear investment in the next few years. Indeed, it now seems not impossible that some countries will find themselves facing an indefinite moratorium on nuclear ordering.

Why this should have happened against a background of growing energy scarcity, and why confidence in nuclear power has survived in some countries but not in others, are questions that we shall be seeking to explain in these pages. However, our main concern in writing this paper is not so much with the energy situation as with the predicament of the civil nuclear industry, and of the reactor industry in particular. Fears have been expressed that, if the environment in which it has to operate does not improve significantly in the next few years, the reactor industry could be severely damaged as a productive entity, diminishing the role that nuclear energy could be called upon to play in meeting global energy needs in the longer term. We wish to examine closely the prospects facing the reactor industries in the principal supplier countries and to consider under what conditions they can be expected to survive in the 1980s as reliable and efficient developers and suppliers of nuclear products and services.

The remainder of this paper contains four sections. In section 2 we consider some of the distinctive internal features of the civil nuclear industry—its technical and corporate structures, and the productive characteristics of the whole and its component parts. Section 3 is addressed to the political and economic environment as it has developed in the 1970s, and may develop in the 1980s, particularly in relation to electricity growth and utility attitudes to nuclear power, to political conditions in advanced countries, to export markets, and to political controversies surrounding development of the back end of the fuel cycle and the fast breeder reactor. Section 4 looks specifically at the predicament of the reactor industry and at the ways endangered firms are trying to cope with the slump in demand. Section 5 contains the conclusions to the paper.

In preparing this report, we have benefited greatly from discussions and correspondence with representatives of nuclear industries, utilities, and governments in Canada, France, West Germany, Sweden, Britain, and the United States. We have been unable to visit Japan and the Soviet Union, so that these countries receive less attention in this report than they deserve. The various countries that are on the way to developing their own nuclear industries (Argentina, India, Italy, Spain, etc.) have also escaped our net, though one cannot ignore the effect they are already having on the international structure of the industry. We have also gone out of our way to seek the opinions of informed persons outside the nuclear communities in the countries we have visited. We owe a substantial debt to all those who have helped us in our task.

We have deliberately taken a very broad approach in our analysis. This is not a subject that can be treated narrowly by considering some aspects to the exclusion of others; one has to try to capture it in its totality. We are well aware of the risk of superficiality in ranging so widely, but we believe it is a risk worth taking. As it stands, we would prefer this report to be regarded as an extended essay — a stimulus to further discussion and analysis — rather than as an attempt at a final statement on the subject.

2. STRUCTURE AND CHARACTERISTICS OF THE CIVIL NUCLEAR INDUSTRY

Some distinctive characteristics of the nuclear industry will be identified in this section, together with structural similarities and differences between countries. It will be seen that the structure of the industry and its relationship to government have important implications for its robustness in difficult and uncertain market conditions.

Choice of Fuel Cycle and Reactor Type

In the earliest days of the nuclear industry — the 1940s and 1950s — technical choices were strongly influenced by three factors: a country's ease of access to natural and enriched uranium, its technical know-how, and its military aims and programs.

After World War II the United States attempted to prevent the spread of nuclear weapons by prohibiting the export of enriched uranium, on which it had a monopoly, and by refusing to share scientific and technical knowledge gained in its weapons program. The three countries that had cooperated, when allowed to, with the United States in wartime nuclear development — Canada, France, and Britain — had to find their own ways of furthering their nuclear programs. Only a handful of other countries (West Germany, Japan, Sweden, the Soviet Union) had the immediate potential to acquire the technical know-how that made nuclear energy attainable, and, of those, West Germany and Japan were politically constrained after the war.

Countries without access to enriched uranium were therefore limited to reactor designs that could be fueled with natural uranium: the British and French gas-cooled graphite-moderated reactors (GCR), or the Canadian (and soon the Swedish) heavy water reactors (HWR). The first British and French reactors were constructed specifically to produce spent fuel that could be reprocessed to yield plutonium for bomb programs. From the start, reprocessing was therefore an essential part of their fuel cycles.

These early European reactors were designed and built to produce plutonium, with electricity generation a secondary concern. In response to fears in the 1950s of energy scarcity owing to inadequate coal production and disruptions to oil supplies after the Suez crisis, nuclear power generation was given higher priority, though in both France and Britain reactor programs remained dual-purpose until well into the 1960s, as, to a lesser extent, did the Swedish reactor program.

The United States also developed and constructed natural uranium fueled reactors to produce plutonium for its military program. But it was led to explore a different technical route, and was enabled to do so by enrichment capabilities, when the nuclear submarine program brought the first demands for a compact reactor design with a high power density to give optimum power output. The design finally implemented, the enriched uranium fueled light water reactor (LWR), was the ancestor of the reactor that became the cornerstone of United States civil nuclear expansion. The uniqueness and scale of this naval program gave the United States an early and clear advantage in this particular line of technical development.

The civil nuclear industry thus grew out of the military programs of the 1940s and early 1950s, and it developed in embryo amid restrictions on the international exchange of materials, equipment, and technical know-how. As far as the United States was concerned, however, the restrictive attitude was largely reversed under the Atoms for Peace program launched in 1953. During the two decades between the mid-1950s and mid-1970s, the connections between civil and military nuclear programs gradually became less pronounced, and trading restrictions were lifted, allowing the worldwide diffusion of nuclear technologies, particularly those originating in the United States. Aggressive marketing of LWR technology by United States reactor manufacturers, their willingness to license it to leading firms in Europe and Japan, and the favorable commercial and political terms attached by the United States government to the supply of enriched uranium combined to establish the LWR and its attendant fuel cycle as the dominant civil nuclear technology in the Western world.

During this period, a number of countries sought to consolidate their civil nuclear industries, while renouncing all claims to nuclear weapons by accepting comprehensive IAEA safeguards, or later by becoming parties to the Non-Proliferation Treaty (NPT). They were prepared to accept a large degree of dependence on existing supplier countries in return for assistance with their civil programs. West Germany and Japan, which had no weapons programs and therefore no substantial fuel cycle experience, took particular advantage of the liberal trading climate and established major civil nuclear industries, based on United States fuel cycle and reactor technologies. France and Sweden also abandoned the reactor designs they had initially developed, in order to gain the commercial advantages offered by United States reactor and fuel cycle systems.

While this favorable climate lasted, countries had comparative freedom in the development of their fuel cycles. Steps were taken to establish enrichment and reprocessing plants in a number of areas, with little political interference from abroad. This was a period during which the United States maintained its postwar dominance of the Western world in political, economic, and military fields, and there was as a consequence little perceived risk attached to the spread of nuclear technologies (though enrichment technology continued to be strictly controlled by its possessors). However, perceptions of these risks changed significantly in the early to middle 1970s. Partly as a result of its diminished authority in the international arena, fears of the consequences of horizontal proliferation led the United States government to attempt to impose new and sweeping restrictions on civil nuclear trade. While continuing to sanction the export of nuclear reactors, it sought to prevent the spread of reprocessing capabilities by attaching restrictions to the supply of enriched uranium, on which it still held a substantial monopoly.

Coincidentally, France and West Germany, having recently established sizable nuclear industries with assistance from United States manufacturers, made a determined move into export markets, offering for sale a complete range of fuel cycle technologies in addition to reactors. In addition, Britain reaffirmed its commitment to commercial reprocessing and offered to reprocess foreign spent fuels at its projected facilities at Windscale.

So, while the United States was endeavoring to restrain trade, other countries, with different perceptions and ambitions, were seeking to promote it, leading to an inevitable clash of opinions between supplier countries, and between supplier and recipient countries, over the terms under which nuclear trade should be conducted.

We shall return to these issues later. But it is worth noting at this stage that military and civilian aspects of nuclear energy were intertwined in the early stages of their development and that their separation—especially as regards the further adoption of nuclear technologies by non-nuclear-weapon states and the open trading that entailed—relied on the establishment and maintenance of international trust against a background of geopolitical stability. The interdependencies promoted in the liberal trading conditions that prevailed in the 1960s and early 1970s made the nuclear industries in many countries vulnerable to any breakdown of that trust and stability.

Productive Characteristics of the Civil Nuclear Industry

In functional terms, the nuclear industry can be conveniently categorized as in figure 8.1. Although its detailed specification depends on the chosen reactor/fuel cycle, it generally has four parts: the uranium mining industry, the fuel cycle industry (front end and back end), the reactor industry, and the fuel cycle equipment supply industry. The nuclear research and development (R&D) industry is essential to its development, in relation to the gradual

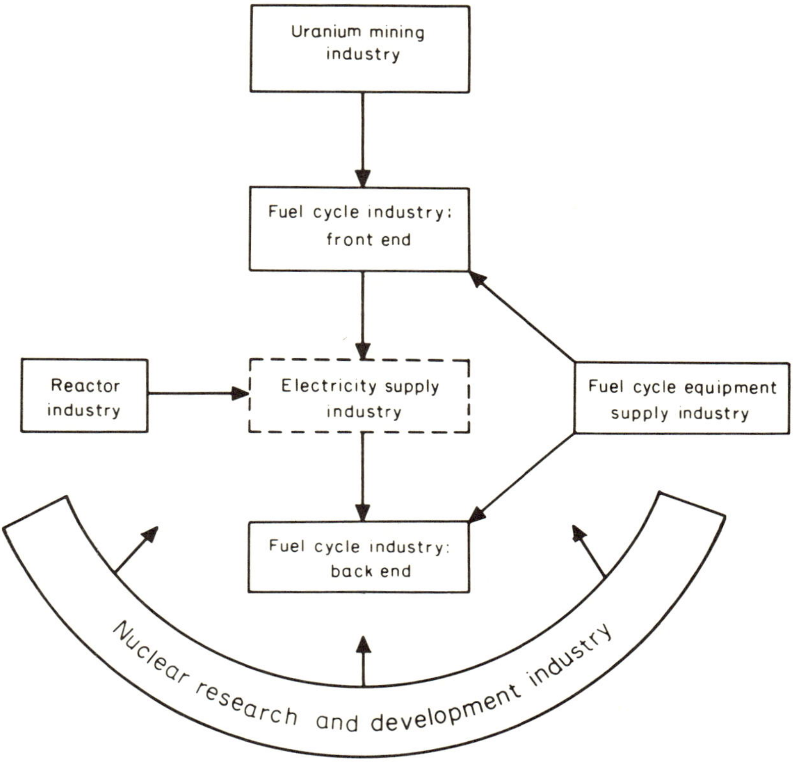

FIGURE 8.1. Functional breakdown of the nuclear industry.

technical improvement of present fuel cycles and reactor designs and, in particular, as the parent of future generations of nuclear technologies. The electricity supply industry interacts with, and overlaps, the nuclear industry but is not strictly a part of it.

Although the nuclear industry has everywhere been conceived on a vast scale, it is concentrated on serving a single market: electricity generation. It is an unusually self-contained, single-minded industry. It lacks major opportunities for product diversification that can bring economies of scale, extend the capital base, and spread commercial risks. It is therefore peculiarly sensitive to conditions within the electricity supply industry, the pivot around which the nuclear industry rotates and in many respects the key decision-making unit.

Another feature of the nuclear industry is the variety of industries with different productive characteristics that it embraces: a mining industry, process/flow industries (the fuel cycle), and capital goods industries (reactor and fuel cycle equipment supply). Its scientific and technological base is similarly varied, embracing mineral extraction techniques, chemicals and chemical process plant, electrical and mechanical engineering, special materials, and so on. The nuclear industry is, in this sense, an ad hoc assembly of industries with their underlying scientific and technical disciplines.

Scientifically and technically, the nuclear industry is also very advanced. It has been developed against a background of intensive scientific research and has relied, to an extent only approached in the aerospace industry, on heavy front-end investments in R&D and technical demonstration. In all aspects of production (uranium mining aside), it also requires the achievement of the highest standards of quality control. As such, this is an industry that benefits greatly from the support of an advanced scientific and industrial infrastructure, as exists in a handful of OECD countries and in the Soviet Union. The exceptional complexity of the industry also requires a high degree of organizational ability. It is not easily transplanted into countries that lack technology, industrial skills and experience, managerial ability, and traditions of quality assurance in manufacturing. Moreover, as in other high-technology industries, the combination of accumulated experience and technical advance tends to reinforce the commercial gap between those possessing and those not possessing developed nuclear industries.

The nuclear power plant is at the heart of nuclear technology. The nuclear-specific part of the power plant is the nuclear steam supply system (NSSS). Typically, nuclear power plant equipment accounts for between 20 and 25 percent of the total cost of a power plant (the remainder being taken up mainly by construction costs and interest charges) and is divided approximately equally by value between the NSSS and nonspecific (although adapted) equipment – turbine generators, switchgear, instrumentation, and so on. There is some interdependence between NSSS and fuel cycle characteristics, making it costly for a utility (or a country) to switch from one reactor type (and hence fuel cycle) to another. However, such costs vary, and they can be exaggerated. While there is considerable "commonality" between some reactor types and their associated fuel cycles (e.g., PWR and BWR), there is less commonality between others (e.g., PWR and HWR). Historic fuel cycle and reactor investments, and the institutional momentum built around them, can therefore tie countries to particular nuclear systems and, it should be added, to particular long-term R&D programs. Licensing requirements developed around individual reactor types and fuel cycles add to the strength of these ties. In this context, it should be noted that the capital cost of a nuclear power plant is roughly one-half of the cost of the fuel cycle services and materials purchased during its lifetime, so that fuel cycle investments are a major consideration in switching between reactor types.

A final characteristic of the nuclear industry worth noting is the exceptional lead times encountered in technical development and plant installation. It needs outlays of large quantities of venture capital, with a view to long-term requirements, and, as such, benefits to an unusual degree from stability of expectations and hence of policies. It also follows that this is an industry that takes a long time (decades) to reach maturity and to bring a satisfactory return on historical investment.

Corporate Structure of the Nuclear Industry: Ownership and Control

In most countries, development of a civil nuclear industry has initially been directed by the state. The industry's military implications, scale, technological novelty, and safety requirements have tended to make the state, not private enterprise, the prime mover in establishing it as a commercial entity, even if the private sector has been involved from the start in nuclear production as a recipient of government contracts and subsidies. Governments have directed and financed initial R&D and demonstration programs and have acted as catalysts in creating and bringing together the many organizations in both private and public sectors that, in concert, can function as a coherent nuclear industry.

Governments exert influence over today's nuclear industries, directly through ownership and the provision of finance and indirectly through legislation, regulation, and a variety of less formal channels. However, the form and extent of this influence differs considerably from country to country. The corporate structure adopted in a given country for nuclear production reflects the compromise struck between national independence and international dependence, the productive characteristics of the industry, the presence or absence of a military nuclear program, and, in particular, inherited traditions of industrial organization.

Patterns of ownership in the nuclear and electric utility industries in the eight supplier countries are displayed in a necessarily simplified and approximate form in table 8.1. These pertain to 1979 and represent the consequences of an evolutionary process that occurred mainly during the period of expansion in the 1960s and early 1970s. As one would expect, the United States comes closest to the private-ownership model (although enrichment is state-owned but privately operated, for historical military reasons), and the Soviet Union to the state-ownership model. Between these two poles there is a spectrum, with Japan and West Germany tending toward private ownership, France and Britain tending toward state ownership. Sweden lies somewhere in the middle, and the Canadian industry has a pattern of ownership that is peculiarly its own. One reason the smaller capitalist countries (e.g., Canada and Sweden) have substantial state ownership of their nuclear industries is that the costs and risks of nuclear production were considered too great, and the domestic market too limited, for relatively small systems of private enterprise to bear. It should be noted, however, that there is significant subcontracting of component manufacture by nuclear plant suppliers to the private sector in all capitalist countries.

TABLE 8.1. Patterns of Ownership in the Nuclear and Electric Utility Industries

	United States	Japan	West Germany	Sweden	Canada	France	Britain	Soviet Union
Enrichment	S	M	P*	–	–	S	S	S
Fuel fabrication	P	P	P	M	P	M	S	S
Electric utilities	M*	M*	M	M*	(S)	S	S	S
Architect-engineering	P	P	P	M			M	
Reactor supply	P				S	M		S
Reprocessing	–	S	P*	–	–	S	S	S
Waste disposal	?	?	S	?	S	S	S	S

S : state (central government) ownership
(S) : state (provincial government) ownership
M : mixed private and state (central and/or provincial) ownership
M*: diversified private and public ownership (e.g., 77 percent of United States electricity generated by private utilities, 23 percent by public utilities)
P : private ownership
P* : private ownership with significant utility shareholdings
? : undecided
– : not relevant

The following points of detail should be noted in relation to table 8.1:

- In the United States, some of the largest utilities (e.g., TVA, Duke Power) perform their own architect-engineering; others rely on the services of private firms (e.g., Bechtel, Ebasco). Elsewhere, architect-engineering is performed mainly by reactor suppliers (where the utility industry is fragmented) or by utilities (where it is large and concentrated).
- The French and British nuclear industries are similar in structure, with state-owned enterprises covering the fuel cycle (Cogema and BNFL) but reactor suppliers that are owned partly by the state and partly by private enterprise.
- In all countries except Britain, reactor suppliers are involved in fuel fabrication.
- Germany has set up ad hoc companies to run each activity within the fuel cycle: enrichment (Uranit), LWR fuel fabrication (RBU), and reprocessing and waste disposal (DWK). Ownership of these companies is shared by utilities and private concerns involved in nuclear production.
- The Canadian nuclear industry has been developed largely within the province of Ontario and has been promoted by AECL (owned by the federal government) and Ontario Hydro (owned by the provincial government). Of the heavy water plants in operation, three are owned by Ontario Hydro and two by AECL.
- Since multinational firms are involved in uranium mining, it was not possible to depict patterns of ownership in the uranium industry within the framework of table 8.1. The uranium mining industry is mainly in private hands (except in Canada), although governments in uranium supplying countries have historically played important parts in trying to

stabilize the industry. Recently there has been a tendency for uranium consumers (i.e., utilities) to inject venture capital into the industry, and exploration has increasingly been sponsored by national governments in producing and consuming countries.

The structure and characteristics of the nuclear reactor industry will be considered in some detail in section 4. By way of introduction at this stage, table 8.2 identifies the main actors in the industry. The market dominance of General Electric and Westinghouse and of their reactor systems, the light water reactors, is self-evident. In addition, table 8.2 shows that, in the United States, West Germany, and Japan, reactor manufacture is in the hands of large diversified companies, all of which entered nuclear markets after long histories of selling power plant equipment to electric utilities, either as suppliers of heavy electrical equipment (turbine generators, etc.) or of steam boilers. Existing patterns of international competition were thereby transferred to reactor markets. In the other major supplier countries, reactor supply is undertaken by consortia set up for this purpose alone (in France, Sweden, and Britain) or by state-owned corporations (in Canada and the Soviet Union), although they are also dominated by large suppliers of power plant equipment.

These enterprises do not represent the totality of the reactor industry in their parent countries. They all depend, to a greater or lesser extent, on supplies of materials and components from surrounding industrial infrastructures. There is also some international trading of reactor components.

The question of coordination and control is considerably more intricate than that of ownership. Patterns of ownership, although very relevant, take one only so far in trying to understand it. We have noted that the nuclear industry is an assembly of industries and institutions with different technical bases, time perspectives, and commercial dynamics. By what means, if any, are the various aspects of nuclear policy coordinated so that the various parts act with some degree of unison?

Although one has to tread warily, it is useful here to make a threefold categorization of countries. In the first category are those countries (e.g., France and the Soviet Union) in which the central government performs the coordinating functions through ad hoc committees, on which utilities, reactor and fuel cycle industries, R&D institutions, and relevant government departments are represented, and by other less formal means within the framework of government. The second category contains those countries (e.g., West Germany and Sweden) in which decision making is also centralized, but by joint action within the private sector with some government involvement. Thus, one finds in Germany that several organizations within the fuel cycle are jointly owned by utilities and industrial firms, and that key individuals hold senior positions in banks, utilities, and industrial concerns, lending an element of cohesion to nuclear decision making. The third category contains, in particular, the United States, in which the "privatization" of aspects of

TABLE 8.2. Principal Thermal Reactor Suppliers by Country

Country	Manufacturer	Ownership, or Main Business of Parent Firm	Reactor Types Supplied	Source of Technology	Nuclear Generating Capacity Installed and Ordered to March 1979 (GW)
		Private Corporations			
United States	General Electric (GE)	Electrical engineering	BWR	In-house	76.5
	Westinghouse	Electrical engineering	PWR	In-house	96.7
	Babcock and Wilcox	Heavy engineering	PWR	In-house	27.8
	Combustion Engineering	Heavy engineering	PWR	In-house	34.1
Japan	Mitsubishi	Heavy engineering	PWR	Westinghouse licensee	8.0
	Toshiba	Electrical engineering	BWR	GE licensee	4.4
	Hitachi	Electrical engineering	BWR	GE licensee	2.3
West Germany[a]	Siemens/Kraftwerk Union (KWU)	Electrical engineering	PWR + BWR + HWR	Former Westinghouse licensee[b]	31.4[c]
		Mixed Ownership and State-Owned Contractors			
Sweden	ASEA-Atom	ASEA (50%) Government (50%)	BWR	In-house	8.1
Canada	Atomic Energy of Canada, Ltd. (AECL)	Federal government (100%)	HWR (CANDU)	In-house	17.5
France	Framatome	Westinghouse (15%) Creusot-Loire (55%) CEA (government) (30%)	PWR	Westinghouse licensee	39.1
Britain	Nuclear Power Company (NPC)	General Electric Company (30%) British Nuclear Associates (35%) UKAEA (government) (35%)	AGR	In-house	10.6

Sources: Westinghouse 1979; NEI 1979.

[a]The West German-Swiss reactor supplier, Babcock-Brown Boveri, which is constructing one reactor in West Germany, is not considered in this report.

[b]The Westinghouse license related only to PWRs; KWU has also constructed BWRs under license from GE, as well as indigenously developed HWRs for export.

[c]PWR only.

nuclear production has, as a deliberate part of government policy, brought about considerable fragmentation of both ownership and control, largely with a view to inducing competition between suppliers. In the United States, except in the area of safety regulation, there is an absence of centralized planning and direction of the nuclear industry in its totality (although not of certain parts), so that its development has tended to reflect the commercial interests of individual private enterprises to a greater extent than in other countries.

In keeping with countries' political and economic traditions, there is thus a basic contrast between those countries in which the state has assumed a large part of the commercial risks involved in nuclear production and has attempted to impose a centralized, coherent, and cohesive structure on the nuclear industry; those countries in which short- and medium-term commercial risks are largely borne by private enterprises and capital markets, with a degree of state risk-sharing, but in which policy is coordinated from within the private sector by a range of institutional devices; and those countries in which commercial risks are borne almost entirely by private enterprises and capital markets and there is only limited coordination of policy by either public or private bodies.

Decisions about the ownership and coordination of nuclear industries were largely taken in the period when market prospects were good and commercial profitability appeared to be just over the horizon. We shall see in sections 3 and 4 that the difficult political and economic conditions that have prevailed since the early 1970s have given a distinct advantage to those countries in which governments exert strong control over and share risks with nuclear industries and, most important, electric utilities, and a distinct disadvantage to those countries in which risks are borne by private concerns and there is limited policy coordination.

Commercial Vulnerability within the Nuclear Industry

The interconnectedness and essential functional unity of the nuclear industry have already been noted. So has its internal diversity. We now turn to the commercial characteristics of the different parts of the nuclear industry identified in figure 8.1, and to the question whether any one part of the industry is inherently more vulnerable in difficult market conditions than the other parts.

The front end of the fuel cycle includes, among other things, enrichment, fuel fabrication, and, for heavy water reactors, heavy water production. Although access to enriched uranium has been the subject of recurrent international dispute, the front end of the fuel cycle has arguably reached a state of commercial and technological maturity that has yet to be attained by other parts of the nuclear industry. And, the present worldwide overcapacity in the uranium enrichment industry notwithstanding, it is insulated to a degree from economic cycles, since, once nuclear power is reasonably established, its business arises out of the routine operation of the installed stock of nuclear power plants. It therefore has a reasonably secure cash flow even when new nuclear facilities are not being ordered.

The same could be said of the back end of the fuel cycle, except that it has yet to be fully established as a politically and economically viable industry. This is especially true of waste disposal, but it also applies to aspects of reprocessing and spent fuel management. While in theory nuclear power can proceed in the short and medium term without the back end of the fuel cycle being established, in practice many of the disputes over nuclear power have emanated from fundamental disagreements over the appropriate techniques, strategies, and time scales for developing it.

The uranium mining industry also derives its cash flow from fueling the existing stock of reactors (although forward contracts are a complicating factor). While there are uncertainties facing individual companies, owing, for instance, to government import/export policies, feed requirements for enrichment plants, and uncertain growth of demand, they are not of such an order as to threaten the survival of the industry. We shall see later, however, that expectations of uranium prices and availabilities have an important impact on other parts of the nuclear industry and a particular influence on policies toward the back end of the fuel cycle and the development of advanced reactor technologies.

The commercial dynamics of the reactor industry differ from those of the fuel cycle and mining industries in one basic respect. In cyclical markets, it lacks continuity of income and capacity utilization, since it depends on present levels of ordering by electric utilities and on the backlog of orders placed over the previous few years. Servicing and maintenance provide some compensation but are a relatively minor, if growing, part of total activity. The fuel cycle industry's income is thus related to the size of the stock of nuclear power plants (although expected stock changes influence the level of revenue coming from forward contracts), while the reactor industry's income is related to the incremental change in that stock. In addition, the reactor industry is essentially a unit production industry — the units being few in number, large in scale, and with long lead times — while the fuel cycle industry is a flow industry. When combined, these features of the reactor industry make it particularly vulnerable to cyclical variations in demand and to uncertainty over future market conditions. It should be noted, however, that the relatively steady and growing income that most reactor suppliers gain from their commercial involvement in fuel fabrication provides some financial cushioning against variations in income from reactor sales.

The fourth part of the nuclear industry identified above — the fuel cycle equipment supply industry — contains, as is implied, a variety of industries reflecting the varied character of the technologies used in the fuel cycle. The process-specific equipment used in an enrichment plant has little in common with that used in a reprocessing plant or in a heavy water plant. The only similarity is that suppliers of fuel cycle facilities tend to be offshoots or divisions of chemical process plant industries. As seen by the very large firms involved in these activities, the nuclear fuel cycle generates a minor part of the sales and tends to be regarded as an occasional market. It is also a considerably smaller market than the reactor market.

Commercially, therefore, the reactor industry is inherently more at risk in uncertain and variable markets than other parts of the nuclear industry. It was to be expected that it would be particularly hard hit by the deterioration in market conditions in several advanced countries that has occurred in recent years.

Sections 3 and 4 will contain extensive discussions of the prospects for the reactor industry and of the ways it is attempting to cope with the present adverse economic and political environment. There will be comparatively little discussion of the lesser commercial difficulties facing the fuel cycle industry.

It should be noted, finally, that the nuclear R&D industry is also vulnerable. In this case, vulnerability is more to do with cycles in the long-term expectations of governments, its main sponsors, concerning demand for future nuclear systems, and those systems' potential viability. It is also to do with the perceived opportunity costs of heavy R&D investment programs, and with the timing of those programs. We shall be giving particular attention to these issues in the discussion of the fast breeder reactor in section 3.

3. CONSTRAINTS ON NUCLEAR EXPANSION

We have already noted that, while nuclear power is being developed with determination in some countries, it is in serious trouble in others. In the Soviet Union and France, reactors continue to be ordered and installed, both front and back ends of the fuel cycle are being developed with relatively little opposition, and substantial fast breeder reactor (FBR) demonstration programs have been launched. Prospects in the United States, West Germany, and Sweden are very different. The de facto moratoria on reactor ordering that have existed in those countries since the mid-1970s could well continue into the early to mid-1980s, and there is intense controversy over the fuel cycle and future nuclear systems. Canada, Japan, and Britain lie somewhere between these extremes, with nuclear power expanding, but at a lesser rate than anticipated and not without internal difficulty.

Overall, then, the expansion of nuclear power has slowed down, with considerable international variations in growth rates and market prospects. To explain how this has happened, we start by looking at the growth of electricity demand and at utility attitudes toward nuclear power, then move on to discuss political conditions in supplier countries, conditions in export markets, and the implications of the controversies over the FBR and the back end of the fuel cycle. It should be emphasized at the outset that the state and prospects of the nuclear industry have been determined as much by the way the various factors discussed in this section have combined and interacted in recent years as by their individual influences.

Electricity Growth and the Role of Nuclear Power

Between 1950 and the early 1970s, rapid growth of demand for electrical energy occurred throughout the Western world, with most countries experiencing growth rates upward of 7 percent per annum. In all supplier countries, nuclear industries were established, and the requisite productive capacity installed, in the 1960s and early 1970s in the expectation that these growth rates would continue at least until the end of the twentieth century. But since 1974 the market for electricity in most developed countries has been relatively stagnant, and it is widely believed that between now and the year 2000 the average per annum rate of growth of electricity demand, while remaining higher than the growth of total energy demand, will be well below the historical average. Only in some developing countries is electricity demand expected to grow rapidly, and then from relatively low levels.

Electricity forecasts have been reduced for two main reasons. First, there is widespread pessimism over the prospects for world economic growth, and hence for growth of energy demand, in the coming decade. In the advanced industrial nations that form the backbone of the world economy, population growth has fallen, markets that grew rapidly in the 1950s and 1960s are approaching saturation, and there appear to be relatively few new technologies and markets around to provide the stimulus for economic expansion. At the same time, there are serious inflationary pressures within the world economy, instabilities in financial markets, and growing confrontation between industrial nations over access to markets and to energy and mineral resources. As a consequence, it now seems likely that the world economy will experience annual growth rates that are low compared with those of the 1950s and 1960s, against a background of considerable turbulence, with successive fluctuations, of unpredictable magnitude and timing, between growth and recession. Some even believe the world economy will be lucky to escape a serious economic depression in the 1980s.

Second, it is anticipated that electricity users will make greater efforts to conserve electrical energy when faced by rises in the real price of electricity, and that there will be improvements in the energy efficiency of electricity-consuming technologies, as industries direct their technical efforts toward developing energy-saving products and processes. The alternatives include more efficient appliances in households, more efficient electric motors in industry, and new and old technologies for savings in space heating (e.g., heat pumps and insulation). Utilities have already been surprised at the degree of price elasticity evident in electricity consumption patterns in the past few years. There is undoubtedly considerable scope for further conservation of energy in general and electricity in particular, though there are many uncertainties. In addition to those of a technical and economic nature (rate of technical change, rate of turnover of capital stock, etc.), a major uncertainty concerns the attitudes of governments toward conservation, as pressures from constituencies opposed to and in favor of energy-saving policies build up.

An additional factor that could reduce the growth of demand for *centrally* generated electricity is the introduction of dispersed forms of electricity production (e.g., cogeneration of heat and electricity, and possibly various solar technologies) and of district heating schemes whereby waste heat from electricity generating stations is used for space heating. Though they are likely to become a major source of competition for centrally generated electricity only if they show distinct economic advantages and if governments promote them actively, the growth of these options on the margin could in some regions further curtail the growth of demand for electricity from centralized facilities, and thereby curtail the growth of demand for large power plants.

There are also, however, factors that may encourage electricity consumption. The use of electricity has expanded rapidly in nonpremium markets in many countries during the last decade, particularly in the space-heating sector. This expansion has been, and remains, a deliberate policy in some countries (e.g., France) and an accepted development in others (e.g., West Germany, Sweden), being justified by the resulting reduction in demand for heating oil. Without the space-heating market, nuclear expansion would have been slower in several countries.

This space-heating market is sensitive to the relative price of electricity versus that of other fuels (natural gas is a particularly attractive alternative in Canada, Britain, and the United States). If, in the long term, the relative price of electricity falls when oil and gas supplies decline (as many forecasts predict), market forces can be expected to give greater momentum to the electrification of space heating. The market is also influenced by various aspects of government policy, concerning relative energy pricing, conservation, promotion of new technologies, and so on. To some extent, therefore, governments are faced by a choice between the pace of nuclear development and advocacy of alternative policies.

The overall picture to the end of the century is thus one of low average electricity growth by historical standards, with considerable uncertainty over the actual trend rate and over the timing and magnitude of fluctuations above and below that trend. In most advanced countries, forecasts of annual electricity growth to the end of the century have varied between 2 and 6 percent. At the upper end of the range are countries that are still incompletely industrialized, or where governments see good reasons for encouraging further electrification of nonpremium markets (e.g., France and Japan); at the lower

TABLE 8.3. Reactor Manufacturing Capacity and Ordering Rates in Noncommunist Countries

Country	Reactor Manufacturing Capacity 1979 (GW/a)	Average Ordering Rate Required by Nuclear Growth Scenarios 1980–89 (GW/a)			Average Ordering Rates 1970–79[a]			Average Loading on Manufacturing Capacity 1970–79	
		IEA		KFA Jülich	Total Orders by Purchaser Country (GW/a)	Domestic Orders by Vendor Country (GW/a)	Export Orders by Vendor Country (GW/a)	(GW/a)	(%)
		Low	High						
Canada	3	1.8	2.0	1.7	0.95	0.95	0.20	1.15	38
France	7	n.a.	n.a.	2.0	3.40	3.40	0.55	3.95	56
West Germany	6–8	2.2	4.2	2.3	1.25	1.25	0.70	1.95	24–33
Japan	6	4.7	6.3	4.9	1.55	1.25	0.00	1.25	21
Sweden	1–2	n.a.	n.a.	0.6	0.55	0.35	0.15	0.50	25–50
Britain	1–2	1.3	1.7	1.0	0.00	0.00	0.00	0.00	0
United States	25–30	12.5	14.4	12.0	11.45	11.45	3.25	14.70	49–59
Others	0	n.a.	n.a.	11.5	4.35	–	–	–	–
Total	49–58	n.a.	n.a.	36.0	23.50	18.65	4.85	23.50	41–48

Sources: IEA 1978; Westinghouse 1979; NEI 1978; Connolly et al. "World Nuclear Energy Paths" (this volume).
[a] Including reactors construction of which has not yet begun or been completed.

end are countries that are already at an advanced stage of industrialization (e.g., the United States), or that have cost-effective alternatives to using electricity in nonpremium markets (e.g., Britain).

In the short term, the result of reduced electricity growth has been that utilities in most countries have been caught with substantial overcapacity in their grids and with large construction programs under way that serve to perpetuate that overcapacity. There has therefore been a tendency to stretch out construction schedules wherever possible, and in some instances even to cancel orders. The present hiatus in ordering of nuclear power plants is partly due to this need of utilities to readjust construction programs with long lead times in the light of lower expectations of electricity demand growth. In the longer term, revised forecasts have led to a reassessment of the rate of growth of nuclear investments over the next twenty years. Forecasts of Western world nuclear capacity of 1,500+ gigawatts (GW) by the year 2000—forecasts that were perhaps more reflections of desired developments than realistic assessments—have been repeatedly revised downward. The World Energy Conference L4 scenario, which is based on 3 percent annual economic growth and high response to real energy price increases compared with other WEC scenarios (although a lower response than has been assumed in a number of recent national studies), proposes an installed nuclear generating capacity outside the centrally planned economies of 970 GW by that same year (WEC 1978). This implies an average ordering rate of about 70 GW per annum (GW/a) in the 1980s, assuming a ten-year lead time in power plant licensing and construction.

More recent studies, especially those taking a disaggregated, country-by-country approach, have come up with even lower scenarios of nuclear growth. Table 8.3 reflects the results of two such studies, the one reporting official figures from members of the International Energy Agency (IEA 1978), the other the scenario-building exercise carried out at the Kernforschungsanlage Jülich for the ICGNE which is reported in the next paper in this volume. The calculations made at Jülich, based on certain assumptions about electricity growth rates and generating mixes in different countries, suggest an installed nuclear electricity capacity of 695 GW outside the centrally planned economies in the year 2000. This would require an average rate of ordering of about 36 GW/a between 1980 and 1989, again assuming a ten-year lead time.

It can also be seen in table 8.3 that the currently installed reactor manufacturing capacity in the seven major Western supplier countries, according to information gained in our interviews with reactor suppliers, is capable of producing between 50 and 60 GW of nuclear generating capacity each year. While inadequate to meet the level of demand implied in the WEC L4 scenario, this level of capacity would, on the face of it, be adequate to supply the demand implied in the IEA and Jülich scenarios, with a fair margin to spare.

TABLE 8.4. Average Reactor Ordering Rates by Vendor Country, 1970–79 (GW/a)

	Domestic		Export		Total	
Country	1970–74	1975–79	1970–74	1975–79	1970–74	1975–79
Canada	0.7	1.2	0.2	0.2	0.9	1.4
France	3.4	3.4	0.4	0.7	3.8	4.1
West Germany	1.7	0.8	0.7	0.7	2.4	1.5
Japan	1.6	0.9	0.0	0.0	1.6	0.9
Sweden	0.3	0.4	0.3	0.0	0.6	0.4
Britain	0.0	0.0	0.0	0.0	0.0	0.0
United States	21.0	1.9	5.3	1.2	26.3	3.1
Total	28.7	8.6	6.9	2.8	35.6	11.4

Sources: As for table 8.3.

However, from past experience, it is unlikely that such a reactor ordering program would be evenly spread over the decade. Cyclical ordering is a well-known phenomenon in heavy electrical equipment markets (given long lead times, small errors in forecasting tend to induce large fluctuations in ordering rates), and the scale and lead times of nuclear power programs make the nuclear industry particularly prone to such variations. Table 8.4 shows a substantial variation in the rate of ordering in the two halves of the 1970s, with more than three times as much being ordered between 1970 and 1974 as was ordered between 1975 and 1979. Cancellations and deferrals in the latter period also smoothed out the ordering fluctuations considerably. While the fluctuations of the 1970s were perhaps extreme, given the stark contrast between rapid economic growth in the early 1970s and recession in later years, this illustrates that some spare capacity might be needed to meet the Jülich ordering rates for 1980–89, since orders would probably again be spread unevenly over that decade.

To be fulfilled, the revised forecasts and scenarios of nuclear growth, most of which appear to converge on ordering rates of between 35 and 50 GW/a in the 1980s, require that utilities and governments find it politically and economically feasible to invest in nuclear power on such a scale. It should be noted in table 8.3 that the relatively conservative Jülich scenario posits a 53 percent higher rate of ordering in the 1980s than was achieved in the 1970s: 36 GW/a as against 23.5 GW/a. Given lower anticipated total power plant ordering (coal + nuclear + others) in the 1980s than in the 1970s, it requires, among other things, that nuclear power have a considerably higher share of the power plant market than in the previous decade.

While this level of ordering is not impossible, it has to be assessed against more detailed studies that have been carried out in the countries in question. Some respected national studies indicate that ordering levels in the 1980s could be closer to, or even below, the average rates of the 1970s, certainly for Canada and Sweden and possibly also for West Germany. Moreover, given present circumstances and government energy policies in the United States, an average 1980–89 ordering level comparable to the 11.45 GW/a achieved between 1970 and 1979 seems high — possibly very high.

In the following pages we shall analyze further the reasons for the setbacks the nuclear industry has experienced since 1974, and we shall assess the feasibility of achieving the rate of ordering implied in these scenarios.

Electric Utilities and the Economic Environment

The past few years have seen a significant change in the climate in which electric utilities operate. The growth and profitability of the utility industry has depended historically on the satisfaction of certain conditions: in particular, the availability of economies of scale in new electricity generating and distribution systems, reasonable and predictable capital costs, and rapid and reliable growth of electricity demand. Until the early 1970s these conditions were generally satisfied, leading utilites into a virtuous circle of declining real electricity prices, growing demand, a high rate of return, readily available finance, and increasing capital intensity.

Utilities have found themselves in recent years in danger of being trapped in a vicious rather than a virtuous circle. The predicament has been most serious for privately owned utilities — particularly those within the United States utility sector — but publicly owned utilities have not escaped the consequences either. Slow growth of electricity demand, high and unpredictable capital costs, rising fuel costs, and diminishing returns to increased scale of equipment have acted together to make the utility industry more averse to taking risks than previously and less able to finance new investments. Furthermore, while the real price of electricity fell regularly and substantially — even the current price did not show a significant upward movement — utilities were relatively free from public influence over pricing policies. With high inflation in some countries, and real increases in electricity prices, utilities have had increasing political difficulty in achieving rate increases to safeguard their profitability and investment programs.

Whereas growth of generating capacity historically led reliably to reduced average costs and increased profitability, for many utilities it now carries the threat of higher average costs and a lower return on investment. With its relatively high capital costs and proneness to long lead times, marginal cost increases have been particularly marked and financially damaging in the case of new nuclear plant. Furthermore, the combined unpredictabilities of demand growth, of capital costs and lead times of new plant, and of future relative prices of primary fuel inputs have made it much more difficult to estimate, and therefore to minimize, future systems costs and availabilities. Confidence in investment plans has decreased as the risks and costs of making mistakes have increased.

The recent accident at the Three Mile Island plant in Harrisburg has added a further twist to utilities' assessments of risks. It is now apparent that equipment failure or operator error can lead to a billion-dollar investment's being severely damaged in a matter of minutes or hours, to heavy bills for replacement power, to enormous claims for financial compensation, to increased insurance charges, and to public acrimony. However unlikely the event, its possibility has nevertheless made utilities — particularly small private utilities with limited resources — more cautious in their attitudes toward nuclear power.

These economic difficulties do not mean that utilities will abandon nuclear power, although in some countries they are tending to reduce the scale of their commitments and expectations. Most utilities still look with favor on nuclear power, for the opportunity it provides for diversification of energy supply inputs and for its low running costs. Once installed, provided a satisfactory load factor can be achieved, nuclear power soon becomes profitable in inflationary periods owing to the effect of historical cost accounting in reducing the real level of annual capital charges. In addition, increases in the risks attached to nuclear investment would not matter if there were readily available alternatives at reasonable cost, which frequently there are not. In Germany and a number of other industrial countries, utilities are placing greater emphasis on coal for base-load electricity, but dependence on coal also carries its risks. Coal is polluting and is subject to inflationary pressures, supply insecurities, and lengthening lead times. The potential impact of those risks will be increased if coal has to compensate for a widespread reduction of nuclear investment. Moreover, there are large uncertainties over the future elasticity of coal supplies. Will the industry be able to produce sufficient coal at competitive prices in the face of a large potential demand — for electricity generation, bulk steam raising, and synthetic fuel production — and of political pressures to reduce the various forms of pollution and environmental degradation associated with its large-scale production and use? Even if only for strategic reasons, electric utilities will try, wherever economically and politically possible, to attain a reasonable balance between nuclear power, coal, and other primary energy forms in their electricity production.

Furthermore, for those utilities that have invested in the past three decades in a pattern of energy supply that has a large oil-fired component, recent oil price rises and supply insecurities have given a strong incentive to diversify the energy inputs to their base-load generating capacity, entailing, in some instances, early retiring of oil-fired plants or, in others, lowering of them in the merit order. Within the power plant replacement market generally, oil and gas are no longer acceptable base-load options, the choice being largely between coal and nuclear power. This market will provide some compensation for reduced demand growth. Indeed, in countries with low expectations of electricity demand growth, such as Britain, it is mainly on the replacement market that power plant suppliers — and the coal industry, where applicable — are basing their plans.

TABLE 8.5. Environment for United States Utility Industry

	Historical (pre-1970)	Today (post-1970)
Load growth	Very predictable (high)	Very uncertain (low)
Project schedules	Short and predictable (3-5 year horizon)	Long and uncertain (8-12 year horizon)
Utility financial status (ability to raise capital)	Good	Weak
Generation mix	Local concern	National concern
Economics of various fuels	Important and predictable – one of major factors	Important but controversial and often overridden by other factors
Public utilities commission environment	Supportive because of rate stability	Unfriendly because of rate increase
Government and regulatory requirements	Not heavily involved Few requirements No energy policy	Major impact on utility business Numerous laws and acts Rapidly changing energy policy
Available economies of scale	Substantial	Limited

The predicament of the utility industry is most serious in the United States, as is illustrated in table 8.5, which is a slightly modified version of a figure shown to us by a major United States reactor manufacturer. There, the risks and costs of investing in any expansion of electricity generating capacity have risen alarmingly. Nuclear power is effectively blocked by political opposition and by utilities' poor financial health; environmental regulation and transport difficulties are pushing up the costs of coal and raising doubts about its availability; inflation is making it politically difficult to obtain rate increases; and oil and natural gas have become uneconomical and, in many instances, illegal for base-load generation.

With marginal costs of all capacity additions on the increase, utilities can absorb the costs and risks of expansion only if electricity prices are allowed to rise substantially. Owing to the politics of and economic approaches to rate regulation in the United States, restraints have been placed on the rise of electricity prices by public utilities commissions in many areas, leading to an erosion of utility balance sheets. This has made utilities unable to finance expansion from internal cash flow, has led to a fall in the market value of their equity, and has forced them to resort increasingly to debt, which they have had difficulty in financing from retained earnings.

In such adverse circumstances, utilities can only respond by expanding supply with great caution, by deferring investment decisions where possible and, increasingly, by trying to restrict demand growth. From being pro-

TABLE 8.6. Distribution of Authority over Electric Utility Industries

Country	Ownership	Generating Mix	External Financing	Rate Regulation	Comments
United States	Mainly private; some state	Utility, but state governments influence decisions de facto	Capital market; state finance for public utilities	Public utilities commissions in each state	Fragmented authority over utilities' financial health; role of federal government very weak
Japan	Private	Utility, but in accordance with central government planning	Capital market	Central government	Central government ultimately controls financial health through rating authority
West Germany	Larger utilities: mixed regional government/ private	Utility, but regional government makes licensing decisions	Capital market; regional government	Federal government	Local, regional, and federal governments all influence financial health through ownership and rating authority
Sweden	State (50%); local government (15%); private (35%)	Utility, but central government makes licensing decisions	Central government for state-owned utility; capital market for others	Almost none; electricity producers compete for distributors and customers	Fragmented authority over utilities' financial health; competition said to hold rates down
Canada	Mainly provincial government	Provincial government, at recommendation of utility	Provincial government; capital market	Provincial government	Provincial governments control financial health
France	State	Central government, at recommendation of utility	Central government; capital market	Central government	Central government controls financial health
Britain	State	Central government, at recommendation of utility	Central government	Central government	Central government controls financial health

ponents of demand growth and supply centralization, some United States utilities are now advocating electricity conservation and supply decentralization in order to defend their commercial integrity. If this becomes more widespread, it could imply a rather fundamental shift in the role of utilities.

Quite apart from the political problems that beset nuclear power in the United States, it therefore looks unlikely at this moment that the utility sector will be able to make substantial headway with nuclear expansion unless the federal government intervenes to give financial relief, whether by providing direct loans and guarantees or by changing the system whereby rates are regulated. Furthermore, it is being questioned whether the fragmented utility industry in the United States will be able to cope with the even more stringent safety and licensing requirements that are the likely consequence of the Three Mile Island accident. Some have even gone as far as to recommend the restructuring of the United States utility industry into a limited number of large generating boards as a necessary condition for the continued expansion of nuclear power in that country.

Utilities relying on private sponsorship in other countries have encountered similar economic constraints, but on a considerably lesser scale. For instance, in West Germany, utilities have had fewer problems in financing nuclear investments, despite long lead times. The reasons they have escaped the crisis that has hit the United States utility industry are unclear but may be linked to West Germany's low rate of inflation, to the West German capital market's historically different interpretation of its financial responsibilities, to a certain amount of state financial support for utilities, to West German utilities' greater ease in adjusting electricity rates, and to substantial federal subsidizing of the West German coal industry.

As a general rule, the greater financial robustness of electric utilities outside the United States owes much to government involvement in providing financial loans and guarantees, approving plans, and regulating electricity prices. Although its manifestation and degree vary from country to country, extensive government intervention in utility affairs in Canada, Japan, and Europe is evident in table 8.6, which seeks to identify key features of utility industries in these countries. Governments in these countries have the responsibility for ensuring that utilities' financial health is maintained so that they are able to supply reliable and acceptably priced electricity to the consumer into the future. While the nuclear programs of state-supported utilities cannot proceed independent of financial constraints and the economic climate, it is also the case that government backing allows utilities – and governments – some freedom to invest in a pattern of electricity supply that reflects perceptions of long-term energy interests, at the risk of some economic wastefulness in the interim, rather than always being prisoners of short-term commercial pressures.

The advantage of close ties between utilities and governments goes beyond the financing and planning of utility investments. In France, the state-owned utility, Electricité de France (EdF), has been an essential policy instrument for the government in developing its wide-ranging strategy of nuclear development, involving the replacement of oil by nuclear power in base-load electricity generation, the electrification of space heating and other nonpremium end uses, and the establishment of a viable and internationally competitive nuclear industry. Through EdF, the government has been able to create the stable and favorable market conditions in which the nuclear industry could be developed (it is notable in table 8.4 that continuity of nuclear ordering was maintained in France throughout the 1970s), thereby allowing it to achieve its goals of substituting for oil in domestic markets and gaining balance-of-payments benefits and political influence from the export of nuclear equipment and materials.

It is an important point in relation to our discussions in section 4 that procurement measures, whereby industries are promoted or sustained by ad hoc government expenditures, are rarely available as instruments of industrial policy to governments with limited influence over utilities.

Besides possible social opposition, the main threat to the nuclear investment programs of state-supported utilities arises from the general economic difficulties that can lead governments to restrain public expenditure, and hence possibly to defer or prune nuclear investment programs. This has already occurred in Canada, where the provincial government of Ontario has placed restrictions on Ontario Hydro's capital spending. One can even imagine a situation in France where inflation, recession, and a slower growth of electricity demand might force the government to slow the rate of nuclear investment. With deflation unavoidable, heavy capital investments in nuclear power would be a natural target for public expenditure cuts.

Before closing this section, we should emphasize one point. The view that a rapid rise in the real price of oil, such as that in the past few months, will be to the nuclear industry's advantage is widely held among proponents of nuclear power. Rising oil prices will improve the economic competitiveness of nuclear power, they argue, will bring home to the public and to governments the risk of forgoing or delaying nuclear investment, and will lead to a lifting of restrictions on nuclear trade. For reasons similar to those outlined above, however, we are skeptical that the ensuing economic and political climate would be particularly favorable to nuclear power—indeed, it might act against it. Inflation and recession can only deter nuclear investment. Thus, while oil scarcity in many respects reinforces the need for nuclear power, it can simultaneously create economic conditions under which that need is extremely difficult to meet. Again, it appears that only in countries where governments are prepared to override these commercial difficulties by giving financial and political support to utilities can one expect any substantial shift to nuclear power in response to oil shortages, though they will also be subject to economic and financial constraints in pursuing their policies of energy diversification.

Political Conditions in Supplier Countries

Although adverse economic conditions and reduced forecasts of electricity growth have lowered expectations of nuclear expansion, it is clear that in most countries the introduction of nuclear power would have been more rapid, and the prospects more favorable, but for the intensification of political opposition in recent years. Indeed, politics and economics cannot easily be separated, since political controversy has raised the economic costs and risks of nuclear investments. In several countries its outcome has been long and uncertain lead times – in plant licensing and construction – and increased environmental and safety regulation, which has made the technology more elaborate and thereby more capital-intensive.

Public opposition has not been universal, but it has been particularly significant in that it has curtailed civil nuclear development, in all its aspects, in two of the three largest and technologically most dynamic Western economies, the United States and West Germany. As heavy consumers of energy, their failure to expand nuclear investment or to implement viable alternative energy policies could have serious implications for the balance of global energy supply and demand, and the frustration of their development plans has placed a greater technological burden on the smaller, less richly endowed industrial countries.

There have been many targets for public concern and opposition. They have ranged across the fuel cycle, from the environmental consequences of uranium mining to reactor safety, the effects of low-level radiation, and the hazards of reprocessing and waste disposal. A peculiarity of the nuclear industry is that problems in any part of it tend to affect the acceptability of the whole. If a process in the chemical industry is considered socially unacceptable, it can be modified or even discontinued without threatening the viability of the entire chemical industry. This is not the case with nuclear power. It is a reflection of both the degree of integration of the nuclear fuel cycle and its political sensitivity that the whole nuclear industry is so vulnerable to specific concerns.

Furthermore, it is apparent that public anxiety relates to today's *and* tomorrow's nuclear commitments. Thus, the political and economic viability of nuclear industries in 1989, 1999, and beyond, indefinable though it may be, is as much at issue as the viability of the industry in 1979. There is a particular fear that political systems will be unable to handle the consequences of large-scale commitments to nuclear power.

The opposition to nuclear power is, however, by no means a homogeneous political movement, whether in terms of its aims, opinions, tactics, or institutional support. Some constituencies are opposed to nuclear power in any form and want past investments dismantled. Others are opposed to new nuclear investment but will accept nuclear plants already installed and under construction. Still others are concerned about particular aspects of nuclear power (e.g., the "plutonium economy") but accept nuclear expansion in principle. Opposition may be expressed by pressure groups, with diverse ideological, community, or commercial interests, or in terms of a general mood, as recorded in public opinion polls. For the nuclear community this means that there tends to be no representative institution or individual to engage in dialogue, which complicates the task of identifying politically acceptable solutions.

While the opposition to nuclear power is not a unified political movement, it is nevertheless evident that industries and governments in a number of advanced countries have been unable to convince large and influential sections of the public that nuclear power is capable of being safe and economical at the same time. They have also been unable to convince significant sections of the public that those with interests in the expansion of nuclear power, whether in government or in the private sector, can be trusted to act in the public interest. There is suspicion in many quarters of the motives and abilities of the nuclear industry, suspicion that has been heightened by a history of exaggerated claims for nuclear power, by tendencies to underestimate the difficulties of such problems as waste disposal, and most recently by the events at Three Mile Island. The accident there demonstrated the susceptibility of nuclear power to accidents, however infrequent they may be, and the potentially drastic political, economic, and psychological consequences of such events. The credibility of regulatory authorities appears to be growing into a major issue, partly as a result of this accident.

Within the nuclear industry, the extreme sensitivity to the possible consequences of nuclear accidents is often viewed with incomprehension. It is argued that very few deaths or injuries can be directly attributed to the worldwide expansion of nuclear power, and that its safety record is superior to that of other energy industries. It is also pointed out that the political and economic costs and risks of forgoing nuclear investment, particularly in relation to any failure to reduce dependence on imported oil, tend to be underestimated by many nuclear opponents. Among advocates of nuclear power, there is suspicion of the motives of opponents of nuclear power, and resentment at being unfairly singled out, as they see it, for public criticism. It is apparent, however, that environmental and safety issues are, to an extent, the tip of the iceberg—the visible parts of a deeper and less easily defined anxiety over the social and political consequences of nuclear expansion. It has also been suggested that a key underlying issue in the controversy over nuclear energy is the distribution of authority to make technical choices in advanced countries.

Whatever lies behind the opposition to nuclear power, it has resulted in a sharp polarization of opinion over its pros and cons, with growing mutual distrust, particularly between those vociferous minorities on either side of the fence that are vying for the support of the "floating vote." Common ground for compromise has been increasingly hard to find, so that confrontation has tended to replace attempts to reach consensus. Whether this tendency will persist is one of the major uncertainties facing nuclear power.

However, there have been wide variations among countries, in the forms public opposition has taken and in the degree to which it has affected nuclear plans. It is striking, for instance, that public opposition in the United States, Sweden, and West Germany has virtually blocked the progress of nuclear power, while in France and Britain it has to date had comparatively little impact. It appears to us that concern over nuclear power is widely felt in advanced countries but that there are certain conditions under which it can develop into a major impediment, while under others it remains dormant. Although it may be the case that public assessment of risks differs between countries, it nevertheless seems clear that the political acceptability of nuclear power in advanced countries is determined to a considerable degree by administrative structures and by the form and extent of democratic involvement in those structures. These factors also affect the extent to which groups opposed to nuclear power can delay and influence nuclear decision making and the extent to which nuclear policymakers are perceived to be accountable for their actions.

It has been noted that establishing a nuclear industry is a large and risky affair, requiring coordinated actions and long-term commitments from manufacturing industry, utilities, regulating bodies, and different departments of government. The distribution of authority to make investment and other decisions in relation to nuclear power varies widely from country to country, as do the possibilities for intervening in the decision-making process through the courts and through political bodies at local, regional, and national levels. Some examples concerning reactor licensing illustrate the fact.

- While reactor safety is almost universally regulated by central governments, authority over siting decisions can be much more fragmented. Such decisions are subject to local or provincial government approval in West Germany, Canada, Japan, Sweden, and the United States (although in several of these countries siting decisions can constitutionally be preempted by central government if the political will exists), but not in France, Britain, or the Soviet Union.
- Regulatory and licensing decisions can be challenged in the courts in West Germany and the United States, but not in Canada, France, Japan, Sweden, Britain, or the Soviet Union.
- Reactor licensing has been made conditional on satisfactory progress with spent fuel handling and/or reprocessing and waste disposal in Sweden and West Germany (and in an increasing number of states in the United States), but not in other countries.
- In most countries there is an overall assessment of the need for a proposed nuclear (or other) power plant after all the different aspects have been cleared in due process. This authority rests with national govern-

ments in some countries (Sweden, Britain, France) and with regional governments in others (West Germany, Canada). In the United States there has been a tendency for individual states to acquire this authority de facto, though as yet it has no foundation in law.

Generalizations are hazardous, but it appears that the complexity of nuclear decision making and the tendency for there to be local opposition if the opportunity is provided make nuclear power particularly vulnerable in countries where authority is fragmented, as in countries with federal structures or judicial systems with broad mandates. The development and introduction of nuclear power appears to benefit from a degree of coordination and centralization in government that runs counter to political traditions and tendencies in several advanced countries.

The general susceptibility of nuclear power to local opposition to plant siting should be emphasized in this context. As with other large-scale investments (airports, oil refineries, motorways, etc.) that generate environmental concern and threaten the value of real estate in the region, there tends to be difficulty in gaining local approval for investment plans. This is one example, among many in the energy field, where the risks and costs of an activity tend to be imposed on local communities while the benefits are distributed on a wider scale. Furthermore, the operation of nuclear power plants is not labor-intensive, so that the long-term employment benefits to local communities are marked only in areas of low population density and high unemployment (e.g., Caithness in the north of Scotland).

Public attitudes are naturally affected by the quality and quantity of information made available by those in positions of authority. It is notable that particular countries, such as France and Britain, where government policy proposals have traditionally been less exposed to wide public debate at early stages of consideration, have had comparatively little difficulty in promoting nuclear power. Still less difficulty has, of course, been encountered in closed political systems such as the Soviet Union. In the short term, there are grounds for believing that, the more visible the issue of nuclear energy is made to the public, the greater will be public misgivings. In the longer term, however, withholding information could carry its own risks and lead to a more serious erosion of credibility regarding nuclear policies, especially if nuclear decision makers were suspected of abusing their authority and ignoring what was deemed to be the public interest. What in the short term might look like a wise policy could therefore turn out, in the longer term, to have been unwise. In countries where there is substantial freedom of information and access to the decision-making process, there is no choice but to conduct an open discussion.

The relatively trivial opposition to nuclear power in Britain, as compared, for instance, with that in Sweden and West Germany, leads one to speculate about the relationship between public concern and rates of development. These latter countries expanded ordering very rapidly in the early 1970s (and had to find several new sites within a brief space of time), whereas Britain started earlier and proceeded much more cautiously. France is, however, an obvious exception to the rule that might imply.

It is evident that the acceptability of nuclear power bears an inverse relationship to the number of present and future energy options open to a country. The existence of these options tends to weaken the case of nuclear proponents and make it more difficult for governments to justify the urgent and large-scale development of nuclear power. Comparison between the United States and France illustrates the point, France having few alternatives to nuclear power and therefore being able to argue that its adoption is inescapable.

The option of continuing to import oil for electricity generation—or at least not taking rapid steps to substitute for that oil—is still significant for those industrial countries that have substantial trade surpluses. Such surpluses give them the foreign exchange to pay for the rising cost of fossil fuel imports. Notwithstanding security of supply considerations, the increased cost of oil imports could in some circumstances be looked on with favor by countries whose trade surpluses are on such a scale as to be politically embarrassing. Generally it is those countries that lack energy options *and* have weak trade balances (e.g., France) that can present the most urgent case for nuclear expansion.

The extent to which nuclear issues interact with, or come to be regarded as symbolic of, broader contemporary political conflicts in advanced industrial societies clearly affects the way nuclear power enters into the political debate. Distrust of big business and central government (United States), regional tensions (Italy, Spain), concern over the social and environmental costs of industrialization (Sweden), fear of terrorism and the police state (West Germany): these are examples of contemporary issues that are recurrent themes in domestic nuclear controversies.

Attitudes of the larger and longer-established political parties toward nuclear power have traditionally been positive. As opposition has grown, parties and leading politicians have generally been led to reexamine their platforms, whether to defend or gain electoral support, for ideological reasons, or from being in charge of a government with responsibilities for nuclear policymaking. Some parties have aligned themselves with particular viewpoints and have remained united in the process; others have been unable to avoid internal dissension. Since many of the issues concerning nuclear energy (and the environment more generally) have been difficult to interpret in terms of traditional left/right political ideologies, there has been a tendency for these issues to inject an element of confusion and uncertainty into party political systems and to upset the frequently delicate balance of power between right and left constituencies in parliaments.

Some patterns have begun to emerge internationally. Conservative parties have on the whole remained pronuclear. Minority parties occupying the middle ground—center parties in Scandinavia and liberal parties in West Germany and Britain—have tended to adopt antinuclear policies, partly in an attempt to attract, or win back, the "green vote." This has been a particularly significant trend, since such parties hold the balance of power in coalition governments in a number of countries. The large social democratic parties in central and northern Europe, which traditionally have been pronuclear, are becoming increasingly split (Austria, Switzerland, Denmark, West Germany, Norway, Sweden). There is also a clear tendency for developments in a party in one country to influence sister parties in other countries, as can be seen particularly in Scandinavia but also in West Germany and Austria. In the same way, the socialist parties in southern Europe have also become more reluctant, at least on the national scale (France, Italy, Spain). The powerful communist parties in these countries, however, remain strong supporters of nuclear power, though they frequently demand nationalization of various parts. There is thus an overall tendency for nuclear energy to be drawn into party politics and electoral campaigns in Europe (and the United States) during the next few years.

Besides raising the economic costs of nuclear power, it is apparent that public opposition has imposed a number of other constraints on nuclear development. First, it has reduced the sovereignty of institutions that formerly had considerable independence in reaching decisions over nuclear plans and investments. For instance, power plant investments can no longer be decided by utilities without lengthy and frustrating negotiations with constituencies at the local, provincial, and national levels. Utilities that cherish their independence are thereby further encouraged to find ways of avoiding nuclear power. Second, it has made every decision a controversial decision, a "mini-Vietnam" as it is called in the United States, delaying and discouraging the development of every aspect of nuclear power. In some cases it has even made it difficult to solve the very problems that have given rise to controversy, for example, by preventing test drilling to find suitable sites for waste disposal. Third, and perhaps most important, public opposition has raised the electoral risks for politicians and political parties who openly advocate nuclear expansion and has generally induced conflict within societies. Government promotion of nuclear power now carries considerable political risk. In some countries, the costs of forgoing, delaying, or slowing nuclear expansion may be considered worth bearing in order to avoid putting the political system under unacceptable strain. Denmark, Norway, and the Netherlands are examples of countries that have shied away from nuclear power for fear of the political stresses it could induce.

Whether domestic political conditions will change for the better or for the worse in the immediate future is unclear. Much depends on public reactions to the oil crisis, on the way nuclear power enters into the West German and United States elections and the outcome of the Swedish referendum on nuclear power in 1980, on the safety record of the nuclear industry over the next few years, and on whether electricity shortages develop and who or what is blamed for them if they do. Many consider public opposition a temporary phenomenon that will evaporate when the costs of forgoing nuclear investments become more visible. We are not so sure. It is not a simple phenomenon, and it seems to reflect deep-seated social trends — fragmentation of authority, distrust of centralized institutions, fear of certain advanced technologies, the advance of communications and education — as much as it reflects concerns over issues that are specifically nuclear. It would be unrealistic not to expect some continued inhibition of nuclear growth in at least the early 1980s owing to social opposition. It will also take time to undo the various procedural knots that some countries have tied around plant licensing and other aspects of nuclear decision making in the past few years. But, as with the economic inhibitions already discussed, the precise extent to which domestic political difficulties will curtail nuclear growth in the 1980s must remain very uncertain.

In the medium term, it is nevertheless clear that reduction of social opposition is a necessary condition for nuclear expansion on any scale in a number of countries — notably West Germany, Sweden, and the United States. In the same way, it is necessary that political conditions do not worsen in a number of other countries if their nuclear industries are to have healthy futures.

One point seems to emerge clearly from this and the preceding section: the greater ease that countries with centralized and relatively closed political systems, and with substantial state control over the utility and nuclear industries, are having in promoting nuclear power. This derives from their ability to avoid both political *and* economic difficulties, at least for the time being. The corollary to this observation is that countries with more fragmented systems for making decisions are vulnerable to the breakdown of political consensus, at local or national levels, and to economic difficulties encountered by private utilities. If political consensus cannot be reestablished and economic difficulties overcome, these countries may have to accept relatively slow and painful nuclear expansion unless they are prepared to embark on changes in their political and institutional approaches to nuclear decision making.

The Back End of the Fuel Cycle and the Fast Breeder Reactor

Development of the back end of the fuel cycle and the fast breeder reactor (FBR) has given rise to some of the most bitter disputes over nuclear power. As these technologies provide points of contact between today's and tomorrow's nuclear systems, between military and civilian aspects of nuclear energy and between domestic and international political issues, the political controversies have been exceedingly wide-ranging, involving a variety of commercial, environmental, strategic, and social concerns.

In this context we can only give the briefest outline of some of the issues involved. It is nevertheless important to give them due attention in this paper, since they have come to have a considerable bearing on the climate in which today's nuclear investments are or are not taking place, and on the form and scale of those investments. This is particularly the case in relation to the back end of the fuel cycle. Failure to resolve national and international disputes in this area has had significant direct and indirect effects on the rate of, and prospects for, thermal reactor ordering in domestic and international markets.

The classic picture of nuclear development from an energy point of view is one of uranium-based thermal reactors creating, as a by-product, stocks of plutonium that can be used to fuel a subsequent generation of uranium-conserving breeder reactors. Thermal reactors reduce dependence on scarce and expensive fossil fuels in the short and medium terms, and breeder reactors reduce dependence on scarce and expensive uranium in the longer term. As such, the FBR and its associated fuel cycle have been regarded as a key long-term option in a world in which nonrenewable oil and uranium reserves are being depleted.

In parallel, there has been the perceived need to develop and establish satisfactory techniques and institutional procedures for treating spent fuel arisings from thermal reactors and for disposing of nuclear waste products. In addressing both the energy goal of supplying plutonium for the introduction of commercial FBRs (and allowing some uranium to be recycled) and the environmental goal of coping with hazardous fission products, reprocessing has therefore been perceived in many countries as performing an indispensable dual function within the civil nuclear fuel cycle.

A third dimension to reprocessing arises from its potential military application to supply weapons-grade plutonium to a nuclear weapons program. We have already noted in section 2 that several of the existing nuclear-weapon states used plutonium as the material for their first nuclear weapons. In some circumstances, military ambitions can therefore provide an added incentive to develop and establish spent fuel reprocessing.

There have also been strong commercial and institutional interests in proceeding with the development and commercialization of the FBR and reprocessing. First, the substantial and politically influential R&D communities that were brought together to develop the first generations of nuclear technologies have had a natural self-interest in the development of future nuclear systems, particularly those containing a high degree of novelty. Second, it has been perceived that successful commercialization of the FBR would bring considerable long-term benefit in terms of industrial employment and foreign exchange earnings from exporting and licensing the technology; hence the rivalry that has existed and still exists between nations developing the FBR to gain an early competitive advantage. In reprocessing, the prospect of commercial gain has been more immediate for those countries – notably France and Britain – with historical investments in the technology and commitments toward its expansion, especially since business risks have been largely carried by domestic and foreign utilities providing forward finance for plant construction. Third, development of these advanced technologies has long carried international prestige and has brought influence in nuclear affairs.

Despite the powerful motive forces behind these advanced technologies, there has been considerable divergence, especially in recent years, between the strategies adopted by the main nuclear supplier countries toward future nuclear development. Countries fall roughly into three groups: France and the Soviet Union, which are proceeding with substantial investment programs with the aims of establishing large-scale reprocessing and demonstrating FBR technology; Britain, West Germany, and Japan, which have similar civil nuclear ambitions but which have encountered some difficulties in achieving them; and Canada, Sweden, and the United States, which either have no plans or have deferred plans to proceed with the commercial demonstration of FBR technology (although the United States continues to maintain a substantial FBR R&D program) and have opted for a once-through fuel cycle for the time being. Sweden is an in-between case. It has negotiated contracts with Cogema to reprocess spent fuel in France since abandoning plans to build an indigenous reprocessing plant, but it continues to give serious consideration to the once-through fuel cycle.

There have been several reasons for these divergent policies. One has been that countries with secure indigenous supplies of uranium (Canada and the United States) face lesser risks in forgoing or delaying the commercialization of reprocessing and the FBR than countries with relatively insecure supplies (Japan and most European countries). Divergent policies have also resulted from what might be loosely referred to as national and international political circumstances, to which we now turn.

Starting with the national context, many of the factors already discussed that have shaped political attitudes toward thermal nuclear expansion have also influenced attitudes toward the FBR and reprocessing: safety, the presence or absence of alternatives (perceived long-term alternatives in this case), availability of information, centralization or decentralization of authority, and so on. The countries that are moving ahead most forcefully with the FBR and reprocessing—France and the Soviet Union—are also those that have continued to expand their thermal nuclear programs in recent years, owing in large part to strong and centralized government advocacy of nuclear power. On the other hand, countries that have experienced significant public opposition to thermal nuclear power development have also experienced opposition to FBRs and reprocessing—not surprisingly, since the issues surrounding present and future nuclear development have become so intertwined in the public mind.

The question of timing has been an important ingredient in controversies over the FBR and reprocessing, in three respects. There have been disputes over the timing of the need for the FBR and its associated fuel cycle, over the timing of the availability of safe and economical technical systems, and over the timing of the availability of institutions and regulatory procedures appropriate for large-scale FBR deployment. The phasing of FBR development and reprocessing plans depends on views held on each of these questions. Given the virtual impossibility of accurately predicting political, economic, and energy circumstances in the time scales involved in FBR development, the assumptions upon which development plans have been based have inevitably been subject to critical analysis and disagreement. Although its early introduction has also been argued on waste management grounds, this has applied in a more immediate sense to the rate of expansion of reprocessing, since its commercialization on an industrial scale has to proceed in advance of FBR deployment in order to gain fuel cycle experience and provide the necessary stocks of plutonium. Furthermore, since FBR programs could be delayed by failure to establish reprocessing on an adequate scale, the possibility that difficult problems could be encountered in expanding reprocessing increases the perceived urgency of pressing ahead with early commercialization.

Recent reductions in expectations of nuclear expansion and increased estimates of the size of economical uranium reserves have altered perceptions of the time scale in which uranium supplies could become a constraint on thermal nuclear growth, whether in relation to the scale and economics of uranium production or to political access to supplies. These developments have made one of the main arguments for urgently preparing the option to deploy FBRs – including establishing reprocessing to provide plutonium for FBRs – considerably more vulnerable to criticism.

Reduced expansion rates of thermal nuclear power also have a bearing on policies toward intermediate nuclear options that do not necessarily require reprocessing, such as the more efficient use of uranium in enrichment plants and thermal reactors and the introduction of advanced converter reactors. While the nuclear expansion rates projected a few years ago would have allowed these options to have relatively little impact on uranium supplies, their potential becomes more interesting as expansion rates are lowered, both as commercial propositions and as technologies that could provide more time for FBR development. Given the thermal expansion rates now considered plausible in many quarters, their effect on uranium supply over the next thirty to forty years could be as substantial as that of the maximum deployment of fast breeder reactors over that time scale. Governments, nuclear industries, and utilities therefore increasingly face policy choices and controversies over the respective emphases that should be given to the development and deployment of these intermediate and advanced technologies.

A further side to the political debate over the FBR concerns the very large scale of investment of venture capital and skilled manpower required to bring this technology and its fuel cycle to commercialization over a long period of time. FBR development is liable to the charge that it diverts R&D resources away from, and therefore leads to underinvestment in, other nuclear R&D options, including some aspects of the development of the present generation of thermal reactors. Within a wider framework, it is also liable to the charge that it takes resources away from other important energy and industrial priorities or more generally puts an undesirable strain on the public purse over a long time frame. It has been partly to reduce these strains on national resources and to encourage a pooling of technical effort that international FBR demonstration projects have been launched, particularly in Europe.

Controversies over the FBR have occurred at different times and with different intensities in the countries involved, but they have on occasion affected thermal reactor ordering by raising the temperature of political debate over nuclear power. In Britain, for instance, the present debate over the FBR and the conflict of political interests behind it are perceived as complicating the task of gaining political acceptance for a sizable new thermal reactor program. More widely, the effect of domestic debates over the FBR has been to intensify the controversy over the back end of the fuel cycle, owing to the requirement for plutonium and consequent implications for the rate at which reprocessing is expanded. It is in these respects that policies aimed at reducing the pressure on uranium supply in the longer term can indirectly influence reactor ordering rates in the short term, since they involve decisions about the rate of development of politically contentious technologies such as the FBR and reprocessing.

The issue of timing apart, domestic political controversies over the back end of the fuel cycle have many shades and angles. In relation to reprocessing, there are the perceived dangers of handling, transporting, and storing plutonium and the risk of its diversion to terrorist or other organizations; there is the question of the safety of the process in all its aspects; there are uncertainties about its economic advantages. Those countries that have, for whatever reason, opted for the time being for the once-through fuel cycle face another varied set of issues concerning its political and economic viability. And final disposal, whether of concentrated high-level waste or of untreated spent fuel, has again given rise to a distinct set of political and institutional debates, with wide variation between countries in their characters and intensities.

So long as there has not been seen to be adequate progress in sorting out the difficulties that have arisen around the back end of the fuel cycle, and so long as there has been a serious lack of political consensus over the policies to be followed, the effect on reactor ordering and on nuclear development in general has been one of inhibition. As previously noted, this inhibition has recently amounted to prohibition in some countries. In Sweden and West Germany, reactor licensing has been made legally conditional on the relevant authorities being satisfied that adequate plans have been developed for disposing of spent fuel and nuclear waste. Similar conditions now apply de facto in a number of states in the United States. Finding acceptable and timely technical, political, and institutional solutions to these problems—even if they are not final solutions but only convincing blueprints for attaining them—has therefore become a matter of urgency and a precondition for further nuclear expansion in some countries.

Turning to the international arena, one finds a further set of issues on top of those already discussed: competition (and in some instances collaboration) between nation-states in relation to the FBR and fuel cycle services, international dependencies in the back end of the fuel cycle, and the vexed issue of non-proliferation. The back end of the fuel cycle is generally an area in which the relationships between nation-states are at their most sensitive. It is also an area in which there is considerable interplay between domestic and international circumstances. Again, the complexity and diversity of the forces at play prevents our attempting a detailed analysis in this context. Some impression of the issues involved may nevertheless be gained in the following observations.

It was noted in section 2 that France, Britain, and the United States developed and established spent fuel reprocessing as part of their weapons programs in the 1940s and 1950s. Development costs were largely borne by these programs. It was a natural strategy in these countries to extend reprocessing to the treatment of spent fuel from power reactors, and, in France and Britain at least, this was achieved without serious political controversy.

France and Britain have persevered with commercial reprocessing, while the United States has abandoned it for the time being because of concerns over proliferation and the "plutonium economy." The United States government has discouraged the expansion of reprocessing in Britain, France, and elsewhere, on the grounds that it encourages the spread of reprocessing to other countries. Britain and France have argued that, by offering reprocessing services to non-nuclear-weapon states, they help to prevent such a development. Recently, there has been a tendency in some quarters in the United States to accept French and British reprocessing. It appears that in return for, and to an extent despite, this acceptance, France and Britain have joined the United States in discouraging the commercialization of reprocessing elsewhere: a move that has not unnaturally been resented in some countries, though others have accepted it as a necessary step in reducing proliferation risks.

Non-nuclear-weapon states have faced the task of establishing reprocessing and other approaches to spent fuel management as specifically civilian activities, subject to normal commercial and political decision-making procedures. In so doing, some, notably including West Germany, have encountered considerable public opposition. Some smaller non-nuclear-weapon states have also felt unable to justify the technical effort and investment costs required to reprocess spent fuel from their limited nuclear programs.

Whether out of choice (e.g., Switzerland), failure to overcome public opposition (e.g., West Germany), or pressure from the United States to defer reprocessing plans (e.g., Japan), utilities in non-nuclear-weapon states in Europe and in Japan have tended to become dependent on France and Britain for the reprocessing of all or part of their spent fuel arisings. They are also investigating schemes for interim storage until the political conditions for reprocessing and its commercial and technical viability become clearer. There has therefore developed in Europe and Japan an "integrated" spent fuel management system, with France and Britain becoming major centers for receiving a significant part of the spent fuel generated in the region. The viability of this system and the extent to which it offers customer countries time to sort out domestic problems in this area naturally depend on the successful construction and operation of British and French oxide fuel reprocessing plants in the 1980s, and implicitly on the continued political and economic support for these ventures in Britain and France. It also, to an extent, depends on the willingness of the United States government to permit transfers to those plants of spent fuel from reactors fueled by enriched uranium of United States origin. In these respects, there is distinct fragility in the web of interrelations that has evolved between these countries in recent years regarding spent fuel management and reprocessing.

While coming to rely on Britain and France for reprocessing services, some non-nuclear-weapon states have shown concern over the political and commercial terms attached to reprocessing contracts, especially in relation to the price of reprocessing and the conditions under which plutonium will be returned to the country of origin. Fears have been expressed, especially in West Germany, that the effective monopoly held by France and Britain has given them an unwelcome degree of influence over dependent countries' civil nuclear policies, particularly in relation to future plutonium-based technologies, and that the offer of reprocessing services could give them a competitive advantage in thermal reactor export markets.

In a number of respects, therefore, the implementation of spent fuel management policies has been complicated by the way national and international political issues and perceptions have interacted. What is perceived to be a domestic issue in one country is frequently regarded by other countries as an issue with strategic significance, requiring international consultation and negotiation. Furthermore, the wide range of considerations that has to be taken into account in formulating policy, and the multiplicity of possible motives for a given policy, open up risks of misinterpretation and unjustified suspicion, as well as providing opportunities for deliberate ambiguity as a form of disguise. This naturally places a heavy responsibility on countries for appreciating the international implications of their domestic actions, for making a particular effort to establish the credibility of these actions in the eyes of other countries, and for understanding the domestic constraints under which their neighbors are operating.

On another level, the international dispute over weapons proliferation has undoubtedly sharpened in the public mind, at least in advanced democracies, a perception that there is some inevitable and unbreakable link between the spread of nuclear power and the spread of nuclear weapons. This has further exacerbated political opposition to the expansion of nuclear power—interestingly enough, in advanced nuclear-weapon *and* non-nuclear-weapon states. International disagreements over reprocessing have also tended to complicate efforts to reach international agreement over acceptable techniques for and institutional approaches to waste disposal, which again have fed back into domestic disputes.

The emergence of various areas of disagreement between the major supplier countries over the back end of the fuel cycle reflects recent tendencies for there to be increasingly pronounced economic and political differences in wider fields such as economic and foreign policy, as well as divergent policies and ambitions in the nuclear field. Whether nuclear disputes will be settled during the coming years probably depends as much on overall political relationships between these countries as on the success of negotiations on specifically nuclear issues.

To summarize, domestic and international disputes over the FBR and the back end of the fuel cycle have placed a number of constraints on nuclear development. The nature and extent of these constraints have varied widely from country to country, and even from locality to locality. Diversity notwithstanding, their overall impact has been the same. They have complicated, and in some instances obstructed, attempts to manage the development of, and transition toward, advanced nuclear reactors and fuel cycles. But, most significantly for our study, they have inhibited the expansion of thermal nuclear power. They have contributed to the reduction of power plant ordering rates in both domestic and international markets and, insofar as many of the disputes are yet to be resolved, have further diminished the prospects for unfettered nuclear expansion in the coming years.

Export Markets

Earlier in this section we considered factors within the frontiers of the advanced supplier countries that have inhibited reactor ordering and discussed how national and international controversies over the back end of the fuel cycle and the fast breeder reactor have also inhibited nuclear expansion. We now turn specifically to circumstances in international markets.

In the 1960s and 1970s it was expected that a substantial part of supplier countries' output of nuclear equipment, materials, and services would be traded internationally. In the liberal trading atmosphere of the time, it was anticipated that, within overall limits set by non-proliferation objectives, nuclear exchanges would develop as "normal" trade according to accepted commercial rules. With this in mind, a margin of spare capacity was installed in most reactor supplier countries to take account of the export trade. The margin chosen was typically in line with firms' experiences of the historical balance between domestic and foreign demand for conventional power plant equipment. Thus, in West Germany, close to 50 percent of extra manufacturing capacity was installed; in the United States it was somewhat less.

Exports of nuclear products and services had an added significance for the smaller supplier countries. If, for instance, France and West Germany were to compete effectively in open markets with the giant United States manufacturers, it was essential to gain economies of scale in production, and these could come in sufficient quantities only by extending the base market through exports. Furthermore, at a time when trade balances in more conventional products were deteriorating in the face of competition from lower-wage economies, high-value nuclear exports provided an opportunity for improving the balance of trade in manufactured goods.

In the event, export markets have proved as depressed and unreliable as home markets. They have turned out to be smaller than originally estimated, and, as noted earlier, international nuclear trade has become the focus for intense political controversy since the early 1970s. On average, 4.85 GW of nuclear capacity were ordered in export markets each year in the 1970s — 20 percent of total reactor orders (see table 8.3) — with over two-thirds of the orders being placed between 1970 and 1974. Of the total for the decade of 48.5 GW, about 5 GW either have been canceled or are in jeopardy, mostly in Brazil and Iran.

The main Western supplier countries accounted for over 60 percent of global electricity-generating capacity in 1976, with Comecon accounting for 22 percent and the rest of the world for 17 percent. Western supplier countries' domestic markets are now all but closed to international trade in nuclear hardware, and they have little or no access to markets within Comecon (Canada's sale of reactors to Romania being the sole exception). Efforts to embark on or expand nuclear programs in most other developed countries, such as Austria, Denmark, Norway, or Switzerland, have met with economic and political difficulties similar to those encountered in supplier countries, so that export orders have been forthcoming mainly from the relatively small, but growing, sector of the world market made up by the larger and more advanced developing countries. Since several of these countries have become the focus of concern over nuclear weapons proliferation, nuclear trade with them has become highly controversial. It has led on a number of occasions to conflict between supplier countries over the conditions of trade (e.g., between West Germany and the United States over German exports to Brazil), and between supplier countries and developing countries that claim they are being denied access to nuclear technologies to which they have a right (e.g., between France and Pakistan).

Developing countries are generally at a disadvantage in relation to nuclear power, owing to the relative immaturity of their industrial and institutional infrastructures. For instance, they tend to be lacking in the skilled manpower, tradition of quality assurance, and regulatory institutions required for the effective construction and operation of nuclear facilities. None of the developing countries that have launched ambitious nuclear power programs has been without its problems. The extreme case is Iran, where the substantial reactor construction program initiated in the mid-1970s has effectively been abandoned. A few of the more advanced developing countries, notably Argentina, Brazil, India, South Korea, and Taiwan, are nevertheless continuing to press ahead with significant nuclear programs, whether in relation to the expansion of nuclear generating capacity or the construction of fuel cycle facilities, though all have been subject to domestic and international constraints of one sort or another and it is unclear to what extent their individual programs will be realized. In the short term, these countries provide established reactor suppliers with potential, though limited, export markets. But their efforts to build up reactor manufacturing capabilities of their own are already having an impact. Import substitution, particularly in relation to component supply, is likely to curtail the relative size of reactor export markets still further in coming years. If successful, these countries will presumably themselves become contenders in world markets in due course, further increasing competitive pressures within a limited market.

As a result of the small potential of export markets, one is forced to the conclusion that the viability of a country's nuclear industry rests ultimately on conditions within the home market. Exports can only be expected to provide a thin and uneven layer of icing on the cake. However, in today's circumstances, where prospects in many domestic markets are poor and where there is excess capacity in most reactor industries, exporting has assumed special importance for supplier countries. It is seen as an instrument for maintaining industrial capabilities and capacity utilization until conditions improve in home markets. In terms of domestic, but not international, politics, it is the simplest form of support for a beleaguered reactor industry that a government can arrange.

There is an added complication to international trade in nuclear reactors. Nuclear trade involves, and has always involved, more than the exchange of hardware and technology. It also involves the supply of fuel cycle services, cheap finance, industrial assistance, manpower training, access to uranium and enriched uranium supplies, and so on. Furthermore, purchasers seek long-term assurances that vendors will help to maintain nuclear facilities over their lifetimes and that the supplier country will generally act as a reliable partner in the development of the recipient's nuclear industry. Nuclear trade can also involve the bartering of products and services for other favors, with, for instance, nuclear supplies being exchanged for oil. To give a specific example, it has been reported that increased Japanese access to Canadian natural resources could be made conditional upon the Japanese purchase of CANDU reactors. For these reasons, the framework within which nuclear sales are negotiated by commercial enterprises tends to be arrived at through bilateral negotiations at the highest level between governments.

Competitive advantage in world markets has thus tended to be determined by such factors as the range of goods and services that can be made available to the customer, the supplier country's reputation for abiding by or breaking contracts and treaties, the manner in which safeguards and other arrangements are interpreted, a government's enthusiasm for promoting nuclear sales, and the recipient's confidence that the reactor supplier will remain active in the development and production of nuclear equipment in the future. Thus, besides the financial incentives that have become a necessary condition for exporting, conditions in home markets, resource endowments, and government policies have a significant influence on export prospects. For instance, Canadian reactors have been made more attractive by their nondependence on enriched uranium and by the possibility of easy access to Canadian natural uranium supplies, while France has been able to offer a full range of fuel cycle services and (where politically acceptable) technologies and has also been helped by its strong national commitment to nuclear expansion. Recently, West Germany has been more constrained in its ability to export by uncertainties in its domestic nuclear program and by the lack of a well-developed fuel cycle (though this did not prevent it gaining the Atucha-2 reactor contract in Argentina). It has also suffered more than other supplier countries from the demise of Iranian nuclear plans and from setbacks in Brazil. The United States industry has lately been most constrained of all in international markets, owing in this case to the United States government's efforts to restrict international transfers of nuclear materials and technologies.

From the above, it can be seen that there is a danger of intensified conflict between the major supplier countries over exports to third markets: conflict that could lead to a further deterioration in the industry's commercial prospects, quite apart from the effect it would have on the international diffusion of nuclear products and processes. Matters are further complicated by signs that the Soviet Union is entering more forcefully into nuclear markets outside the Communist bloc (witness the trade deals with India, Iraq, and Libya). Thus, one has a situation in which the Canadian and German industries, eagerly seeking outlets for sales overseas to compensate for depressed home markets and to support endangered reactor suppliers, are competing with one another and with the French (and Soviet) industries, which are in their turn seeking to exploit their positions of international competitive advantage, while all of them are risking confrontation with the United States over the conditions on which nuclear trade is conducted. While these countries have shown their readiness to exploit to the full any available commercial and political advantages in competing for export contracts, it is clear that they are now exercising great restraint in the negotiation of nuclear trade agreements in order to avoid political confrontation with other suppliers over non-proliferation aspects of trade. There remains the danger, however, that some countries may feel themselves forced to risk confrontation if adequate orders are not forthcoming in domestic markets and if other countries are considered to be trying to impose excessive restrictions on nuclear trade. In this respect, ordering prospects in domestic markets have a significant bearing on international nuclear relationships.

Summing Up: Reactor Ordering Rates during the 1980s

We have discussed in this section a range of factors that have acted together to inhibit the growth of nuclear investment in the Western world in the second half of the 1970s. What deductions can be drawn from this analysis regarding the probable rate of ordering of nuclear power plants in the 1980s?

Studies cited above tended to arrive at average reactor ordering rates outside the centrally planned economies of at least 35 GW/a in the 1980s — rates well above those achieved in the 1970s. However, the scenarios developed in these studies are largely "demand-driven," in a conventional sense, in that they are built upon a limited number of assumptions concerning the growth of electricity demand, generating mixes, the availability and relative prices of other energy supplies, and conservation responses. Excluded from the calculations are several other factors on both supply and demand sides that we have discussed in earlier parts of this section and that can significantly influence nuclear growth: the availability of finance for large-scale nuclear development, the technical performance of nuclear plant, utility behavior in the face of supply and demand uncertainties, social attitudes toward nuclear power, institutional structures, and so on.

We have attempted to weigh these additional factors, as they seem likely to bear on major Western reactor-manufacturing countries and on their export markets during the 1980s. The result is set out in table 8.7. As that shows, rather than a minimum average of 35 GW/a of nuclear plant being ordered in the noncommunist world during the 1980s, we believe that a much lower average total ordering rate, in the region between 14 and 25 GW/a, represents a contingency that demands serious consideration by policymakers in government and industry. While persuaded that the most probable ordering rate lies within that range, we fully recognize the elements of uncertainty involved and do not in any way exclude the possibility that pressures of circumstance or policy may produce an average ordering rate either above 25 GW/a or below 14 GW/a. We nevertheless believe that the 14–25 GW/a range constitutes a more realistic basis for planning in relation to the reactor industry than the higher-growth scenarios that have gained currency in recent years.

An average ordering rate between 14 and 25 GW/a would be on the low side in relation to currently installed manufacturing capacity (50–60 GW/a) in the reactor industry. It might also be considered an inadequate response to

TABLE 8.7. Estimated Average Reactor Ordering Rates in Noncommunist Countries, 1980–89 (GW/a)

Country	Rate
Canada	0.5–1
France	3–4
West Germany	1–2
Japan	3–4
Sweden	0–0.5
Britain	0.5–1.5
United States	2–7
Others	4–6
Total	14–25

growing oil scarcity. Against that, it could equally well be argued that it is on the high side when recent experience and present adverse conditions facing nuclear power in a number of countries are taken into account, and that attaining it would therefore require considerable improvement in these conditions. The estimates in table 8.7 reflect, for instance, an assumption that the present de facto moratoria on nuclear ordering in the United States and a number of European countries will end relatively soon, except possibly in Sweden. It is also implicit in our assessment that growing oil supply problems will reinforce the drive in industrial countries to conserve and diversify energy supplies and will lead to some reassessment by electorates of the acceptability of nuclear power in countries where its development has been strongly opposed in recent years. At the same time, we assume that the outcome of such a reassessment will fall short of an unqualified acceptance of rapid nuclear expansion.

The estimates in table 8.7 represent our personal judgment, and not all the factors entering into that judgment are quantifiable, or even tangible. It may help to illuminate our argument if we set out, even briefly and impressionistically, some of the considerations we have taken into account in relation to individual reactor-manufacturing countries. (We assume ten-year lead times for plant licensing and construction throughout.)

France. Substantial momentum has been built up in and behind the French nuclear program, involving major efforts to expand the share of nuclear power in electricity generation and of electricity in overall energy consumption, the construction of large-scale enrichment and reprocessing facilities, and the commercial-scale demonstration of a fast breeder reactor. The extent of this political and economic commitment is publicly justified in France by reference to the country's heavy dependence on foreign fossil fuel supplies, to the potential shortage of generating capacity in the electricity supply system, and to anticipated industrial and foreign exchange benefits.

These factors are likely to remain strong motives for continued nuclear expansion in the 1980s. At the same time, while there is some public opposition to nuclear power at both the local and the national levels, there is little sign that it will constitute a significant obstacle to France's nuclear program in the near future. For these reasons we consider that an ordering rate of 3–4 GW/a—close to that anticipated by EdF and the reactor industry—is a plausible outcome. It should nevertheless be noted that the French nuclear program could prove vulnerable if electricity growth in the 1980s were less than expected, if serious difficulties were encountered in the expansion of oxide fuel reprocessing, or if there emerged an unanticipated degree of political opposition, particularly from within the trade unions.

Japan. Like France, Japan lacks indigenous supplies of fossil fuels. The government and utilities are therefore promoting nuclear power as a means of obtaining greater national energy security through diversification. The pressure to find alternatives for oil in electricity generation is reinforced by anticipated growth rates of electricity demand that are higher than in most other industrial nations. There are, however, potential restraints on nuclear expansion. Power plant licensing and construction lead times have been lengthening, gaining site approval has proved increasingly difficult, and the operating performance of installed reactors has been somewhat disappointing. This may prompt changes in reactor types that could introduce delays for institutional and other reasons.

The Japanese nuclear program is also peculiarly vulnerable to setbacks in the United States, owing to its present dependence on United States reactor designs and on United States supplies of enriched uranium. Spent fuel management is a further area that could be increasingly problematic for Japan in the 1980s, especially if international political factors continued to impose effective restrictions on the development of the back end of the fuel cycle in Japan or if technical or political difficulties were encountered in reprocessing Japanese spent fuels in France and Britain.

In these circumstances, the 4.7–6.3 GW/a ordering rate given in the IEA *Steam Coal* study (IEA 1978), which is roughly in line with the most recent official Japanese projections, appears to be on the high side. A 3–4 GW/a ordering rate makes some allowance for the national and international obstacles that have yet to be overcome.

Britain. Unlike France and Japan, Britain has substantial, though not unlimited, fossil fuel reserves, anticipates relatively low growth in electricity demand, and at present has considerable excess capacity in the electricity supply system. The scale of investment in thermal reactors in the 1980s will depend to a considerable degree on policy toward the domestic coal industry, toward the operation of the large oil-fired generating stations that are near completion, and toward replacement markets generally.

After a lengthy period of hesitancy in government over civil nuclear development, the present government and the utilities are firmly committed to nuclear expansion, ostensibly involving a higher rate of reactor ordering. Subject to the outcome of a public inquiry, resources may also be committed to the demonstration of a commercial-scale fast breeder reactor. Owing to the unsatisfactory outcome of the AGR construction program, a PWR design may be developed as an alternative option for the next reactor orders. There is, however, a perceived risk that efforts to launch a major nuclear program at this time could trigger substantial public opposition and lead to politicization of nuclear issues along party lines, neither of which has occurred in Britain to date. Given the continuing risk of deepening economic recession in the early 1980s, there is also a real possibility that electricity demand may grow even more slowly than now anticipated.

The IEA *Steam Coal* study suggests an average ordering rate of 1.3–1.7 GW/a for the British domestic market in the 1980s (IEA 1978). While agreeing that the latter figure represents a plausible maximum, we consider that the lower bound may be too high, and that 0.5 GW/a as a lower bound would more accurately take account of the impediments that could arise in the British market.

Canada. There are some similarities between the overall energy situations in Canada and Britain. In Canada there is also excess capacity in the electricity supply system, a variety of short- and medium-term energy options, including major natural gas and hydropower resources, and a relatively low rate of electricity demand growth anticipated in the 1980s.

Policies toward nuclear ordering are largely formulated by utilities within constraints applied by their owners, the provincial governments. Any significant market improvement can come only from an expansion of ordering by Ontario Hydro, since other provinces have abundant fossil fuel or hydropower resources. However, further nuclear investment by Ontario Hydro is made difficult by financial constraints imposed by the provincial government of Ontario, by the anticipated low growth of electricity demand in the region, and by the presence of excess generating capacity.

For these reasons, a lower rate of nuclear ordering is expected in the 1980s than in the 1970s, when, on average, two nuclear power plant units (a little over 1 GW) were ordered each year. In addition, there are signs of growing political opposition to nuclear expansion in Canada, which could increase further if opposition in the United States does not abate. This could result in some inhibition of nuclear investment in the early 1980s at least.

Particularly in the light of the recent reports by the Royal Commission on Electric Power Planning in Ontario (1978) and by the Stanford Research Institute (n.d.), the IEA *Steam Coal* study figures of 1.8–2 GW/a in Canada in the 1980s (IEA 1978) seem too high; 0.5–1 GW/a seems a more realistic assessment.

West Germany. A de facto moratorium on nuclear ordering has existed in West Germany since the mid-1970s. This moratorium has arisen out of a range of legal and regulatory problems and from the political divisiveness that has recently plagued nuclear power in West Germany at both federal and regional levels. The recent deferral of plans to build a large reprocessing plant at Gorleben was a setback to the nuclear industry but could open the way to political compromise over the course of nuclear development. There are signs within the ruling coalition that such a compromise is being actively sought, possibly involving a reduced rate of nuclear investment, expansion of spent fuel storage, the investigation of alternative sites for reprocessing, and the attachment of increased priority to conservation and coal in energy plans. It remains to be seen whether such an approach will be acceptable to a wider public and whether it will enable legal restrictions on plant licensing to be removed. In the short run, much depends on the way nuclear power enters into the federal election in 1980 and on its outcome.

There seems little doubt that West German utilities would like, and have the financial resources, to resume nuclear ordering if they could gain construction licenses. West Germany remains a dynamic economy, with relatively high growth rates of energy consumption by international standards in recent years. It is heavily dependent on imported oil and high-cost coal. If electricity shortages were to develop in the next few years, pressures to grant currently blocked construction licenses, and possibly to resume reactor ordering, would naturally mount.

The IEA *Steam Coal* study specifies an average ordering rate in the region of 2.2–4.2 GW/a in West Germany in the 1980s (IEA 1978). It is reported that a recent (August 1979) paper prepared by the federal Ministry of the Interior for a discussion between federal and regional governments cites figures close to the lower end of this range. Other studies have reached still lower figures, depending on assumed growth rates, degrees of promotion of other energy options, and assessments of other variables. There is, in fact, particularly high uncertainty over the rate and timing of nuclear ordering in West Germany during the 1980s. In light of the present search for a political compromise over the role of nuclear power in West Germany in the 1980s, an average ordering rate in the region of 1–2 GW/a seems a reasonably balanced expectation.

Sweden. The immediate fate of the Swedish nuclear program will be determined in a referendum early in 1980. Given the present stances of the major political parties, it seems probable that the result will be a continued moratorium on nuclear ordering but a decision to complete and fuel the six reactors at present under construction or awaiting operating licenses. There remains the possibility that public opinion will become more favorable toward nuclear power later in the 1980s, for which reason we specify an upper bound of 0.5 GW/a for the 1980s taken as a whole.

United States. There has been an effective moratorium on reactor ordering in the United States since 1976, recently reinforced by the reactor accident at Three Mile Island. This moratorium has been caused by a complicated array of political and economic circumstances, with social opposition leading to licensing difficulties and discouraging government from active promotion of nuclear investment, lack of progress on spent fuel management adding to public concern, the financial plight of many utilities encouraging discrimination against nuclear power in investment plans, and so on. It is possible that this moratorium could be legalized, pending improvements in safety regulation procedures, as a result of the various inquiries under way into the Three Mile Island accident.

Obstacles to nuclear ordering seem more intractable in the United States than in any other country—apart, possibly, from Sweden. As in West Germany, the political climate will have to change substantially for there to be any significant resumption of nuclear plant ordering. In addition, it seemed evident in our discussions earlier in this section that the prospect of such a resumption is conditioned by financial and structural problems facing the electric utility industry, the solution of which may depend upon restructuring and the development of closer ties with state and federal governments. More generally, the prospects for restoring political confidence in nuclear power and its associated institutions and for reducing the risks confronting utilities when embarking on new nuclear investment may depend upon the willingness of central government to give a stronger lead in a number of areas. Whether it will become politically feasible for the federal government to take such an initiative remains an unanswerable question at this stage.

There remains the question whether the alternative energy options to nuclear power that are being advocated in the United States at present will prove viable. There are many uncertainties here. All that can be said is that, if widespread energy scarcity is encountered in the United States in the 1980s, public perceptions of the risks of investing or not investing in nuclear power may change.

The IEA *Steam Coal* study suggests an ordering rate in the United States of 12.5–14.4 GW/a in the 1980s (IEA 1978). The annual average ordering rate in the 1970s was 11.45 GW/a. Given the circumstances described above, it seems unlikely that any similar rate will be achieved during the next decade. Indeed, it may be rash to assume any significant ordering during the first half of the 1980s. Thereafter, if the de facto moratorium were lifted, one might expect a cautious resumption of reactor ordering, at least by those utilities that have large and positive experiences of nuclear power. When the whole decade is considered, an average ordering rate of between 2 and 7 GW/a appears reasonably plausible.

Conclusion. Our comments on individual countries serve to emphasize the extent to which any estimate of future ordering rates for nuclear power plants is subject to considerable uncertainty, if only because it involves such diversity of variables. It is not impossible that failure to overcome problems now affecting nuclear power in several major countries will result in an average total ordering rate during the 1980s of below 14 GW/a. Nor is it impossible that, if the majority of economic, social, and political obstacles are overcome, the average rate will rise toward the 35+ GW/a implied by other studies. The recognition of such possibilities does not, however, offer a basis for prudent planning by those who must make policy decisions at the national or industrial level. It is with the need for such a planning basis in mind that, having taken a wide range of factors into account, we would argue for assuming an overall average ordering rate in the noncommunist world of 14–25 GW/a in the 1980s.

4. STRUCTURE AND PROSPECTS OF THE REACTOR INDUSTRY

After an early experimental period, the reactor industry was established, and capacity installed, during the period of economic expansion and relative political stability in the 1960s and early 1970s. It was set up on the assumption that these conditions would continue into the 1970s and beyond. Since the early 1970s, notwithstanding the threats to economic and strategic security posed by growing oil scarcity, the economic and political climate surrounding nuclear power has deteriorated in a number of important countries, and several reactor suppliers now find themselves having to cope with substantial excess manufacturing capacity.

Reflecting this change in climate, expectations of nuclear growth, whether in terms of reactor ordering rates or other indicators, have repeatedly been revised downward. Yet there has been little visible change in the scale and structure of the Western world's reactor industry, which still has the capacity to produce between 50 and 60 GW of nuclear generating capacity each year. If our assessment of ordering prospects in the 1980s proves justified, it seems unavoidable that the reactor industry should undergo some structural adjustment. In this section we consider the form this adjustment might take. In particular, we consider the predicament of those suppliers that have received few if any reactor orders since the mid-1970s and may have to face a further period of inactivity in the 1980s. What options are open to these suppliers in seeking to maintain an active capability to supply nuclear generating plants if and when the market revives?

The present overcapacity in reactor manufacturing in the Western world is undesirable for a number of reasons. It brings large financial losses to the industry. It is damaging to corporate morale. It encourages attempts to use political leverage to gain domestic and export orders. It creates difficult problems for governments. Furthermore, the division of a limited number of spoils between a number of oversized suppliers carries the risk of an erosion of productive abilities (economic efficiency, quality control and safety standards, design expertise) in the industry taken as a whole.

In an "ideal" competitive system, excess capacity would be overcome by the strong eliminating the weak, leading to concentration and relocation of industrial production internationally. The survivors would be those with the lowest costs of production and the most successful technology. It could be argued that, in the 1960s and early 1970s, such factors influenced the emerging international structure of the reactor industry. United States manufacturers owed their market supremacy even more to financial and technological strength and to a large home market than to the persuasive power of the United States monopoly on enriched uranium. If conventional market forces had continued to drive the industry in the later 1970s, the international structure of the industry might have evolved rather differently. For instance, the European reactor industry might have followed the course taken by the aircraft industry, with an international division of labor being arranged within Western Europe to counter United States competition.

As it has turned out, the United States reactor manufacturers have been unable to press home their commercial advantage. Difficulties at home and the politicization of international markets have allowed other suppliers to gain in market share at their expense. In the absence of strong commercial competition and open markets, the tendency toward increasing concentration that has been a characteristic of many industries in recent years, and of the nuclear industry in the 1960s, has been lacking; indeed, the tendency has if anything been toward greater fragmentation and reduced commercial efficiency.

There are other reasons why it is unlikely that a more "rational" international division of labor will develop in this industry. For reasons of sovereignty and security of supply, countries, for right or wrong, place self-sufficiency before economy and resist increased foreign influence over reactor supplies. In all major industrial countries, the trend in thermal reactor production has been toward trying to reduce foreign involvement – in some contrast to collaborative trends in the nuclear fuel cycle (e.g., Urenco, Eurodif) and in FBR R&D and demonstration (e.g., the Serena agreement). Different reactor licensing requirements are a further barrier to close integration, making it difficult to achieve common designs and encouraging utilities to maintain their loyalties toward domestic reactor suppliers that have accumulated experience of these requirements. Suppliers are also reluctant to become embroiled in political controversies over reactor licensing in other advanced countries.

We therefore do not consider it likely that the present difficulties in the reactor industry will result in significant mergers or international collaboration between the established NSSS suppliers. However, at the level of LWR component manufacture (pressure vessels, pumps and valves, tubing, etc.), which is subcontracted in varying degrees by NSSS suppliers, there may be some scope for international rationalization. A consequence of reduced market demand could be that agreements will be reached whereby individual countries or firms will specialize in the production of certain components and will cease production of others, in order to promote an international division of labor that is better adapted to market conditions. Such rearrangements are even more likely to take place nationally, in cases where there are several domestic suppliers competing for a given component market, as in the United States, though the way any such development occurs will be influenced by national policy and legislation, for example in relation to antitrust restrictions.

Structural change could also occur in the next decade if some reactor manufacturers abandon their present technologies and move to different reactor systems. For instance, it is possible that Britain will adopt a PWR design as an alternative option for its next reactor orders. The possibility that Japan might adopt CANDU reactors, though more remote, could be even more significant, since it would give this reactor system a degree of international recognition it has lacked to date and thereby make it a more formidable competitor to the LWR in industrial and semi-industrial markets. But, unless there is a general move away from LWRs if safety concerns are not allayed, the industry's international structure is unlikely to be affected in a major way by such shifts in reactor type.

Thus, in considering the future evolution of the thermal reactor industry, one has to look principally within national frontiers, at least as far as the 1980s are concerned. Changes seem most likely to take place as a result of national efforts to restructure or to adopt different reactor designs, or as a result of individual firms' adjusting in one way or another to difficulties in domestic markets. We have already observed the critical role domestic markets play in supporting reactor industries. It would be true to say that restructuring of the reactor industry is at present being driven by conditions in the home markets of individual producers, not by the forces of international competition, and is the responsibility of national, not international, policymakers.

In relation to capacity utilization and market prospects, we can divide the world's major reactor suppliers into three groups:

- France and the Soviet Union, which are moving ahead with major nuclear investment programs, with industrial capacity reasonably well utilized;
- Canada, Japan, and Britain, where there are prospects for reactor ordering, although the timing and scale of that ordering are uncertain;
- West Germany, Sweden, and the United States, where there is substantial overcapacity and short-term market prospects appear to be very limited.

The first group of countries faces no urgent need to adjust manufacturing capacity, although table 8.3 shows that some overcapacity could develop in France in the 1980s. In the second group, Japan and Canada will probably need to adjust capacity to be more in line with a lower growth of demand for nuclear power, whereas Britain will have to reestablish capabilities after a long period of inactivity in the reactor industry. Barring a possible reduction in the number of Japanese reactor manufacturers, it would be surprising if any of the producers in these first two groups encountered difficulties in reactor markets so serious as to make them consider withdrawal, although the Canadian industry may have been brought closer to this position by recent failures to secure export orders. The most difficult problems are faced by suppliers in the third group of countries: how to reduce capacity more sharply (and to what level) and, in the immediate future, how to maintain an adequate manufacturing capability in a period of low or zero ordering. For these suppliers, withdrawal or some form of tactical retreat could be difficult to avoid if market prospects do not improve and if satisfactory policies cannot be devised for staying in a "holding pattern" while orders are insufficient or are not being placed. How long can they justify staying in such a holding pattern if the timing and scale of demand growth and eventual profitability (or losses) remain very uncertain?

Before considering the positions of individual suppliers, it is necessary to look at a general level at the problems reactor manufacturers face in uncertain markets and at the options available to them in seeking to cope with these circumstances. We start by looking at a problem that all reactor suppliers face when orders are not forthcoming in sufficient quantities or on a regular basis: how to keep design and engineering teams together.

As we indicated earlier, reactor manufacture involves the design and construction of the NSSS: the pressure vessel, reactor internals, piping, pumps and valves, instrumentation, and other hardware that make up the heart of the nuclear power plant. This manufacture requires investment in sophisticated capital equipment, some of which has been specially developed but most of which is of a type to be found on shop floors in other branches of engineering (machine tools, welding equipment, gantry cranes, etc.). More than anything, it requires the assembly and training of teams of highly skilled workmen. On the one hand, there are the engineering and design teams whose skills and technical knowledge have been developed rather specifically for nuclear production; on the other hand, there are the teams of skilled workers occupied mainly on the shop floor and on the construction site (e.g., machine operators, welders, die and jig makers, fitters, supervisors, and project managers). The former are the critical, nuclear-specific elements in the reactor business. The size of these teams is not large in relative terms. According to one reactor supplier (ASEA-Atom), they account for less than 10 percent (2,000 man-years) of the 25,000 man-years that are typically involved directly in nuclear power plant manufacture and construction. Their importance is out of all proportion to their size, however, though they can add up to a few thousand people in the large firms. The latter categories of skilled workers are less nuclear-specific. But they contain skilled and experienced workers who are in much demand in the marketplace and are therefore as easy to lose as they are difficult to recruit.

Two further points should be made about the requirements of nuclear manufacturing. The first is that nuclear design and production require the maintenance of a level of quality assurance that is absent from most other areas of manufacture. This is not only a question of skills and equipment. It requires effective and diligent management, confident teamwork, and a political structure within a firm that ensures that there are adequate checks and balances to prevent negligence or oversight. While no manufacturer has approached the ideal in this respect, most have achieved a level of expertise that they can ill afford to lose.

The second point is that design and engineering teams are largely employed at the front end of nuclear reactor projects, while the skills on the shop floor (and the capital equipment) are employed in the middle of projects. Given a ten-year lead time, design teams are typically most fully loaded in the first two to three years after the order is placed, and the shop floor in the middle four to five years, the remaining years being taken up by construction and assembly on site. This means that fluctuations in demand affect design teams and the shop floor at different times. Design teams are the first to suffer from a fall-off in demand and the first to benefit from an increase in demand.

It is for these reasons that reactor manufacturers regard their design and engineering teams as the parts most vulnerable to reduced ordering. They fear they will be unable to prevent many highly skilled and experienced workers from leaving the business altogether, especially those who are in their thirties and forties and are still relatively mobile. The age profile could then move out of balance, with a loss of younger personnel who could take over senior positions in the coming years. Once they start leaving, it is feared there could be an avalanche, with design and engineering capabilities being severely eroded. Manufacturers are consequently putting much effort into finding ways of keeping these key people on their payrolls. How can they achieve this?

It is ironic that the very factors that are hindering the recovery of reactor industries in some countries are also helping it to survive in the short term. Long lead times allow manufacturers to maintain a certain level of activity in their factories and on construction sites, though they do not keep the designers and engineers adequately employed. Work has, however, been created for them by the technical problems that have been encountered in all reactor designs. Even the Three Mile Island accident—disastrous for the industry in every other respect—will probably lead to revised safety regulations and a consequent need to redesign and refit parts of reactor systems. This raises a wider issue. Besides extending lead times and raising capital costs, increased safety regulation over the past decade has brought an increase in the numbers of designers and engineers required per unit of output. Manufacturers could afford to cut down on their design overheads, and thus their vulnerability to fluctuating demand, if greater stability in power plant design and licensing could be achieved.

As the number of reactor installations has risen, servicing and general maintenance have also provided opportunities for occupying some designers and engineers and have brought some continuity of activity to production generally. For reactor manufacturers involved in the nuclear fuel cycle, an increasing stock of operating reactors provides similar opportunities within the expanding fuel cycle business. Otherwise, the only options open to reactor suppliers are to diversify into nonnuclear markets or to buy time by moving key personnel to R&D projects. Prospects for diversification are, however, limited. Many other markets have been hit by recession. Furthermore, there are few areas of production that have the level of skill intensity and quality assurance of nuclear manufacturing, so that diversification can provide employment for only a small proportion of the skilled labor force and can do little to maintain quality standards. It is said, for instance, that for each designer or engineer required in the supply of conventional power plants — an obvious direction to move in — ten are required in the design and erection of nuclear power plants.

R&D also has its limitations. Expanding R&D to provide temporary employment is more or less equivalent to providing a pure subsidy, since there is no short-term return on the outlay. With many reactor divisions running at a substantial loss, boards of directors need some convincing that such expedients are justified. It is here, however, that governments can intervene to give financial assistance, though at the risk of creating competition for resources between government nuclear R&D establishments and reactor suppliers.

Taken together, the existing backlog of orders, technical problems with installed reactors, new regulations, servicing and fuel operations, diversification, and R&D provide some opportunity for maintaining an adequate manufacturing capability when markets are depressed, though the extent to which these options are available and effective varies considerably from manufacturer to manufacturer. Suppliers claim they can be effective only for a limited time — useful for a few years after the last orders have cleared the design and engineering stage. Much depends on the prospects for future ordering. If prospects are good, the labor force remains confident in its longer-term security of employment and is willing to put up with temporary diversions. If prospects remain poor, those with ambition and family responsibilities will make every effort to move elsewhere.

Without visible improvements in market conditions, affected manufacturers are clearly anxious that they will find themselves unable to retain an adequate productive capacity beyond those few years. Besides the loss of skilled personnel, they fear a general loss of expertise in their design and manufacturing divisions: in effect, that they will begin to move rapidly back up the learning curve, losing the tradition of quality assurance developed over many years. Hence the urgency that is expressed regarding the need for these firms to gain new orders. There may be short-term measures they can enforce to maintain capabilities, but there is no effective substitute for designing and building reactors.

Orders have an added significance. They help to sustain those parts of the industrial infrastructure that feed components and materials to the reactor supplier. This brings us to a second area of vulnerability in the reactor business: component supply.

We have mentioned that the degree to which component manufacture is subcontracted by reactor vendors varies widely from vendor to vendor and country to country. At one end of the spectrum there are suppliers such as Westinghouse and the Japanese firms that manufacture the majority of components in house; at the other end are suppliers such as KWU, AECL, and ASEA-Atom that subcontract large parts of NSSS component manufacture to other industrial enterprises. There are advantages and disadvantages attached to subcontracting. On the plus side for the main supplier, it externalizes some risks and costs and lessens the financial burden of both building and maintaining a supply capacity. It also potentially spreads the benefits of high-technology production more widely within a country's industrial infrastructure and allows the main contractor to shop around for the best supply arrangements in the case of components that are not nuclear-specific. On the minus side, it lessens the control exerted by a reactor manufacturer over its component supplies and requires that a high degree of confidence and unanimity be maintained between the parties concerned.

In periods of low or zero ordering, subcontracting can lead to particular difficulties. If component manufacture is undertaken in house, the reactor supplier can adjust capacity, relocate labor, and inject subsidy where necessary. The main supplier generally retains control over manufacturing operations, even though the internal risks and costs incurred can be considerable. If there is subcontracting on a large scale, there is a risk that external component supplies will become insecure as component manufacturers reassess their prospects in reactor markets.

For those component suppliers that are themselves large diversified companies, sustaining relevant manufacturing divisions during downturns in reactor markets can be achieved without serious difficulty, especially if the production facilities are not "dedicated" and can be diverted into serving other markets. There are two sets of suppliers that may be particularly at risk: suppliers of very large components (e.g., pressure vessels, steam generators) that have made substantial nuclear-specific investments of human and physical capital, and small companies, admittedly few in number, that have specialized in supplying nuclear-specific components to reactor markets.

Thus, reactor manufacturers face two particular structural problems in periods of low ordering – the one of maintaining adequate teams of designers and engineers in their own organizations, the other of ensuring that supplies of bought-in component parts remain available. The former is arguably the more serious problem, since skilled manpower is the critical resource in any high-technology industry. There is also greater room for maneuvering in relation to component supply. For instance, if a component supplier leaves the market, a reactor vendor has the option of replacing it by creating (or purchasing) its own in-house capability or by finding an alternative source of supply, at home or abroad. We have already noted that there may be scope for rationalizing component supply industries internationally, leading to a more marked exchange of component parts between suppliers and between supplying countries. It is clear, however, that, wherever possible, reactor vendors prefer to have reliable sources of component supply within their own spheres of influence and fear a loss of sovereignty in their operations if large parts of their reactor systems have to be imported.

Before turning to the positions of individual reactor suppliers, four further general points should be mentioned, influencing as they do the responses of suppliers to difficult market conditions. The first is that, as we noted in section 2, most reactor suppliers are also involved in the manufacture of fuel elements for the installed stock of nuclear reactors. Since this is expected to grow into a substantial, reliable, and profitable business, firms' attitudes toward losses incurred in reactor supply are influenced by the anticipated gains to be made from fuel fabrication. In general, a reactor manufacturer's involvement in fuel fabrication, though on a smaller scale than reactor supply itself, raises the opportunity costs of abandoning reactor manufacture and gives it a degree of financial resilience when reactor markets are depressed.

The second point is that most reactor suppliers are themselves large, diversified companies for which nuclear activities constitute a minor part of turnover. On the one hand, this brings a degree of security to the industry, in that the firms in question have the internal resources to support their nuclear divisions. On the other hand, losses in reactor markets can constitute a significant drain on resources, especially if the overall profitability of the firm is unsatisfactory, and will be tolerated only if market expectations justify a subsidy. Some of the firms at present having to find ways of absorbing these losses are in the private sector. Their policies toward the nuclear market will be based mainly on their own commercial interests, not on an assessment of the longer-term national interest, even if they are under political pressure to take that into account.

The third point is in qualification to the second. The relationships between reactor industries, on the one hand, and utilities and governments, on the other, differ substantially within and between countries. Some governments have adopted paternalistic policies toward domestic reactor suppliers (e.g., in France and Britain), as have some large electric utilities (e.g., RWE in West Germany, Tokyo Electric Power Company in Japan). They have regarded the establishment and maintenance of a viable reactor industry as being in the interest of the nation or utility in question and have therefore devoted effort and finance to those ends. As a consequence, while some reactor suppliers, particularly in the United States, are more or less free agents in deciding strategies toward reactor markets and have more limited access to external sources of political and economic support, other suppliers are subject to utility and government influence in their policymaking, and may be able to gain commercial support in the form of finance, technical assistance, and expanded ordering programs.

The fourth point is that many reactor suppliers have contractual agreements with utilities extending well into the 1980s, and possibly even beyond, that, if breached, could result in substantial claims for financial compensation being made against them. These agreements may relate, for instance, to the completion of nuclear power plants with long lead times, the repair and maintenance of installed power plants, or the supply of fuel elements. There could therefore be substantial costs attached to leaving, or even declaring a future intention to leave, reactor markets. Thus, in formulating policies toward reactor markets when ordering prospects are poor, the costs of staying in the market have to be weighed against the possible costs of leaving it.

It is apparent that, in descending to the level of individual reactor manufacturers or supplier countries, one is confronted with substantial diversity that makes generalization hazardous. There are no two manufacturers that are equivalent in, for instance, their internal structures, patterns of diversification, corporate goals, supply commitments, or utility relationships, so that responses to difficult market conditions are bound to vary widely. Nevertheless, all reactor vendors faced with extended periods of low or zero ordering and with uncertain future prospects ultimately face the same choices: whether to remain in or withdraw from the market or to aim for intermediate positions whereby they can remain on the periphery of the market, meeting remaining commitments and keeping open the option for reentry for a relatively extended period.

This intermediate position is worth considering in a little more detail. In this context it is useful to draw a distinction between what may be termed a reactor manufacturing *capability,* whereby a manufacturer can bid for a contract and feel confident in its, and its component suppliers', ability to supply the required services and equipment to an acceptable price and time schedule, and what may be termed a manufacturing *potential,* whereby a supplier possesses the core of expertise required to enter or reenter the market but will, in the latter case, be faced by costs and delays in reconstructing a full manufacturing capacity and in reestablishing the levels of quality assurance required in nuclear production. The difference is essentially between having an active and a latent reactor industry. Although there are many uncertainties here, and one has again to acknowledge diversity among reactor suppliers, it is our impression that the latter position — the maintenance of a manufacturing potential, as defined above — is relatively easy to hold in the medium term. This is so because, in the worst case of an apparently indefinite absence of orders, the suppliers concerned can opt to become service organizations, supplying fuel fabrication, repair, and maintenance services to utilities with installed nuclear power plants. They may also be able to hold on to some manpower with nuclear skills and experience by moving it into R&D and other high-technology areas of production. These courses of action would preserve some design and manufacturing activities, though they would fall short of those required for active reactor manufacture. In addition, some transferable expertise would be retained in utilities with installed nuclear facilities, in R&D institutions and, where applicable, in submarine reactor programs. The advanced industrial infrastructure on which reactor manufacturers depend for supplies of materials and components would also provide a continuing launching pad for future reactor manufacture.

Consequently, it may be unnecessarily alarmist to think in terms of the possible extinction of a firm's or a country's reactor manufacturing capability in the face of an excessively low rate of, or an extended moratorium on, nuclear power plant ordering. It is probably more correct to regard the main risk to manufacturers in these situations as that, beyond a certain point, they will be unable to maintain their capability to supply the market effectively and in good time and will therefore be forced to beat a tactical retreat into a position that will imply reentry costs and delays if the market revives. Such a position could obviously also imply a loss of international competitiveness and credibility, at least for the time being.

With this as background, let us now consider the situations of the major reactor suppliers in the Western world. Their present and historical levels of activity may be inferred from table 8.8. This table is accurate as far as announcements of orders are concerned, but it may overestimate the number of orders that are pushed through to completion and underestimate lead times.

France. A high rate of ordering has been maintained in the period 1976–80, ensuring that the reactor industry will be well occupied until the middle to late 1980s. EdF plans to continue ordering nuclear power plants in the 1980s, though with the rate declining toward the end of the decade. Framatome's markets therefore seem reasonably assured into the 1990s, provided the political and economic climate in France continues to favor nuclear expansion at the given rate. If the average ordering rates of 3–4 GW/a posited in section 3 are achieved, Framatome will have little need to adjust capacity, particularly if export orders are forthcoming.

Canada. It is apparent in table 8.8 that there is sufficient work in the pipeline to keep AECL and most of its component suppliers occupied

TABLE 8.8. Reactor Units Ordered from and Completed or Due for Completion by Major Manufacturers, 1951–95 (as of October 1979)

Manufacturer		1951–55	1956–60	1961–65	1966–70	1971–75	1976–80	1981–85	1986–90	1991–95	Total
United States											
General Electric	A	1	4	11	37	32	0	—	—	—	85
	B	0	1	5	7	21	15	32	4	0	85
Westinghouse	A	0	4	7	39	51	6	—	—	—	107
	B	0	0	3	8	19	29	31	17	0	107
Babcock and Wilcox	A	1	0	0	13	11	2	—	—	—	27
	B	0	0	1	0	6	4	8	8	0	27
Combustion Engineering	A	0	0	0	12	15	4	—	—	—	31
	B	0	0	0	0	5	3	12	7	4	31
Japan											
Mitsubishi	A	0	0	0	4	3	3	—	—	—	10
	B	0	0	0	0	3	2	4	1	0	10
Toshiba	A	0	0	0	3	2	1	—	—	—	6
	B	0	0	0	0	0	4	2	0	0	6
Hitachi	A	0	0	0	1	1	1	—	—	—	3
	B	0	0	0	0	1	1	1	0	0	3
West Germany											
KWU	A	0	0	2	9	17	2	—	—	—	30
	B	0	0	0	4	5	7	8	6	0	30
Sweden											
ASEA-Atom	A	0	0	1	3	6	1	—	—	—	11
	B	0	0	0	1	3	4	3	0	0	11
Canada											
AECL	A	0	2	3	7	12	6	—	—	—	30
	B	0	1	0	1	6	6	10	6	0	30
France											
Framatome (and predecessors)	A	1	3	2	3	19	20	—	—	—	48
	B	0	2	2	3	1	14	26	0	0	48
Britain											
NPC (and predecessors)	A	4	16	8	8	0	6	—	—	—	42
	B	0	8	10	6	2	1	7	8	0	42

Sources: Westinghouse 1979; NEI 1979.
Note: A = Units ordered in period.
B = Units completed or due for completion in period.
Orders are allocated here to dates of announced commitment to specific NSSS vendors; data therefore differ in some cases from those in tables 8.3 and 8.4.

some way into the 1980s, even if at less than full capacity. However, there is expected to be substantially less nuclear ordering within the domestic market than in the 1970s. For this reason, the fortunes of the Canadian reactor industry in the coming decade will depend to a considerable degree on capturing a significant share of export markets. The contract to supply two reactor units to Romania notwithstanding, the Canadians have suffered setbacks in export markets in recent months. The failure to secure orders in Argentina and Japan has been a major disappointment to the industry.

If prospects do not improve, serious underutilization of capacity could develop in parts of the Canadian reactor industry in the early to mid-1980s. The Canadian Nuclear Association has recently expressed concern that there could be several casualties among component suppliers if orders are not soon forthcoming.

Japan. The Japanese reactor industry is young in comparison with other major reactor suppliers. Reactor manufacture is undertaken under license from GE and Westinghouse by three large industrial conglomerates: Hitachi, Mitsubishi, and Toshiba. The Japanese government and utilities are committed to an expansion of nuclear power in the 1980s, even if at a somewhat lesser rate than earlier anticipated. If their intentions are realized, there should be an adequate level of capacity utilization in the reactor industry. Nevertheless, given the problems that have been encountered with existing reactor designs—particularly the BWR—and the presence of three suppliers in a possibly limited market, it is open to question whether a lasting structure has yet been arrived at for reactor production.

It has been suggested that a feature of Japanese manufacturing that helps to sustain production capabilities in the face of peaks and troughs in ordering is the tradition of lifetime employment in Japanese firms. There is little fear that technical and managerial skills will be eroded during the troughs by manpower leaving the industry; they will be shifted temporarily into other areas of production and kept available to meet increased demand when it arises. Hitachi, Mitsubishi, and Toshiba also manufacture a high percentage of nuclear components in house and are therefore less vulnerable to the loss of component supplies than some other manufacturers.

Britain. Although first into the market, Britain has had difficulty in establishing a viable reactor industry owing to conflict between sponsoring organizations in and out of government, general industrial inefficiencies, design problems with AGRs, and a lack of continuity in demand. Between 1967 and 1978 no reactor orders were placed, leading to loss of manpower, deterioration in the age profile of design and engineering teams, and some loss of confidence in manufacturing and construction capabilities. In some respects, the problems encountered by the British reactor industry in the late 1960s and 1970s prefigured those now being encountered by other suppliers.

Although its effectiveness may be in doubt, the continued survival of reactor design and manufacturing abilities in Britain after a long period of underutilization provides some support for our earlier argument that a manufacturing potential, but not necessarily an active capability, can be maintained over an extended, if not indefinite, period.

Britain is now faced with the task of reestablishing its reactor industry to meet plans for an ordering program (possibly based on a PWR design) that may be launched in the early 1980s. Provided no significant political or economic barriers are erected in the way of such a program, the British reactor industry could find itself in the exceptional position of having insufficient capacity to meet demand.

West Germany. Kraftwerk Union (KWU) was set up in the 1960s to be a major producer of nuclear and conventional power plants, with a capacity of six to eight nuclear units per year. Since 1975 it has received only two firm reactor orders (including the recent reactor order from Argentina), and some of the seventeen plants ordered in the previous five years have been canceled or are being delayed in construction. Export markets have been as problematic as the home market, with Iranian contracts canceled and the large Brazilian construction program delayed.

KWU and its component suppliers have been placed under considerable strain by the de facto moratorium in the West German market and by difficulties encountered overseas. Recent press statements have indicated that some shedding of manpower has already taken place and have highlighted the predicament of Gutehoffnungshuette, the principal supplier of heavy nuclear components to KWU, and of some smaller specialist component suppliers whose survival is threatened.

We suggested in section 3 that an average ordering rate of 1–2 GW per annum may be the level achieved in the West German market in the 1980s. If this is the case, the West German reactor industry will no doubt survive intact at reduced capacity. If, however, the present de facto moratorium in West Germany continues far into the 1980s, the industry's future status must be in doubt.

Sweden. It seems probable that ASEA-Atom will face an indefinite moratorium on new plants once the twelve reactors installed or under construction are fully operational. If no export orders are forthcoming by 1985, ASEA-Atom's involvement in nuclear markets may be limited to supplying nuclear services and fuel. Efforts are being made to diversify into other markets.

If political conditions did change, however, the Swedish government has the powers to stimulate demand and increase ordering, so that nuclear expansion could take place with comparatively little delay, provided the reactor supply industry could be reconstituted. If this proved difficult, the price to be paid for further nuclear expansion could be greater import dependence.

United States. United States reactor manufacturers have been faced with an effective moratorium on new reactor ordering in the United States since the mid-1970s and with a deteriorating commercial position in external markets. General Electric is least well placed, having received no nuclear power plant orders since 1975, partly, it appears, because the operating performance of installed BWRs has been unsatisfactory. GE might retreat from reactor markets in the next few years unless there is an abrupt change in market conditions and utilities show renewed interest in the BWR. However, it would probably keep in touch with the market through the provision of fuel fabrication, repair, and maintenance services. If conditions do improve and the BWR remains uncompetitive, it is an interesting question whether GE might consider promoting an alternative reactor design.

Babcock and Wilcox's reputation has suffered from the accident at Three Mile Island. Its future position in nuclear markets is now uncertain. It may choose to concentrate on supplying nuclear components (pressure vessels, etc.) to other reactor vendors and withdraw from its role of main contractor for nuclear reactor systems.

Combustion Engineering and Westinghouse are the United States suppliers in the strongest positions. They have work in hand to take them into the 1990s, though they could face the usual difficulties of sustaining design and engineering teams from the early to mid-1980s. Westinghouse, with a relatively high percentage of turnover coming from reactor sales and substantial overseas interests, is the most heavily committed of United States producers to staying in the market. It is also more involved in FBR R&D than other firms.

It is unclear whether and how the de facto moratorium on nuclear ordering in the United States will be broken. We have suggested that ordering rates in the range of 2–7 GW/a could be all that will be achieved in the 1980s. If we are proved correct, major structural changes in the United States reactor industry, which at present has a capacity to produce 25–30 GW/a, seem unavoidable.

Let us try to summarize the positions of the leading reactor manufacturers, as we see it. The early 1980s will be critical for a number of reactor suppliers, notably ASEA-Atom in Sweden, Babcock and Wilcox and General Electric in the United States, KWU in West Germany, and, possibly, AECL in Canada. If they fail to gain significant new orders at home or abroad, their reactor manufacturing operations will be placed under severe strain. The loss of manpower and component supplies, and the curtailment of investment in reactor development and production that would unavoidably accompany an extended absence of orders, would place limits on these firms' ability to respond quickly and effectively to renewed market demand if and when it eventually occurred. However, the adjustments forced on such manufacturers would not necessarily imply complete withdrawal from the market or the total loss of reactor manufacturing capabilities. By becoming service organizations and adopting other supportive policies, and by completing remaining backlogs of orders, they could probably ensure that sufficient nuclear expertise was retained to enable the reconstruction of their supply capacities, at some cost and with some delay, if they so desired.

In the event that the de facto moratorium on nuclear ordering in the United States continues, the other United States reactor vendors would also begin to face similar difficulties in the mid-1980s. The other major suppliers – Framatome, NPC (or its successor), and one or more of the Japanese suppliers – seem to have their futures reasonably assured into the 1990s, barring major political and economic reversals in their home markets.

It may be considered unlikely that market conditions in the 1980s will be so severe as to force more than one or two reactor suppliers to retreat from the market. It should also be recalled that the financial risks of leaving the market could be considerable for some reactor vendors, in the light of possible claims for compensation for broken contracts. A wider retreat of reactor suppliers from the market is nevertheless a possibility that cannot be ignored.

On an international scale, the nuclear option could probably withstand the withdrawal, temporary or permanent, of some reactor manufacturers. If, by way of example, the four or five reactor manufacturers most threatened in present circumstances were all to retreat from the market, between 30 and 40 percent of reactor manufacturing capacity in the Western world would be lost, or at least would no longer be immediately and reliably available. This would still leave sufficient capacity to meet a demand higher than the 14–25 GW/a we have suggested as a plausible outcome in an earlier context, given the 50–60 GW/a of currently installed capacity in the international reactor industry. It would also raise the potential market share of remaining suppliers and improve their capacity utilization, encouraging them to stay in the market. However, if the erosion of manufacturing capacities went much further than this, say with the withdrawal of the remaining United States suppliers, the risk of irreparable damage would be considerably greater, and an adequate response to any increases in demand in later years could no longer be assumed to be forthcoming from the manufacturing sector.

The consequences of a limited withdrawal of manufacturers from reactor markets would be felt mainly at the national level. The Swedish political system seems likely to arrive at a consensus against further ordering of nuclear power plants and in favor of exploiting other energy options, so that the withdrawal of ASEA-Atom from active reactor manufacture would have minor political repercussions. The United States nuclear program could survive the loss of, say, Babcock and Wilcox and GE. The loss of manufacturing capabilities in Canada and West Germany would have more serious political and economic consequences, at home and abroad. In both cases, governments may find themselves in difficult positions in the next few years, fully recognizing the desirability of keeping their reactor industries alive and internationally competitive for energy and industrial reasons, yet, for political, institutional, and economic reasons having their hands tied, whether in relation to promoting new orders or arranging financial support.

We should again emphasize in this context the distinction between loss of active manufacturing capability and loss of the potential to enter or reenter reactor markets. While the former is undoubtedly threatened in some instances, the latter need not occur. There is no inevitable coupling between the two. Consequently, one should hesitate to conclude that a loss of reactor manufacturing capabilities will lead to an inevitable loss of the nuclear option by a country that finds it politically unacceptable, as seems to be the case in most instances, to import from abroad such a strategic and controversial technology. Although the economic and technical difficulties involved should not be underestimated, one could envisage the reestablishment of an indigenous reactor industry at a later date if the necessary steps are taken to ensure that the potential survives.

5. CONCLUSIONS

We have seen that the civil nuclear industry is a relatively young, large-scale, high-technology industry, with long lead times in technical development and plant construction. It was developed in the 1950s, 1960s, and early 1970s in a generally favorable political and economic climate. There has been a pronounced change in this climate in the second half of the 1970s, leading to reduced expectations of nuclear growth in a number of countries. This has been principally due to a conjunction of the following circumstances:

- Economic recession and higher electricity prices have reduced electricity growth rates and left many utilities with overcapacity in their supply systems and large investment programs on hand. Lengthening lead times, cost inflation, and a hostile regulatory climate have also discouraged some utilities from increasing their investment in nuclear power.
- Public confidence in nuclear power has been eroded in several countries, owing to concerns over safety and the social and political consequences of nuclear expansion. In addition, certain technical and institutional difficulties have proved more serious than anticipated. All of this has led to disruption of nuclear plans and increases in their economic costs.
- International controversy has arisen around the conduct of nuclear trade, following concern, first in the United States and then in several other countries, that the spread of nuclear power and transition to a

"plutonium economy" could increase the likelihood of a horizontal weapons proliferation.

Only in France, Comecon, and a few developing countries have nuclear programs survived these circumstances intact. This has been due to a significant degree to extensive government control over, and support for, nuclear and utility industries in those regions.

In sections 3 and 4 we discussed at some length the difficulties being faced by reactor industries. We noted that the problems confronting nuclear policymakers, and the conditions under which they might be overcome, vary considerably from country to country. While recognizing this diversity, we have argued that the reactor industry, as a whole, may have to adapt itself to average ordering rates outside the centrally planned economies within the range of 14–25 GW/a in the 1980s – well below its present capacity to supply 50–60 GW/a – although we are aware that there is a large margin of uncertainty that could lead to a higher or lower outcome.

More immediately, a number of reactor suppliers face serious deterioration in their commercial positions because they have received few, if any, reactor orders since the mid-1970s. Besides incurring financial losses, they run the risk that losses of skilled manpower could lead to an erosion of expertise in the design, engineering, and project management teams which, though small in size, are the critical elements in reactor manufacturing. There is an added risk, faced particularly by those reactor suppliers that rely heavily on subcontracting, that some external component suppliers could leave the market.

It is important here to make a distinction, at both national and international levels, between the viability of the nuclear option and the viability of the nuclear reactor industry, particularly in its present size and structure. A further distinction should be drawn between the *capability* of a reactor industry to make an effective and timely response to market demands and the *potential* to enter or reenter reactor markets. It is the former industrial condition that is particularly threatened in periods of low ordering. Involvement in the supply of fuel elements and repair and maintenance services, together with other measures available to manufacturers (e.g., diversification and R&D), provide a continuing background of activity against which to sustain a reactor manufacturing potential over a relatively extended period. The presence of an advanced industrial infrastructure also ensures that supplies of many components and materials could be reinstated. While the costs and delays involved in reconstructing a full manufacturing capability from such a potential should not be underestimated, the feasibility of reconstruction means that the nuclear option itself is not seriously threatened by the present difficulties being encountered by reactor industries.

Nevertheless, the deteriorating commercial positions of several reactor suppliers significantly complicate the political problems being faced by governments in relation to nuclear power. Until recently, governments have weighed the implications of low or zero ordering mainly in terms of energy diversification and economics. Governments and utilities in a number of countries are now coming under growing pressure from reactor industries to remove, or at least reduce, barriers to reactor ordering. Industrial implications of low ordering therefore must increasingly be taken into account, particularly since having a viable reactor industry is seen to bring significant benefits outside the energy field, including national prestige, the promotion of high-technology manufacture, and balance-of-payments improvements.

In addressing this situation, there are three considerations that we believe require particular attention by governments, utilities, and nuclear industries. The first concerns the nuclear reactor industry in its various parts. Where an indigenous reactor industry faces an uncertain and depressed home market and where, for one reason or another, it is considered important to maintain it, attention will have to be given to finding ways, besides ordering, of sustaining a sufficient manufacturing capability in the short term or a potential to reenter reactor markets in the longer term if market prospects do not improve. Moreover, if it is anticipated that the average rates of ordering in the 1980s will fall well short of currently installed capacity, which we believe is a likely outcome in most supplier countries, attention will have to be given to how to manage, and even encourage, an orderly reduction of manufacturing capacity to more sustainable levels. This might require substantial restructuring of reactor manufacturing, nationally and internationally. Ordering rates of 14–25 GW/a should then provide a sufficient basis for a viable reactor industry in the 1980s, even if some of today's suppliers are forced to retreat wholly or partially from the market.

The second consideration concerns export markets. We perceive a serious risk of increased confrontation between supplier countries over the terms on which reactor trade is conducted, if domestic markets remain depressed and reactor manufacturers are threatened. This risk will be heightened if exporting comes to be widely regarded as a valid means of avoiding necessary but painful adjustments to domestic manufacturing capacities and of gaining time in which to resolve domestic disputes over nuclear power. The problem of containing pressures to compete for contracts by offering terms that further undermine confidence in international nuclear relationships requires urgent attention.

The third consideration concerns improvement in the political climate surrounding nuclear power, which we believe may be a necessary condition for maintaining the nuclear option in a number of countries. This climate may improve by itself, or in response to changing energy circumstances and assessments of risks, but it would be imprudent for governments, utilities, and nuclear industries not to proceed on the assumption that determined efforts will be required to increase the overall political acceptability of nuclear power.

There are several approaches to policy that could contribute to this improvement, though all entail considerable uncertainty and are more or less appropriate depending on the national context. Within an overall energy perspective, the acceptability of nuclear power could be enhanced by a clear demonstration by governments and utilities that other options, including conservation, are receiving as strong political and institutional support as nuclear power. In relation to nuclear power itself, there are obviously a number of problems that require urgent attention, including the safe design and operation of reactors, regulatory procedures, and waste management. However, opposition to nuclear power has many dimensions, and one cannot be entirely confident that urgent attention to these problems will by itself restore a favorable political climate. Additional approaches may have to be explored.

A question we think it important that nuclear policymakers consider in this context is whether reduced plans and expectations for nuclear growth in the 1980s, problematic though they may be in many respects, could provide an opportunity for developing a more durable consensus over the course of nuclear development. For instance, a slower than expected rate of ordering in the 1980s could ease pressures on uranium supply, strengthening the case for proceeding more cautiously with advanced or contentious technologies like the FBR and reprocessing and for giving a higher priority to other uranium-conserving technologies that could even further reduce pressures on uranium supply. In addition, political acceptance of a slower rate of expansion, by providing a more cautious background against which to address the technical and institutional problems that face nuclear power, could increase the probability that broadly acceptable solutions to these problems can be found and generally help to reduce the temperature of the political controversy surrounding nuclear power.

If the political climate could be improved, the outcome could only enhance the prospects for the nuclear option. Attempts to promote more substantial reactor ordering programs without improvements in the present climate carry risks of further deterioration in the political and economic conditions facing utilities and reactor industries, perhaps even to the point of jeopardizing the nuclear option itself.

REFERENCES

International Energy Agency. 1978. *Steam coal: Prospects to 2000.* Paris: International Energy Agency.

Nuclear Engineering International. 1978. Power reactors 1978. *Nuclear Engineering International,* suppl., July 1978.

______. 1979. Power reactors 1979. *Nuclear Engineering International,* suppl., July 1979.

Royal Commission on Electric Power Planning. 1978. *A race against time: Interim report on nuclear power in Ontario*. Toronto: Royal Commission on Electric Power Planning.

Stanford Research Institute. N.d. *Long-range electricity forecasts for Canada: A methodology*. Stanford: Stanford Research Institute.

Westinghouse Electric Corporation. 1979. *Nuclear market acceptance and comparative experience*. Pittsburgh: Westinghouse Electric Corporation.

World Energy Conference. 1978. *World energy resources 1985–2020: Executive summaries of reports on resources, conservation and demand to the Conservation Commission of the World Energy Conference*. Guildford/New York: IPC Science and Technology Press.

9

World Nuclear Energy Paths

Thomas J. Connolly, Ulf Hansen,
Wolfgang Jaek, and Karl-Heinz Beckurts

1. OVERVIEW AND CONCLUSIONS

As this report is being written, the nuclear power option is seriously at risk. Questions of nuclear safety are of political concern in the aftermath of the Three Mile Island accident. There is international disagreement about the relation between nuclear power expansion and the proliferation of nuclear weapon capabilities, leading to divergent opinions on reprocessing and to restrictions on nuclear trade and technology transfer. And there is wide confusion about adequate and acceptable means of nuclear waste disposal. A de facto moratorium on reactor orders exists in the United States, Sweden, and West Germany, and there are delays in nuclear power programs in most of the other countries. Only France and the Comecon countries appear to be moving ahead in line with their original plans to develop nuclear power as a large-scale energy source and as protection against rising prices and decreasing stability in the supply of mineral oil. In these circumstances, what will the world nuclear future look like?

Nobody can answer this question with any certainty; most of the prognostic attempts over the past decade have failed, usually through having overestimated the speed of nuclear expansion. We simply do not have a good historical precedent in the energy technology field to permit us to judge whether the present political difficulties of nuclear power will severely constrain this energy source for the indefinite future or will turn out to be merely a transitory phenomenon. In this study we assume that:

Date of origin, December 1979.
We should like to thank Dr. Gerhard Kolb, KFA Jülich, for performing the computer calculations. We should also like to thank those members of the ICGNE, and in particular its chairman, Mr. Ian Smart, who participated in discussions at various stages of the study. This interplay was a test for our ideas, and their critical review and constructive suggestions have, we hope, given the study its proper balance.

- the world economy will grow somewhat more slowly than in the past, leading to reductions in electricity demand growth rates;
- national and international political impediments to the deployment of nuclear power will gradually disappear over the next few years;
- the further development of nuclear power will proceed steadily, without serious interruption but with realistic lead times for the introduction of advanced technologies.

Given these assumptions, this paper attempts a study of possible world nuclear energy developments, disaggregated on a regional and national basis.

We have used the scenario technique. Several alternative fuel cycle scenarios have been developed. Each is an internally consistent model of technically and economically feasible paths to the further development of nuclear power in an aggregate of individual countries and regions of the world. The main purpose of this modeling exercise has been to gain some insight into the probable international location of reactors and other nuclear facilities, the future requirements for uranium and fuel cycle services, and the problems of spent fuel storage and waste management. The study is also intended to assess the role nuclear power might actually play in meeting future world energy demand.

Any nuclear scenario has to be considered in a context of some wider view of energy supply and demand, and thus in a general economic context. Rather than develop a new general energy scenario, we decided to refer to the work of the Conservation Commission of the 1977 World Energy Conference (WEC) (WEC 1978*b, c*). At the time we started our study, we judged this to be the most comprehensive and well-documented energy supply/demand analysis for the period up to 2020—a time span sufficiently long to permit some market penetration by new technologies, but not so long that analysis becomes obscured in speculation. Out of the various cases considered in the WEC study, the so-called L4 scenario, based on low economic growth, a high price response of energy demand, and constraints on oil supply, was selected as the reference case for this work. This L4 scenario projects a total world primary energy demand of some 830 exajoules (EJ) in 2020, compared with 320 EJ in 1979, and thus assumes a much more modest rate of growth than most other forecasts. Electricity growth rates in the L4 scenario vary regionally and with time, but they roughly average 3 percent per annum for OECD countries, 4.5 percent per annum for the centrally planned economies (CPEs), and 5.5 percent per annum for developing countries (LDCs).

The total electricity demand in the WEC L4 scenario has been used as a starting point for the study. However, a completely independent assessment has been made of the contribution nuclear energy might, in fact, make to meeting this electricity demand. For that purpose it has been assumed that, up to 2020, nuclear power will be used on a large scale only for electricity production. While we firmly believe in a considerable long-term potential for

TABLE 9.1. Projected Distribution of Nuclear Generating Capacity (GW)

	1985	2000	2020
OECD	235 (13)	610 (22)	1,380 (22)
LDCs	15 (9)	85 (23)	390 (39)
CPEs	30 (6)	200 (10)	780 (10)
Total	280 (28)	895 (55)	2,550 (71)

Note: Numbers of countries using nuclear power are shown in parentheses.

nuclear process heat technologies, it appears premature to speculate on their worldwide market penetration during the next few decades.

A general set of rules has been established to calculate, on a disaggregated level, the possible contribution of nuclear power to the L4 electricity demand. It is assumed that, between now and 1995, nuclear capacity additions will be delayed and that, in 1995, only those plants that are now operating, under construction, or firmly planned, with sites selected, will be in service. From 1995 to 2020, however, nuclear expansion will be determined by policy choices yet to be made. On that basis it is assumed that one-third, one-half, or two-thirds of incremental electricity generation will be nuclear, depending respectively on whether a country has high, intermediate, or low fossil fuel resources. This is admittedly an arbitrary assumption, but it seems to be supported by experience. Many countries, although rich in fossil fuels, have, in fact, installed nuclear power. On the other hand, even a nation without fossil fuel resources would not be likely to have more than two-thirds of its electricity generated by nuclear fission, since this would make it necessary to operate nuclear power plants at uneconomically low load factors in order to carry intermediate loads.

It is further assumed that some minimum size is required of an electrical grid before nuclear power plants can reasonably be introduced: an assumption that naturally represents an impediment to the deployment of nuclear power in a small and less developed country. That threshold has been set at 17 terawatt hours a year (TWh/a), corresponding to a grid size of some 3 gigawatts (GW) at a 65 percent capacity factor. Once this threshold is passed, a 500 megawatt (MW) nuclear plant can be introduced, and the subsequent deployment proceeds according to the earlier rules.

The most important aggregated results are given in table 9.1. A number of conclusions can be drawn immediately from those figures:

- nuclear power growth, as determined by our assumptions and criteria, would be slower than in almost all other studies so far published;
- nuclear power, by 2020, would produce 43 percent of the world's electricity and represent roughly 20 percent of its primary energy supply, as projected in the L4 scenario;
- although its relative share may appear small, the absolute contribution of nuclear power in 2020 would nevertheless be impressive: 160 EJ of

TABLE 9.2. Natural Uranium Consumption and Spent Fuel Arisings (Once-through Path)

	Annual Uranium Consumption (10^3t/a)		Cumulative Uranium Consumption (10^3t/a)		Spent Fuel Arisings (10^3t/a)
Year	Reference	Improved	Reference	Improved	Reference
1985	41	41	305	305	7
2000	130	120	1,420	1,380	25
2020	350	275	6,090	5,350	75

primary energy a year, or approximately twice today's world coal production;

- the role of nuclear power in developing countries would remain very limited for some time but would increase rapidly after 2000.

It can also be concluded that, at least for the rest of this century, the rates of nuclear reactor ordering in OECD countries will be considerably below the announced capacities of the nuclear industry, and that export markets will also be limited.

What are the fuel cycle implications of this overall nuclear power scenario? Three basic fuel cycle paths have been considered:

1. a once-through path, in which a mix of current thermal converter reactors operate in an "open" or once-through mode (i.e., without spent fuel reprocessing) and, as a subcase, a similar path but with evolutionary improvements in uranium utilization;
2. a fast breeder path, in which fast breeder reactors (FBRs) are added to this mix according to a well-defined development and market introduction strategy;
3. a modified fast breeder path, in which some recycling of plutonium and uranium in light water reactors is also carried out.

The utilization of the world's thorium resources in current or advanced reactors could also be a promising future path. However, in view of the present state of national programs, it is difficult to construct a realistic deployment scenario that would have a major impact before 2020.

For the once-through path, it is assumed that most power plants would be based on a reference pressurized light water reactor (PWR), but some contribution by heavy water reactors (HWRs) and advanced gas-cooled reactors (AGRs) is taken into account. Compared with the standard "reference" case, the modified or "improved" subcase that has also been studied, involving progressively better fuel utilization, leads to an incremental 1 percent per annum reduction in the natural uranium consumption of all reactors completed after 1985. Some aggregated world results are given in table 9.2, which contrasts the effects of the two cases on uranium demand.

The cumulative spent fuel figures for the reference case in table 9.2 are 50,000 metric tons (t) (containing 250 t of plutonium) by 1985, 250,000 t (containing 1,350 t of plutonium) by 2000, and 1.2 million tons (Mt) (containing 6,350 t of plutonium) by 2020. It is obvious, therefore, that a once-through path, even if rather modest nuclear growth is assumed, leads to problems with uranium demand and the accumulation of spent fuel. The cumulative uranium requirements would exceed present world reserves by soon after the turn of the century. To meet that demand, there would have to be additions to reserves and new mine openings at a high and sustained rate. Evolutionary improvements in the natural uranium utilization of once-through converter reactors would reduce this rate somewhat. However, their main achievement would be to shift the demand figures forward by, at best, five years.

As a next step, we have investigated closed fuel cycles, including spent fuel reprocessing and the recycle of uranium in thermal reactors and plutonium in fast or thermal reactors or both. As is well known, there are different motives in different national circumstances for closing the fuel cycle by reprocessing fuel discharged from thermal (converter) reactors:

- as a prerequisite for safe nuclear waste management;
- as a prerequisite for more efficient uranium utilization by recycling plutonium in fast breeder reactors;
- as a hedge against high uranium prices or other uranium supply difficulties.

The second motive is generally recognized, since FBR operation, and thus substantial improvements in uranium utilization, cannot be achieved without reprocessing. The first objective is also considered very important in some countries, including the Federal Republic of Germany. However, although an early introduction of reprocessing capacity would be advantageous for several reasons, a more cautious route has been chosen in the scenarios, limiting the build-up of reprocessing capacity to the rate required for the introduction of fast breeder reactors.

A stepwise FBR strategy, involving consecutive demonstration, commercialization, and deployment, has been assumed, in recognition of the technical complexity of this reactor system. Thus it is assumed that, as a first step, a full-scale demonstration FBR begins operation five years after the nuclear capacity of a country reaches 25 GW. This threshold exists because plutonium from spent thermal reactor fuel has to be recovered in order to deploy an FBR, and economies of scale in spent fuel reprocessing require a substantial plant capacity. The threshold also reflects to some degree the maturity of the nuclear infrastructure required before a country can embark on a complex FBR system. For some countries, however, specific dates have been set for the start-up of demonstration FBRs, in the light of announced plans or our judgment of the earliest likely date: that is, France, 1985; Soviet Union, 1990; United States, 1995; West Germany, Japan, and Britain, 2000.

The second step — the operation of a first commercially ordered FBR — is assumed to occur fifteen years after the start-up of the first demonstration plant, since it appears unlikely that commercial orders could be placed before some experience in operating demonstration plants had been gained. It is assumed, moreover, that one or two additional demonstration plants will be started, at five-year intervals, during the interim period.

The subsequent steps imply a gradual transition of nuclear capacity over a further period of fifteen years, until FBRs represent two-thirds of all new reactor completions (unless limited by plutonium availability). In this scenario, six countries would operate FBRs by 2000, with a total capacity of 17 GW. The numbers would increase to sixteen countries by 2020, with a total FBR capacity of roughly 400 GW: that is, 15 percent of total world nuclear capacity. These figures are much lower than those generated by many other studies, but they are presented here as the highest likely. It would certainly be physically possible to have a more rapid global expansion of FBR capacity. However, we believe that, in view of technical, commercial, and public acceptance problems, our scenario is more realistic. It also appears unlikely that the established thermal reactor infrastructure would be abandoned quickly, even when the FBR is established.

It is assumed that thermal converter reactor fuel reprocessing capacity is built up in the countries concerned in accordance with the need to fuel FBR programs. This would lead to a total reprocessing capacity of about 3,600 t/a in five countries by 2000, and about 36,000 t/a in sixteen countries by 2020. In addition, the capacity for reprocessing spent fuel from FBRs would have to be built up, mainly in the lead countries.

Although the effect of FBR introduction on uranium needs will hardly be noticeable by 2000, a 25 percent reduction in annual uranium requirements, to 260,000 tons of uranium a year (tU/a), can be achieved by 2020. (Separative work requirements for enrichment are similarly reduced, to 210,000 separative work units a year [SWU/a].) Cumulative consumption by 2020 would then be 5.23 million tons of uranium (MtU): a global conservation effect comparable to the effect of improving the once-through fuel cycle. These figures are based on the assumption that recovered uranium is recycled in thermal reactors, a measure that accounts for half the uranium savings indicated. However, two facts must be borne in mind. Expansion of the FBR introduction strategy beyond 2020 would lead to a very rapid decrease thereafter in annual natural uranium demand and to the leveling-off of cumulative requirements – effects that could never be achieved by any once-through fuel cycle improvement. Also, it should be noted that the impact of FBR introduction is considerable in countries where the system is introduced early. In France, for instance, cumulative natural uranium savings would amount to 33 percent by 2020, and the annual natural uranium requirement in that year would have fallen to 45 percent of the once-through value. Detailed values for the leading FBR countries are presented in Appendix C.

We have also considered a modified closed cycle path in which, in addition to introducing the FBR and recycling uranium in thermal reactors, some plutonium recycling in thermal reactors would take place. The countries in which reprocessing then occurs are the same as in the pure FBR case, as are the size and timing of the first reprocessing plant and the scheduling of FBR plant construction. However, reprocessing capacity is expanded somewhat faster (to 6,000 t/a in 2000 and 44,000 t/a in 2020), and essentially all the plutonium not immediately needed for FBRs is manufactured into mixed oxide (MOX) fuel to be recycled in thermal reactors. Although this could be done in "dedicated" plutonium burners, we have assumed that, in those reactors where recycling occurs, only 30 percent of the fuel is MOX. In the period 2000–2020, about 250 1 GW reactors would operate in this mode. As a result, annual natural uranium needs would drop to 235,000 tU in 2020, and cumulative consumption to 4.91 MtU. Effects on spent fuel inventories would be somewhat more noticeable: while the world inventory is 1.2 Mt by 2020 in the reference once-through case, it would fall to 0.66 Mt in the FBR case and to 0.51 Mt in the FBR/plutonium recycle case. Again, the effects are more important in countries that close the fuel cycle early; in some cases, as shown in Appendix C, spent fuel inventories can essentially be eliminated by 2020.

A major part of our study is devoted to a discussion of the future world uranium situation and of the impact on it of the various reactor strategies. As input data on uranium resources, we have compiled information from several sources, including the joint NEA/IAEA study *World Uranium Potential* (NEA/IAEA 1978). In the cost category up to \$50/lbU_3O_8 (\$130/kgU), we have assumed values of:

- 5 MtU of "known" resources ("reasonably assured" plus "estimated additional") in WOCA countries;
- 1 MtU of "known" resources in CPE countries;
- 10–20 MtU of speculative or "unknown" resources.

We have then chosen to represent the cumulative natural uranium requirements of the different scenarios at three different levels. The first (level I) is the actual mine/mill production required up to a certain time. This is essentially the same as the cumulative natural uranium requirement of reactors, as given above, corrected to take account of the two-year lead time allowed for enrichment and fuel element manufacture. The second level (level II) we call "cumulative reserves required for mine openings." For a uranium mine to be opened, it is necessary that the decision maker be satisfied that the uranium ore body is well defined. We have therefore used a simple model, in which it is assumed that a mine operates at its rated capacity for fifteen years and that the uranium ore in question must have reached "reserve" status five years before mine production actually begins. The third level (level III) is "cumulative commitments implied by reactor orders." This is the total amount of natural uranium needed to fuel all reactors built, building, or ordered up to a given time for the whole of their thirty-year lifetime, assuming a ten-year lead time between ordering and start-up.

Figures 9.14–9.17 show the evolution over time of cumulative uranium requirements, at these three levels, for the reference once-through and FBR cases, and compare these with uranium resources. It is obvious that, for the once-through path, while total known resources seem sufficient to support actual mine production (level I) over the time range considered, substantial resources must change from the "unknown" to the "known" category soon after the turn of the century in order to justify new mine openings (level II), which points to the need for intensive exploration. Meanwhile, cumulative commitments (level III) reach the limit of known resources as early as 1995. This kind of information is required by nations or utilities in setting up reactor programs, and it may be important for long-term uranium price developments. The degree of certainty needed may be less than assumed in level II; however, general comparison of the long-term level III curves with the range of total world resources indicates a clear need to develop and introduce more uranium-efficient reactors. It is important to note that a partial extension of our model up to 2070 has shown that, in the once-through case, commitments rise appreciably above the band of estimates of both "known" and "unknown" resources, whereas, with a strong FBR deployment strategy, they level off at about 17 MtU. To put that comparison into a wider context, however, producing the same amount of electricity in nonnuclear power plants would consume, by the end of the next century, all the proved reserves of coal, oil, and gas recorded in the WEC analysis of world resources (WEC 1978*c*).

Although we have devoted only limited attention to the following questions, some supplementary observations are also worth making.

- Currently operating and planned uranium production capacity appears to be sufficient up to 1990 (fig. 9.12); an increase of annual production to a maximum of about 300,000 tU/a in the period immediately after 2020 would, however, be required, with a decrease thereafter only in the FBR and FBR/plutonium recycle scenarios. There is a question as to how difficult it would be in practice to achieve this production level.
- Currently operating and planned enrichment capacity seems to be sufficient up to 1990 (fig. 9.13). All our calculations of separative work requirements are based on a tails assay of 0.2 percent, which we believe to be economically justified by present uranium and separative work prices. We have studied the potential for increasing uranium utilization by lowering the tails assay or by reworking tails inventories and have concluded that a gradual decrease of tails assays, as part of the scenario for evolutionary improvements in the once-through cycle, is the most likely path.
- The timely availability of fuel fabrication capacity does not pose a problem.
- A model for the future development of uranium costs and prices has been developed within this study. It is based both on figures for the cumulative reserves needed for mine openings, as indicated by our

scenario, and on more speculative figures for the long-run marginal cost of exploiting uranium resources. This model leads to the important conclusion that a real price increase of 2–3 percent a year should offer a sufficient profit inducement for the necessary investments in exploration and mine opening. That, however, assumes a well-functioning world market, with fair price setting and a minimum of political interference.

We have given some attention to the problems of spent fuel and waste management on a more disaggregated level. Of particular interest is the regional distribution of spent fuel generation for the once-through path, as well as for paths involving reprocessing (cf. fig. 9.26).

Whether or not reprocessing is carried out, there will be very large inventories of spent fuel to be taken into account. The question arises whether to store this at reactor sites or at "away-from-reactor" (AFR) facilities. A plausible criterion seems to be that, for the first ten years after discharge, storage will be at the reactor site, and that central storage will subsequently be used in those cases where the total amount of spent fuel in a country exceeds 1,000 t. Applying this criterion, we find, for instance, in the once-through path that thirty countries would hold a total inventory of 90,000 t of fuel older than ten years in 2000 (out of a total of 255,000 t), of which 83,000 t would be in AFR storage facilities. The corresponding figure for 2020 would be 610,000 t (out of a total of 1.2 Mt), including 600,000 t in AFR storage facilities. Accordingly, construction of many AFR storage facilities on a national basis seems highly probable. We have not investigated the concept of international spent fuel storage, but our results do not seem to produce economic or technical arguments in favor of this approach, given that larger nations may prefer to establish their own AFR stores while smaller ones would more probably choose to extend storage facilities at their reactors. Closing the fuel cycle would, of course, strongly reduce the incentives for AFR storage, since most of the older fuel would be selected for reprocessing. The actual strategies of individual countries on central or reactor-site storage of spent fuel will ultimately depend on many factors, including costs (which at present appear to favor reactor-site storage), licensing conditions, and political pressure to demonstrate and carry out permanent spent fuel disposal. It is unlikely that the world will see a uniform approach to the spent fuel problem.

One other figure of interest is the age distribution of the plutonium contained in accumulated spent fuel, since age is an indication of its accessibility, and conversely of the difficulties facing an attempt to divert it. In the once-through path, taking into account the radioactive decay of Pu241, there will be 3,000 t of plutonium in 2010, out of which 500 t will be older than twenty years. The corresponding figures for 2020 are 6,000 t, with 1,200 t older than twenty years.

What are the economic implications of our future nuclear scenarios? To reach the level of market penetration by nuclear power indicated in our report, some kind of economic advantage must exist over other kinds of electricity generation. By and large, this is now the case when comparing nuclear and fossil fuel costs per kilowatt hour (kWh) in many countries of the world. However, the structures of nuclear and fossil fuel power costs are quite different, with a much higher share of capital costs in the former. Future trends will see some increase in real capital costs, owing to the evolution of safety, environmental, and licensing requirements for both nuclear and fossil fuel stations, together with some increase in nuclear fuel costs and a more substantial increase in fossil fuel costs. On the whole, therefore, we expect some relative improvement in the competitive position of nuclear power.

A much more complex question concerns the economic aspect of the fast breeder reactor. The specific capital costs of the FBRs at present envisaged are much higher than those of present thermal converters, and, with present uranium prices, the overall costs per kWh are also much higher. A question thus arises concerning the uranium price level at which the FBR would become competitive. We have considered the same point from a different direction by investigating the specific capital cost differential between a thermal converter and an FBR that is needed in order to break even as uranium prices develop in a certain way. Assuming a 2 percent per annum real uranium price increase, a 5 percent discount rate, and a thirty-year plant lifetime, we arrive at a value for this differential of $265/kW, which corresponds to about 30 percent of specific LWR plant costs based on 1979 figures. In other words, it would be economically rational to order the FBRs called for in our scenario provided their capital cost was no more than 130 percent of that of the same amount of LWR capacity. This does not seem to be completely out of reach if some future decrease in FBR costs is assumed. However, one must also take into account the considerable cost of developing FBRs and the costs of demonstration plants and fuel cycle facilities. All things considered, it seems somewhat doubtful that the market introduction of the FBR can be based on purely microeconomic arguments. The decision to proceed with the system is more likely to be a political decision, taken by a government as insurance against possible shortages in uranium supply. It is obvious that the complex nature of the breeder system, which includes the full fuel cycle, leads to a high commercialization threshold. For smaller countries, the breeder may never prove a suitable system, for reasons having nothing to do with technological development. For developing countries, the diffusion of breeder technology has its obvious difficulties. But it is very important to note that all countries could profit from a significant deployment of the breeder, since this could keep uranium prices in check and thus make other reactor systems accessible in countries and situations to which the breeder is not well suited.

A related question concerns the economics of plutonium recycling in thermal reactors, especially in LWRs. There have been major controversies over this question in the past few years, with a general tendency to downgrade the economic incentives for recycle in view of the rising cost of reprocessing. We have investigated this question but have not come to a definitive conclusion in view of the large uncertainties in the cost data. It is unlikely that thermal recycling in present LWRs can be justified, in isolation, solely on economic grounds. However, if a country has decided to carry out reprocessing in any case, as a prelude to eventual waste management or to prepare for a later FBR program, there may be a high economic payoff from thermal recycle. The situation on thermal recycling could also change considerably if future improvement of the LWR produced a higher conversion efficiency. We have not, however, investigated that contingency.

Although it is not considered in our scenarios, we have given some thought to the feasibility of further developing advanced thermal converter reactors, such as the heavy water reactor (HWR) or the high temperature gas-cooled reactor (HTR). These systems would give very good fuel utilization if operated in a closed thorium-uranium fuel cycle. During their initial stage, they would be operated in a once-through mode and therefore lend themselves to an evolutionary strategy in which the reactor system could be introduced without the concurrent establishment of reprocessing facilities. This compares favorably with the "all-or-nothing" approach of the FBR strategy, which requires both the introduction of the FBR and the simultaneous closure of the fuel cycle. Although large nuclear countries are more committed to the FBR route, an evolutionary strategy based on an advanced converter and the thorium cycle may be advantageous for other, mainly smaller, countries. Even for the larger countries, this route may be an attractive complementary strategy. We strongly believe that development of this option should be pursued.

A study of nuclear futures must proceed along one main road or a few of them, but it inevitably opens up many tempting views into side roads. We have, by and large, resisted the temptation to enter such side roads, of which we have seen quite a few attractive ones, as a thorough reader of this report will find. Many of these passing views might justify more in-depth study.

One particular and major problem we have deliberately not touched upon is the implication of our scenarios for the proliferation of nuclear weapon capabilities. Again, we leave this question for a more profound analysis. We feel, however, that the fairly slow spread of nuclear power, and in particular of sensitive fuel cycle facilities, indicated by our study should allow more time for the emergence of a stable, worldwide non-proliferation regime.

What, then, are the main lessons we have learned? The most important one is probably that, to remain available as a long-term and large-scale option, nuclear power must be developed and deployed steadily. Even the comparatively modest nuclear power capacity figures in our study—which fall short of the nuclear share of electric energy projected in the WEC L4 scenario—can be reached only if the present infrastructure is kept intact, particularly in the Western industrialized countries. If the complex fabric of people, skills, and equipment within the nuclear industry is not maintained, nuclear capacity might fall well below the levels projected in our scenarios. The crucial question would then be whether the political will to rebuild a nuclear industry could be developed. Continuity and confidence are also required for the development of a major uranium mining industry over the next forty years.

Even if nuclear expansion proceeds as slowly as in our scenarios, the need to develop advanced uranium-conserving reactors, in particular the FBR, is urgent. It is extremely important to recognize that, if uranium is used only with the efficiency of thermal converter reactors operating in a once-through mode, the world's total uranium resources, in energy terms, are of the same order of magnitude as its petroleum resources. Nuclear power would be an ill-planned episode if a steady transition were not foreseen to a stage where uranium utilization was greatly improved and resources became practically inexhaustible. Our study shows that this transition is possible and could well occur about the middle of the next century. However, given its very complex character, the transition will occur only if the leading industrial countries proceed now. A sustained and well-funded program to develop and deploy the fast breeder reactor in these countries will open the way. Studies that reach the conclusion that the breeder reactor is "not needed" until 2000 to 2020 offer little basis for complacency; the danger of trading away too much time before making a strong commitment must be clearly understood. Limited research and development efforts on the part of lead countries, with the ostensible purpose of "keeping the option open," do not fulfill the great responsibility of those countries. The level of financial and political support needed really to open up the FBR option and to keep it open may hardly be less than that needed for a wholehearted commitment.

We recognize that there is not a single global nuclear future, but rather a sum of national ones. The differences between countries are great, and the introduction of nuclear power and of advanced nuclear technologies will come not from a universal agreement but rather as the result of decisions in individual countries. But, even in a national context, nuclear power is unlikely to make a substantial and developing contribution unless there is:

- a wide measure of international consensus on the ways and means of solving the problems of nuclear power;
- a stable set of international rules, in the context of which nuclear trade and technology transfer can proceed, with adequate regard to the questions of security of supply, environmental protection, and nonproliferation;
- a strong effort of international collaboration, cost sharing, and technology transfer in the development of advanced reactor systems.

It is therefore desirable that national nuclear energy development paths be coordinated, so that they may interlock in a cooperative world nuclear energy future. Many nations carry a heavy responsibility here, but the main responsibility is with those large lead countries that have considerable resources of

TABLE 9.3. Energy Projections for WOCA to 2000

		WEC L4 Scenario			WAES D-7, D-8 Cases		
		1972	1985	2000	1972	1985	2000
Economic growth	%/a	3.4	3.1		3.2	2.8	
Electricity growth	%/a	3.7	3.9		4.4	4.0	
Primary energy demand	EJ	192[a]	245	350	176[a]	251	364
Electricity demand	EJ	38	61	107	37	64	115
Nuclear capacity	GW	34	280	970	34	291	913–1,524[b]

Sources: WEC 1978*b, c;* WAES 1977.
[a] Difference due to varying evaluation of non-commercial fuels.
[b] Strong coal and strong nuclear cases respectively.

their own and possess the technological capability to promote a reasonable world path. Despite the uneven present and future distribution of nuclear power capacity in the world, all nations would benefit from its introduction. Whether or not an amount of primary energy equivalent to twice today's coal production, or more than the present oil supply, will be available in 2020 is an issue of vital importance to meeting the energy needs of smaller countries, particularly in the developing world. If the world of the future is to receive these potential benefits of nuclear energy, time must not be wasted.

2. WORLD SUPPLY AND DEMAND PROJECTIONS

Demand Scenarios

There has been no shortage of energy forecasts in recent years, and we will use some of these to provide a general background before attempting to sketch our nuclear paths. In energy studies, most forecasts take the shape of scenarios, in order to allow a representation of various sets of assumptions. From each set of assumptions about economic growth and energy utilization efficiency, there results a specific energy demand. From scenarios published recently by three international working groups—the World Energy Conference (WEC 1978*b, c*), the Workshop on Alternative Energy Strategies (WAES 1977), and the Organization for Economic Cooperation and Development (OECD) (NEA 1978)—we have compiled a simplified comparison of some estimates for the world outside the centrally planned economy area (WOCA) in table 9.3 and for the OECD region in table 9.4.

Since these research bodies had already presented a well-documented range of plausible future developments, it seemed more appropriate to choose one of their scenarios that met the specific needs of this study than to develop our own. The scenario we used had to cover a long period and include the whole world, but also allow us to view the individual development of the main countries. These criteria are best fulfilled by the scenarios of the Conservation

TABLE 9.4. Energy Projections for OECD to 2020

		WEC L4 Scenario				OECD "Present Trend"			
		1972	1985	2000	2020	1975	1985	2000	2020
Economic growth	%/a		3.2	2.9	2.6				
Electricity growth	%/a		3.3	3.6	2.9		5.7	3.7	2.1
Primary energy demand	EJ	150	183	259	370	169	245	372	510
Electricity demand	EJ	34	52	88	154	45	77	133	199
Nuclear capacity	GW	33	280	860	2,053	68	278	1,000	1,980

Sources: WEC 1978*b, c;* NEA 1978.

TABLE 9.5. World Primary Energy Demand by Fuel Type: WEC L4 Scenario

Fuel Type	1972		1985		2000		2020	
	EJ	%	EJ	%	EJ	%	EJ	%
Coal	66	25	77	22	125	24	276	33
Oil	115	43	147	42	159	31	135	16
Gas	46	17	50	14	67	13	84	10
Nuclear	2	1	22	6	83	16	212	25
Hydropower	14	5	23	7	34	7	56	7
Others	26	10	30	9	48	9	70	8
Total	269	100	349	100	516	100	833	100

Source: WEC 1978*b*.

Commission of the World Energy Conference (WEC 1978*b, c*), from which we have chosen as a frame of reference a "low economic growth" projection, assuming world economic growth at an average of 3 percent per annum up to 2020 and thus implying a doubling of present per capita GNP by the end of that period. Within this economic projection, we have selected scenario L4, which is based on a high energy price elasticity and has the lowest primary energy growth of all the WEC scenarios. In the year 2020, world primary energy demand in this scenario reaches 833 EJ, or 29 billion tons of coal equivalent (Gtce)—nearly three times present energy consumption. The L4 projections are summarized in table 9.5 and figure 9.1.

As we see it, the L4 scenario represents a middle-of-the-road future, between a zero-growth economy and an extrapolation of the vigorous expansion in the past decades. In this sense it provides an apt setting for assessing the magnitude of the various nuclear paths we have set out to explore.

At the end of the period, in 2020, world primary energy demand in the L4 scenario is met by 33 percent coal, 16 percent oil, 10 percent gas, 25 percent nuclear energy, and 15 percent renewable resources (including hydropower). In this scenario, coal and nuclear power thus become the main pillars of energy supply in the next century. The present dominance of oil is lost as it reaches a peak at the end of the 1990s and then declines. The gas fraction falls slightly relative to other fuels, but gas consumption increases in absolute

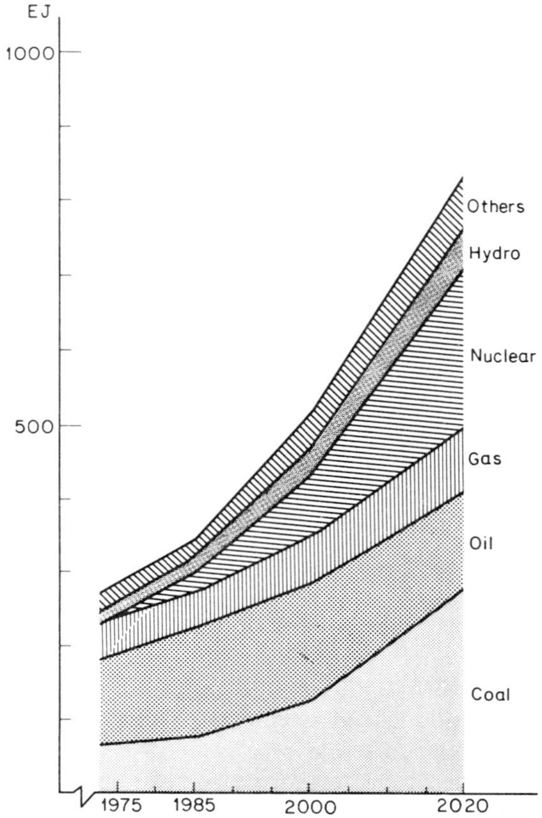

FIGURE 9.1. Future world energy demand: WEC L4 scenario.

terms. Increased use is also foreseen for renewable energy resources, grouped under "hydro" and "others," and the magnitude of their contribution exceeds that of gas in 2020. The use of noncommercial fuels remains important in many developing countries and is estimated to account annually for roughly 26 EJ worldwide, today and over the time frame of the study.

A breakdown of the WEC L4 projection is shown in figure 9.2, in which energy demand is given for the main world groupings: the OECD members, the less developed countries (LDCs), and the centrally planned economies (CPEs). For comparison, figure 9.2 also shows actual consumption figures for 1979, which include 26 EJ of noncommercial fuels to make them compatible with the WEC projection. The CPE average growth rate happens almost to coincide with the global one (2.6 and 2.7 percent per annum respectively, based on 1976 consumption), but the OECD growth rate reaches only 2 percent per annum, while the LDC annual growth rate runs at 4 percent. This is explained, on the one hand, by different rates of population growth and, on

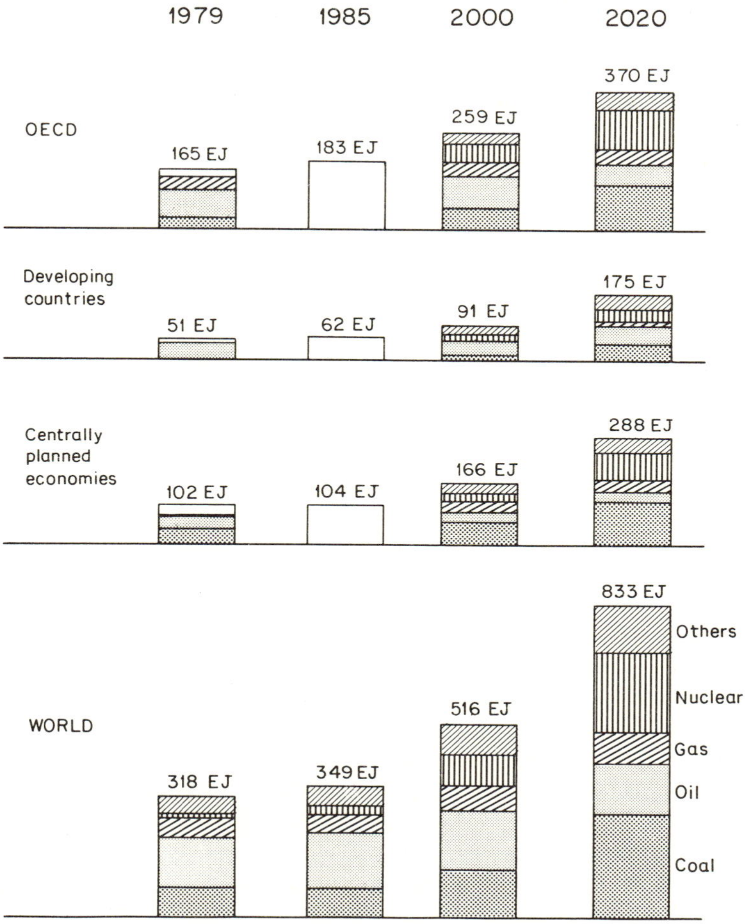

FIGURE 9.2. Primary energy demand in the world and main regions: WEC L4 scenario.

the other, by the stronger energy saving efforts and the saturation effects expected in the industrialized countries.

Fuel Supply

The reference demand scenario has been chosen without specifically considering the world's available energy resources. The resource base normally comprises two categories. One is reserves, consisting of those occurrences that are well defined and can be produced with today's technologies at competitive prices. The other, termed either additional or potential resources, is more speculative, since these are not well delineated and may only be inferred from

TABLE 9.6. World Fuel Supply Status (1977)

Fuel Type	Annual Production (Gtce/a)	Proved Reserves (Gtce)	Additional Resources (Gtce)
Coal	2.95	636	9,400
Oil	4.36	135	250
Gas	1.66	85	275
Uranium	0.56[a]	48[a]	50[a]–1,500[b]

Sources: WEC 1978*c;* NEA/IAEA 1978.
[a] Based on LWR once-through cycle.
[b] Based on advanced converter reactors (e.g., HTR or HWR) and including speculative resources at costs of less than $50/lb$U_3O_8$.

general geological knowledge. In many cases, production technology and costs for resources in the latter category are still uncertain. Knowledgeable mining experts nevertheless regard them as potential reserves, which can be worked economically given the time and money to develop them. The current status of commercial fuels in the world, based on WEC and NEA/IAEA estimates, is summarized in table 9.6 in terms of billions of tons of coal equivalent (Gtce) (WEC 1978*c;* NEA/IAEA 1978). Reserves are a dynamic quantity, which historically has continuously expanded. The relative magnitude of reserves and additional resources, on the one hand, and levels of consumption, on the other, justify the statement that the problem of future energy supply is likely to be not geological scarcity but the timely availability of production capacity.

Renewable energy resources are not globally available or conveniently transportable resources, but rather a source of energy that may be exploited to varying degrees in different countries. Their use depends on geographical location, the development of appropriate technologies, and competitive costs. In a sense, the situation is comparable to that of "resources in the ground": the limitation is the timely availability of economic utilization technologies.

Coal. Resources of hard and brown coal are very large, amounting to more than 10,000 Gtce or 293,000 EJ, of which 636 Gtce are currently regarded as technically and economically recoverable reserves. Nearly half of these are situated outside the centrally planned economies. In the L4 scenario, annual coal demand in 2020 rises above 9 Gtce. Compared with the production level in recent years, this implies a doubling time for demand of some twenty-five years. Given lead times of ten years or more, achieving the output to match that rise calls for a sustained worldwide effort by the coal mining industry.

A further point is that the geographical distribution of coal reserves does not always coincide with centers of demand. The present annual world trade in coal (223 Mt in 1978) is not large, judged by oil industry standards, and a substantial increase will be necessary if coal is to fulfill the role attributed to it in the WEC scenarios. The likelihood that coal will achieve this goal will not be discussed in our report.

Oil. The WEC experts estimate the ultimately recoverable oil resources to be between 175 and 350 billion tons of oil equivalent (Gtoe), with an average value of 260 Gtoe (approximately 10,000 EJ). Somewhat more than a third (90 Gtoe) is classified as assured reserves. To maintain a reasonable forward position in the period between now and 2020, some 3 Gtoe will have to be added to the reserves, on average, each year. With present exploration around the world, it appears that the oil industry is capable of achieving that target.

However, oil supply is not only a geological and technical problem but also a political one. Considerable reserves are concentrated in a few countries, some of which have announced possible limits to oil production in order to extend the life of their resource. In view of such developments, oil experts believe that oil production may peak in the 1990s and continuously decline thereafter. It is expected that the resulting upward trend of prices will bring liquid fuels from other sources into the market. Alternative sources include heavy oil deposits, petroleum recovered from the deep sea and from polar zones, hydrocarbons extracted from oil shales and tar sands, and hydrocarbon liquids produced by coal liquefaction.

Gas. Proved reserves of natural gas amount to 2,500 EJ, and the remaining undiscovered resources have been estimated at 8,100 EJ. Generally, the situation with natural gas is viewed more optimistically; WEC experts believe world production could be raised to twice the present level and sustained for more than fifty years. The contribution from gas is perhaps not so much a question of resource availability as of developing infrastructure and the gas market. As with oil, unconventional resources such as the gasification of coal may add to the supply of gas.

Uranium. Proved reserves, taken as "reasonably assured resources at costs less than $\$50/lbU_3O_8$," were reported by the Nuclear Energy Agency of the OECD and the IAEA to be 2.6 MtU for WOCA at the beginning of 1979 (NEA/IAEA 1979). Allowance for reserves in CPE countries might bring this value to 3 MtU for the world. Further resources in the "estimated additional resources" category are of the same order of magnitude. Estimates of speculative resources exploitable at costs less than $\$50/lbU_3O_8$ have been made, leading to a total figure for the ultimately recoverable resources of uranium in the range of 15–25 MtU. The energy equivalence and lifetime of uranium resources depend greatly on the type of reactor in which they are used. A detailed discussion of uranium is contained in section 6.

3. DEPLOYMENT OF WORLD NUCLEAR ELECTRICITY CAPACITY TO 2020

Alternative Futures

Our work was done at a time when projections of installed nuclear capacity in the year 2000 and beyond were falling with a half-life measured in months. In modeling, there is a temptation to follow the trend just for the sake of the company. At the same time, it is hard to deny the political realities that are at the base of these declining projections.

The successful demonstration of nuclear power technology in the 1960s led to a rush of orders in the late 1960s and early 1970s. In the United States and many West European countries, most power plants ordered or planned were nuclear. This shift proved too sudden. The nuclear industry, licensing structures, and processes of political acceptance had insufficient time for orderly evolution. Nuclear energy could not step in and substitute for conventional fuels at the pace and scale envisaged. In the wake of the economic recession following the 1973 oil crisis came cancellations of orders and delays in licensing and construction, accompanied by public reaction to reactor safety, environmental impacts, and a host of perceived problems.

Today's situation in the power industry is characterized by two effects. First, nuclear energy has been introduced into the electricity supply system of most of the industrialized countries, and, as a result of the spate of early orders, a large number of stations are at an advanced stage of construction. Second, very few new orders have been placed during the past four years. As a number of old projects have either been halted by legal action or deferred by utilities, the medium-term prospect for nuclear power appears to be one of stagnation. In the longer term, one can choose a nuclear future from a spectrum of possibilities. At one end of the spectrum, nuclear power fails to regain the political acceptance of former years and simply atrophies, with the electricity demand of nations being met by various alternatives. Toward the other end is a future in which, for a variety of possible reasons (failure of alternatives to live up to expectations, satisfactory progress on nuclear waste disposal, a good safety record in nuclear power plant operation), public confidence is restored and utilities begin again to place orders for nuclear power plants.

For those who believe events will unfold in consonance with a nuclear rejection scenario, analyses of nonnuclear futures obviously hold the greater interest. For the purpose of this analysis, however, we have chosen to construct a reference scenario nearer the latter end of the spectrum. As the starting point for our simulation, we selected the estimates of world and regional electricity generation contained in the WEC L4 scenario discussed in section 2. Those regional growth rates of electricity demand to 2020 are shown in table 9.7. The electricity production trajectory for each country was then

TABLE 9.7. Regional Electricity Growth Rates: WEC L4 Scenario

WEC Region	Average Annual Growth (%/a)	
	1972–2000	2000–2020
1. North America	3.4	2.9
2. Western Europe	3.7	2.7
3. Japan, Australasia	4.1	2.7
4. Soviet Union, CPE Europe	5.1	4.0
5. China, CPE Asia	7.5	4.8
6. OPEC members	7.2	5.5
7–11. Non-OPEC LDCs	5.5	4.8

Source: WEC 1978*b*.

developed by taking United Nations statistics for electricity generation in 1971–75 (UN 1977) and expanding them from 1973 onward according to the assumed growth rates for the region in which the country in question was situated.

Rules for World Nuclear Capacity Expansion

Having thus disaggregated the WEC L4 electricity scenario, we next assigned values to the nuclear share of electricity generation country by country. This was done according to rules of our own formulation and represents a clear departure from the World Energy Conference's work. The rules were formulated for three time stages:

Rule 1 In 1985, all nuclear plants currently in operation, plus all those under construction with scheduled completion not later than 1 January 1984 (as in Kempken 1978, 1979*a, b*), will be in full operation at a 65 percent capacity factor.

Rule 2 In 1995, all plants now under construction with scheduled completion later than 1 January 1984, plus all for which firm plans (i.e., with site already selected) have been announced, will also be in full operation. (Because the quality of information regarding nuclear plans in different countries is so variable, it was necessary to make some country-by-country judgments to obtain this value.)

Rule 3 After 1995, nuclear capacity expansion is assumed to be policy-driven. Nuclear power has a share of incremental electricity generation in countries that exceed an annual generating threshold of 17 TWh (3 GW at a 65 percent capacity factor). This share is one-third, one-half, or two-thirds, depending on the classification of the indigenous resource base of conventional fuels in the particular country as high, intermediate, or low. A nation is classified as having high fossil fuel resources if its combined reserves of coal, oil, and gas (as in WEC 1978*a* and Parent and Linden 1977) exceed seventy-five years of its current annual primary energy consumption and if its current ratio of primary energy production to consumption exceeds 0.5. A nation is classified as having low fossil fuel resources if its combined reserves of coal, oil, and gas represent less than ten years of its current primary energy consumption. A nation falling into neither of these categories is classified as intermediate.

These rules are somewhat arbitrary, but they have their rationale. Even a nation with very low fossil fuel resources would probably not have more than two-thirds of its electricity generated by nuclear power, because this would mean that a large share of its nuclear plants would have to operate at low capacity factors in order to provide for intermediate loads. Even at the two-thirds level, some of the nuclear capacity would probably have to be load-following. This is one reason for our selecting the relatively low capacity factor of 65 percent for nuclear plants. The assignment of a one-third nuclear share of new electricity output even in countries richly endowed with fossil fuel resources is conversely justified on the basis of the demonstrated policy of several such countries to develop nuclear power.

It should be emphasized that we are using the term "share" in this context to mean a proportion of the actual electricity generated, as distinct from the nominal power rating of the generating equipment. For example, if the nuclear share of incremental electricity generation is a half, then the share in terms of rated capacities would be less than 50 percent, because the non-nuclear portion would include intermediate and peak-load plants that have capacity factors lower than 65 percent.

A question to which we have found no simple answer concerns the role of nuclear energy in smaller countries, mostly found in the developing world. If nuclear power is to be an important part of the world's future energy supply, then it seems that its role must extend to the developing countries. However, nuclear power is a complicated technology, requiring a large corona of supporting activities. These activities can be economically justified only when the total electricity supply system has reached a certain scale. There is thus some threshold for nuclear power introduction, related to the size of the electricity grid transmission system. The unit size of nuclear reactors is also relevant, given the need for a prudent distribution of generating capacity. In this respect, recent trends toward larger reactor unit sizes have not been favorable to their early introduction into small grids. In this study we have assumed that 17 TWh/a is the threshold level and that, once this level of electrical output has been reached, a 500 MW nuclear power plant can be introduced. This implies that nuclear power produces about 14 percent of electricity in a grid with a total equivalent capacity of 3 GW, assuming that all plants operate with the same 65 percent capacity factor as the nuclear station. (In reality, as pointed out above, the total installed capacity would probably be higher.) After the initial 17 TWh/a threshold is passed and the first 500 MW reactor has been installed, nuclear deployment in smaller countries is assumed to take place according to the 1/3:1/2:2/3 rule we have applied elsewhere. This rule tacitly presupposes that all plants are connected to one national grid

TABLE 9.8. Regional Projections of Nuclear Electricity Production (TWh/a)

WEC Region	2000		2020	
	WEC L4	KFA Jülich	WEC L4	KFA Jülich
1. North America	2,600	1,700	6,260	3,270
2. Western Europe	1,690	1,290	3,910	3,080
3. Japan, Australasia	670	460	1,710	1,520
4. Soviet Union, CPE Europe	1,930	1,030	6,510	3,670
5. China, CPE Asia	250	120	1,170	770
6. OPEC members	85	50	430	280
7-11. Non-OPEC LDCs	560	440	2,590	1,930
Total	7,785	5,090	22,580	14,520

and neglects the fact that, in many developing countries, a fragmented supply system exists for reasons related to geography and different regional development. On the other hand, it is possible that smaller countries may link their grids to form a larger intercountry electricity supply system.

Nuclear Capacities to 2020

The nuclear electricity projections that follow from the application of these rules to form our own (KFA Jülich) reference scenario are shown in tables 9.8, 9.9, and 9.10 and in figures 9.3 to 9.6. Nuclear electricity production by region is displayed in table 9.8 and compared with the WEC L4 values (WEC 1978*b*). The primary reason for the different results is the much severer constraints our rules place on the nuclear share of total electricity production. Our simulation shows nuclear electricity expanding at an annual rate of 8 percent in the period 1985–2000 and 5 percent in the period 2000–2020. At the beginning of 2020, nuclear power plants contribute 43 percent of the total 33,700 TWh/a net electricity production of the WEC L4 scenario. In terms of primary energy, nuclear power represents roughly one-fifth of the L4 total of 833 EJ. Measured against today's energy production, nuclear energy would then contribute the energy equivalent of twice the world's present output of coal. The distribution among individual countries of the generating capacity responsible for that contribution is shown in table 9.9.

Table 9.10 gives our projection of nuclear capacities to the beginning of 2020 in rounded figures, disaggregated into conventional political groupings (OECD, CPEs, and LDCs). (Incidentally, it should be noted, looking back to table 9.8, that, because we use a constant value of 65 percent for the capacity factor of all nuclear plants, there is a constant ratio of installed capacity in GW to electricity production in TWh/a of 1:5.694.) In figure 9.3, it can then be seen that our values for WOCA fall below the "present trend" values of the recent OECD study (NEA 1978). The discrepancy is most obvious in the medium term, where we have adjusted for the prevailing hesitant ordering position of utilities.

TABLE 9.9. Country Projections of Nuclear Generating Capacity: KFA Jülich Reference Scenario

Country	Fossil Fuel Resource Category	Nuclear Generating Capacity (GW)						
		1980	1985	1990	1995	2000	2010	2020
North America								
United States	High	51.8	115	150	200	270	370	485
Canada	Intermediate	6.3	9	12	15	29	56	90
Subtotal		58.1	124	162	215	299	426	575
Western Europe								
Austria	Low	0	0	0	2	4	10	17
Belgium	Low	1.7	5.5	5.5	5.5	8	14	20
Britain	High	8.1	13	14	15	24	42	65
Denmark	Low	0	0	0	0	1	4	8
Finland	Low	1.1	2	3	5	7	10	13
France	Low	8.4	35	42	50	62	81	96
West Germany	Intermediate	8.7	19	29	36	52	80	110
Greece	Intermediate	0	0	0	0.5	1.5	3	5
Italy	Low	1.4	1.5	3.5	5.5	16	39	68
Ireland	Low	0	0	0	0.5	1	2.5	4
Luxembourg	Low	0	0	0	1	1	1	2
Netherlands	Intermediate	0.5	0.5	0.5	0.5	3.5	9.5	18
Norway	High	0	0	0	0	2.5	8	16
Portugal	Low	0	0	0	1	2	4	6
Spain	Low	1.1	7.5	11	15	20	31	39
Sweden	Low	3.8	7.5	8.5	9.5	15	25	36
Switzerland	Low	2	4	4	5	7	12	18
Subtotal		36.8	95.5	121	152	227.5	376	541
Japan, Australasia								
Australia	High	0	0	0	0	3	9	18
Japan	Low	14.5	17	27	37	76	150	240
New Zealand	Intermediate	0	0	0	0	1.5	4	8
Subtotal		14.5	17	27	37	80.5	163	266
Soviet Union, CPE Europe								
Bulgaria	Intermediate	0.8	2	2	3.5	5.5	10	17
Czechoslovakia	Intermediate	1	2	3	6.5	12	24	40
East Germany	Intermediate	1.4	3.5	5.5	9	16	33	55
Hungary	Intermediate	0	1	2	2	4	8	12
Poland	High	0	0	1	3	8.5	21	39
Romania	Intermediate	0	0	0.5	2	7	17	33
Soviet Union	High	10.8	20	37	68	125	260	440
Yugoslavia	High	0	0.5	0.5	2	3	6	9
Subtotal		14	29	51.5	96	181	379	645
China, CPE Asia								
China	High	0	0	0	2	17	50	105
North Korea	Intermediate	0	0	0	0	4	14	30
Subtotal		0	0	0	2	21	64	135

TABLE 9.9. Country Projections of Nuclear Generating Capacity: KFA Jülich Reference Scenario *(continued)*

Country	Fossil Fuel Resource Category	Nuclear Generating Capacity (GW)						
		1980	1985	1990	1995	2000	2010	2020
OPEC members								
Algeria	High	0	0	0	0	0	1	2.5
Ecuador	High	0	0	0	0	0	0	0.5
Indonesia	High	0	0	0	0	0	1	2.5
Iran	High	0	1.5	2.5	3.5	5	8.5	15
Iraq	High	0	0	0	0	0	1	3
Kuwait	High	0	0	0	0	0.5	1.5	4
Nigeria	High	0	0	0	0	0	1	2.5
Saudi Arabia	High	0	0	0	0	0	0	1
Venezuela	High	0	0	0	1.5	3.5	9	18
Subtotal		0	1.5	2.5	5	9	23	49
Latin America								
Argentina	Intermediate	0.3	1	1.5	3.5	6	13	24
Brazil	Intermediate	0	3	6	9.5	16	32	58
Chile	Intermediate	0	0	0	0	1	3	6.5
Colombia	Intermediate	0	0	0	0	1	4	8.5
Cuba	Low	0	0	0	0	0.5	2.5	6
Jamaica	Low	0	0	0	0	0	0	0.5
Mexico	High	0	1.5	1.5	3.5	6	13	24
Peru	Intermediate	0	0	0	0	0.5	2.5	5.5
Puerto Rico	Low	0	0	0	0	2	6.5	14
Uruguay	Low	0	0	0	0	0	0	1.5
Subtotal		0.3	5.5	9	16.5	33	76.5	148.5
Middle East, North Africa								
Egypt	Intermediate	0	0	0	0.5	1	3.5	7
Israel	Low	0	0	0	0	1	4.5	9.5
Morocco	Low	0	0	0	0	0	0	1.5
Turkey	Intermediate	0	0	0	0.5	1.5	5	10
Subtotal		0	0	0	1	3.5	13	28
Africa south of Sahara								
Ghana	Low	0	0	0	0	0	0.5	2.5
South Africa	High	0	2	2	2.5	5.5	12	20
Zaire	Intermediate	0	0	0	0	0	0.5	2
Zambia	Low	0	0	0	0	0	1	3
Zimbabwe	High	0	0	0	0	0.5	1.5	3.5
Subtotal		0	2	2	2.5	6	15.5	31
Southeast Asia								
Hong Kong	Low	0	0	0	0	1	3.5	7
South Korea	Intermediate	0.6	2	4	6	8	12	18
Malaysia	Intermediate	0	0	0	0	0	1.5	3.5
Philippines	Low	0	0	0	0.5	2	6.5	13
Singapore	Low	0	0	0	0	0	0.5	2.5
Taiwan	Intermediate	1.2	3	5	7.5	10	15	24
Thailand	Low	0	0	0	0	1	3.5	7.5
Subtotal		1.8	5	9	14	22	42.5	75.5

TABLE 9.9. Country Projections of Nuclear Generating Capacity: KFA Jülich Reference Scenario *(continued)*

Country	Fossil Fuel Resource Category	Nuclear Generating Capacity (GW)						
		1980	1985	1990	1995	2000	2010	2020
South Asia								
India	High	0.6	1.5	4	6	11	25	48
Pakistan	Intermediate	0.1	0.5	0.5	1	2	5	8
Subtotal		0.7	2	4.5	7	13	30	56

Note: All figures are for the beginning of the relevant year.

TABLE 9.10. Installed Nuclear Generating Capacity (GW): KFA Jülich Reference Scenario

Year	OECD	LDCs	WOCA Subtotal	CPEs	World Total
1970	14.3	0.4	14.7	1.5	16
1975	63.2	1	64.2	6.8	71
1980	115	5	120	20	140
1985	235	15	250	30	280
1990	310	25	335	50	385
1995	405	45	450	100	550
2000	610	85	695	200	895
2005	780	135	915	305	1,220
2010	965	200	1,165	445	1,610
2015	1,130	285	1,415	595	2,010
2020	1,380	390	1,770	780	2,550

Note: All figures are for the beginning of the relevant year.

The number of nuclear countries in our scenario is shown in figure 9.4, and the geographical distribution of nuclear capacity in figures 9.5 and 9.6. OECD and CPE countries have more than 90 percent of the capacity in 2000. It is therefore clear that most nuclear investment will take place in the industrialized world during the next two decades. By 2020, however, the LDCs are estimated to increase their share of world capacity to 15 percent. After the turn of the century, moreover, the number of nuclear countries increases only in the developing world. With the assumption of a total electricity production of 17 TWh/a as the nuclear threshold, reactors are operating in thirty-nine LDCs by 2020. Significantly, with an alternative threshold criterion of 34 TWh/a and subsequent introduction of a 1,000 MW reactor as the first nuclear power plant, figure 9.4 also shows that the number of nuclear countries in 2020 would decrease by ten. (It should be added that two of the smaller European countries would not have nuclear generating capacity, according to our other rules, had they not already announced firm projects.) However, applying that alternative criterion, with that effect, would reduce total installed nuclear capacity in 2020 by only 12 GW. Thus, world nuclear capacity is hardly affected over this term by the criteria concerning national

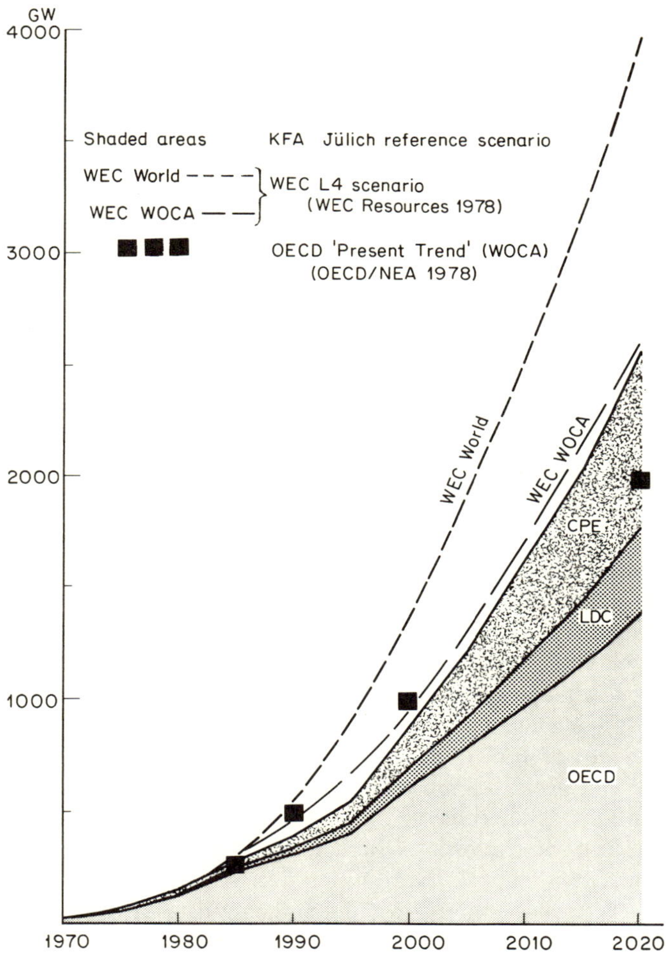

FIGURE 9.3. Nuclear generating capacity projections.

grid size and the unit size of reactors, though the number of countries using nuclear power is affected more strongly.

Table 9.11 shows the new reactor plant completion rate implicit in our reference scenario and furnishes some insight into the magnitude of the undertaking. The average value of about 25 GW completed each year in the 1980s derives from reactors currently under construction, so that this rate should fall comfortably within the physical capacity of the world manufacturing infrastructure to produce pressure vessels, turbine generators, and other critical path items. The reference scenario then shows a sharp increase of plant completions in the second half of the 1990s, to an average value of some 70 GW/a. Since the present aggregate construction capacity in the

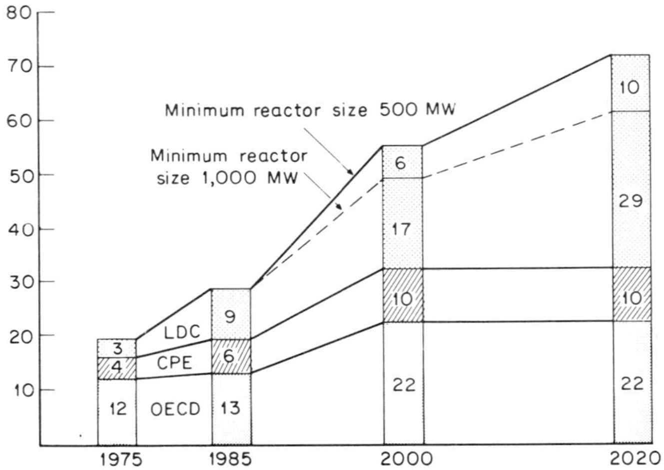

FIGURE 9.4. Number of countries operating nuclear power plants: KFA Jülich reference scenario.

world is 50–100 GW a year, this completion rate also seems to present no industrial problem. However, since most of the relevant capacity has not yet been committed, the scenario completion rate in the 1990s, given a ten-year lead time, will require a corresponding rate of ordering in the late 1980s. This would mean a considerable increase from the present rate.

Probably the most serious question about the realization of the scenario is whether the rate of utility orders will increase in the near future. If this increase does not occur, then not only will the market penetration of nuclear energy be shifted back in time but, at least in the Western industrialized countries, there is also some question whether reactor manufacturing capacity can be maintained. If it cannot, the resulting decrease below the nuclear energy supply in the reference scenario could be much greater than that represented by a simple shift of capacity values five or ten years back in time.

Comments on the Nuclear Capacity Scenario

Before proceeding to the development of nuclear fuel cycles to meet the nuclear electricity production schedule we have projected, we should take detailed note of its limitations. The country-by-country disaggregation that we made for the purpose of the analysis (table 9.9) should not be interpreted as an attempt at prophecy. The values in that table should be viewed as the result of applying our rules, not as a prediction for any given country. The scenario assumes that a country will proceed with the nuclear investments allotted to it under these rules. This implies that nuclear power is universally accepted and that none of the larger countries rejects it on political or other grounds.

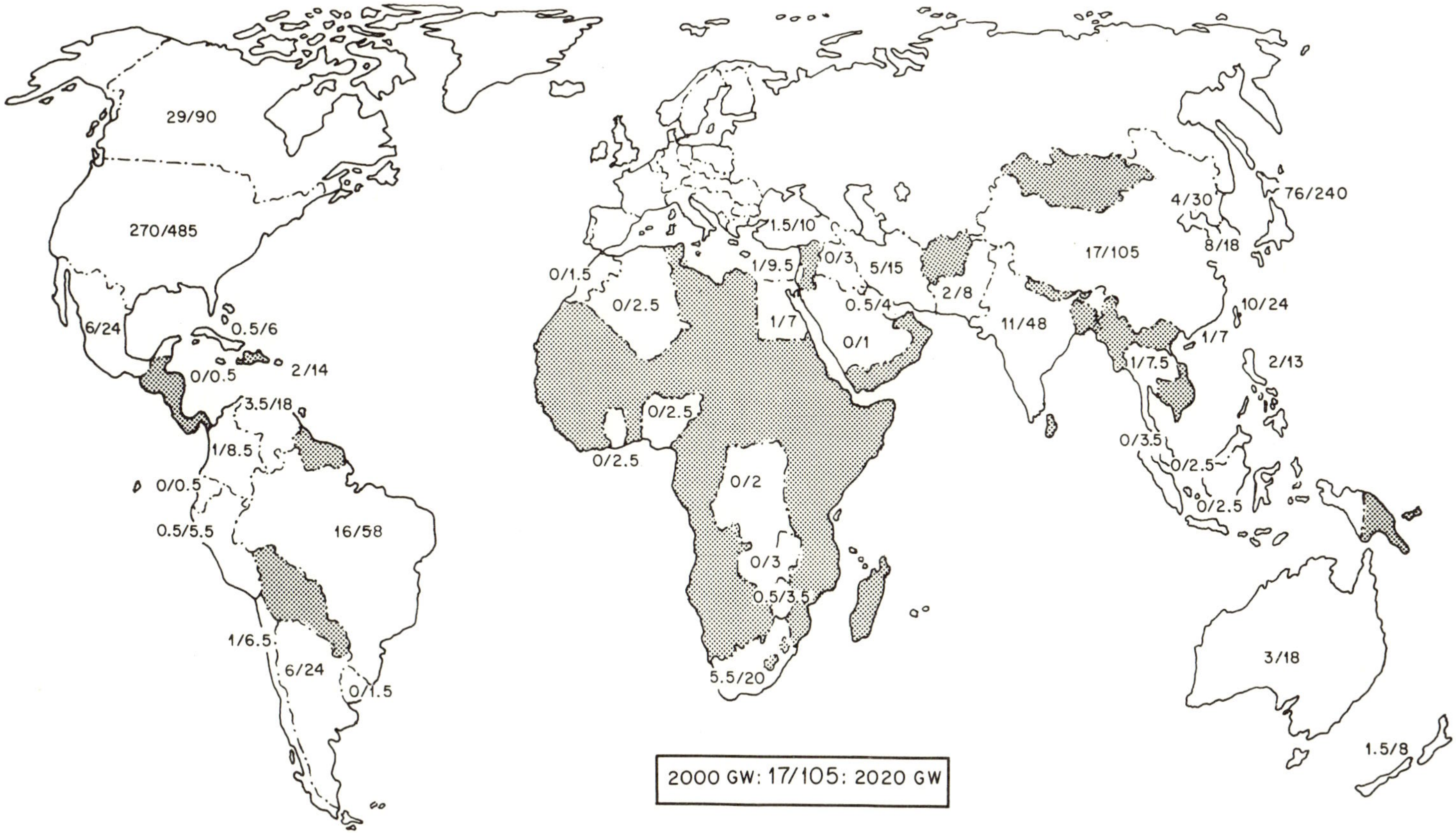

FIGURE 9.5. Distribution of nuclear generating capacity outside Europe in 2000, 2020 (GW): KFA Jülich reference scenario.

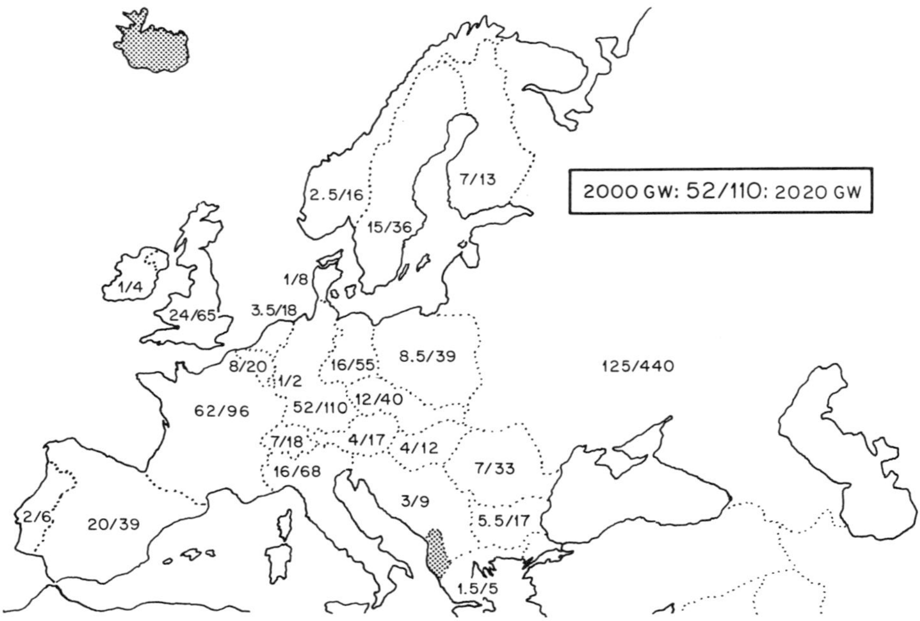

FIGURE 9.6. Distribution of nuclear generating capacity in Europe in 2000, 2020 (GW): KFA Jülich reference scenario.

The world and national nuclear capacity expansion trajectories developed here are coupled directly to the total electricity projections of the WEC L4 scenario. These projections represent a growth rate markedly reduced from the past two or three decades. Although we found the assumptions behind the L4 scenario, as discussed in section 2, to be in broad conformity with our own perceptions, it is plausible that electricity growth over the next forty years will lie closer to the pattern of the past forty years than to that assumed in the L4 scenario. If some such higher total electricity growth projection had been chosen for this study, then it is clear from our rules that nuclear capacities would have been higher in the post-1995 world.

Within the L4 electricity projections, we accept responsibility for adjusting the nuclear share of electricity downward on the basis of two judgments: that nuclear installation will be severely constrained through the rest of this century, and that there are many countries well-endowed with fossil fuel resources that will follow a policy of limiting the nuclear share of electricity. In assigning national nuclear capacity values, we deliberately chose not to use announced national plans as references. Such plans have not proved historically to be very good forecasts. Furthermore, no worldwide set of national plans made at the same date and on a consistent basis is available. In order to keep subjective judgment of country-by-country nuclear growth to a

TABLE 9.11. World Nuclear Generating Capacity Expansion: KFA Jülich Reference Scenario

Year	Period	Installed Capacity (GW)	Average Expansion Rate (GW/a)		
			Net Additions	Replacements	Total Completions
1965		5			
	1966–70		2	0	2
1970		16			
	1971–75		11	0	11
1975		71			
	1976–80		14	0	14
1980		140			
	1981–85		28	0	28
1985		280			
	1986–90		21	0	21
1990		385			
	1991–95		33	0	33
1995		550			
	1996–2000		69	2	71
2000		895			
	2001–5		65	11	76
2005		1,220			
	2006–10		78	14	92
2010		1,610			
	2011–15		80	28	108
2015		2,010			
	2016–20		108	21	129
2020		2,550			

minimum, we have simply applied the WEC L4 growth rates to all countries within a given region. Since there will obviously be variations within regions, it should be clearly understood that we do not offer the distribution in table 9.9 as a forecast.

It seems important to stress that the rigid application of our rules, without adjustments for intraregional variations, has led in a few cases to unrealistic nuclear capacities. It is not necessary to identify countries by name to acknowledge that there are some countries whose nuclear programs are more aggressive than we have indicated and others whose nuclear expectations at present are much lower. There is therefore nothing normative about the values in table 9.9; they are not presented as preferred alternatives to countries' plans. We believe that the test of modeling and its associated analysis lies in the value and correctness of the overall insight provided rather than in the ultimate accuracy of the projection for a particular country or year.

Implicit in our scenario construction is the judgment that some forms of nuclear electricity generation will be economically advantageous throughout the world over the time span of this study. Finally, we should point out that our scenario includes nuclear energy for the production of electricity only. We have regarded its application in the production of other energy forms (e.g. district heating, coal gasification) as outside the scope of our analysis, without prejudice as to its merits. Obviously, however, any investment in such applications would increase demand for uranium.

Fuel Cycle Options

In principle, the nuclear electricity capacity represented in the reference scenario could be met in a great many ways, in terms of reactor types and fueling modes. In practice, the options are severely limited. In any case, the objectives of this study can be met by an analysis of a very limited number of separate fuel cycle strategies. We have therefore limited ourselves to analyzing a mix of current thermal converter reactors operating in an open or once-through mode, the introduction of fast breeder reactors into this mix, and the additional effects of recycling plutonium or uranium or both in light water reactors. The effect of potential evolutionary improvements in the natural uranium consumption of present-day reactors is also analyzed.

None of these paths, described in detail in the following sections, is intended to represent our perception of the probable combination of reactors and fuel cycles likely to develop over the years to come. The paths are highly simplified models, selected to isolate the major effects of the most important fuel cycle decisions. A "most probable" future undoubtedly would contain a synthesis of our paths, as well as others not considered.

We have not analyzed the introduction of the thorium cycle because, within the time frame of this study (1980–2020) and with a realistic introduction rate, it would have little effect on the results. There does not appear to be any national program that could lead to a closed thorium fuel cycle operating commercially much before 2010. This omission, as well as the omission of other fuel cycle options, does not constitute a judgment that these options are not worthy of development. We shall discuss incentives for developing them later in this report.

The reactor types referred to in this study are described in Appendix B. Tables 9.A.1 and 9.A.2 give the fuel cycle numerical data used in the calculations. Table 9.A.3 expresses the uranium resource utilization of the different reactor/fuel cycle combinations, in terms of the natural uranium and separative work required over the reactor lifetime to produce a fixed quantity of electric energy.

4. ONCE-THROUGH PATHS

Reactor Types and Fuel Flows

All commercial reactors today operate on a once-through fuel cycle in which fresh fuel replaces old fuel and no use is made of the fissile material in the discharged fuel elements. This is caused more by force majeure than by

TABLE 9.12. Once-through Path: Installed Nuclear Generating Capacity by Reactor Type (GW)

Year	LWR	HWR	AGR	Total
1970	11.5	0.5	4	16
1975	62	3.5	5.5	71
1980	125	7	8	140
1985	255	12	13	280
1990	353	18	14	385
1995	508	27	15	550
2000	815	55	25	895
2005	1,110	80	30	1,220
2010	1,460	110	40	1,610
2015	1,810	150	50	2,010
2020	2,285	200	65	2,550

intention, since most light water reactors were designed to accept batches of recycled fuel if available. The problems that have led to this state of affairs are well known and may persist for some time to come. In this sense, the once-through scenario maps a possible boundary path into a nuclear energy future.

There are three main types of thermal converter reactor employed in the world on a commercial basis: the advanced gas-cooled reactor (AGR), the heavy water reactor (HWR), and the light water reactor (LWR). AGR installations are limited to Britain, and this limitation is applied throughout the analysis. For simplification, Magnox reactors, also deployed in Britain, are included under AGRs. Four countries—Argentina, Canada, India, and Romania—have announced intentions to use the HWR in their nuclear capacity expansion programs. The natural uranium fueled version (CANDU) of the HWR was assigned to these countries in the scenario. All other countries were assumed to deploy LWRs. There are two types of LWRs, the boiling water reactor (BWR) and the pressurized water reactor (PWR). To reduce the number of calculations, we have used the characteristics of a PWR to represent all LWRs. For the kinds of results we are seeking, we do not believe this generalization introduces significant errors. Given these national assignments of reactor types and the expansion programs in table 9.10, the resulting nuclear capacities by reactor type, over time, are shown in table 9.12.

In this once-through version of our scenario, no discharged spent fuel is reprocessed. In practice, spent fuel will be stored in various places: at the reactor, at some interim storage facility serving several reactors, at the site of a potential disposal facility, or at the site of a potential reprocessing facility. No attempt is here made to assign the spent fuel to these different kinds of storage. The analysis is limited to tracking simple quantities, such as the current and cumulative flow of spent fuel, identifying spent fuel by country of origin and reactor type, classifying it by age (time from reactor discharge), and so on. We do not attempt, therefore, to draw distinctions between nations whose policies call for the ultimate disposal of untreated spent fuel ("throwaway" mode), those whose plans call for eventual reprocessing ("stowaway" mode), and those in which the choice between these two alternatives has not been made. For many years to come, the distinctions may be primarily ones of policies and plans rather than ones of physical facilities and operations. Even in the case of a firm policy of final disposal of unreprocessed spent fuel, the spent fuel will probably be stored for many years, both to permit decay of its radioactive thermal energy and to permit design, construction, and testing of the final disposal facility. It is also likely that, for a number of years after placing spent fuel in a final disposal site, the operator will be able to retrieve it.

The important operations in the once-through path are the mining and milling of uranium, conversion of uranium to the hexafluoride form (UF_6) required for enrichment, isotopic enrichment of uranium, fabrication of uranium oxide or UOX (UO_2) fuel elements, and the transportation and storage of spent fuel. In the case of HWRs loaded with natural uranium, the conversion and enrichment steps are obviously not required. The annual requirement for fabricated fuel and the annual discharge of spent fuel, on a gigawatt year (GW/a) basis, assuming no differences in the thermal efficiency of reactors, are inversely proportional to the average burnup of the discharged fuel. Table 9.A.1 in Appendix B indicates, therefore, that the rate of discharge of spent fuel from an HWR in tons of heavy metal a year (tHM/a) per GW is 111.2, or 4.9 times as great as the 22.6 tHM/a from an LWR, just as from an AGR it is 29.9 tHM/a per GW, or 1.3 times as great.* There are also large differences in the radioactivity and plutonium content of spent fuels from these different reactor types. The radioactivity or fission product content is, as a first approximation, proportional to the burnup, so that the radioactivity in LWR fuel is 32/7.3 (or 4.4) times as great as in HWR fuel, on a unit weight basis. Since radiation intensity and its associated heat generation are the most important characteristics affecting the transportation, storage, and disposal of spent fuel, it is clear that tonnage alone does not give an efficient measure of the storage and disposal problem. Table 9.A.1 also gives the fissile plutonium content of spent fuel. Although HWR spent fuel contains a lower concentration of plutonium than LWR spent fuel, the HWR produces about twice as much plutonium per GW/a.

* Throughout this report, the unit kgU or tU refers to the mass of uranium atoms, whereas kgHM or tHM refers to the mass of heavy atoms in cases where one or more of the other heavy elements (e.g., plutonium or thorium) is present in addition to or instead of uranium. As a bookkeeping convention, a given quantity of reactor fuel is assigned the same kgU/tU or kgHM/tHM value on leaving the reactor as spent fuel as it contained when entering the reactor as fresh fuel. It is clear from tables 9.A.1 and 9.A.2 that there are changes in both mass and composition, but the convention removes any ambiguity as to what is meant. A spent fuel reprocessing capacity of 1,500 tHM/a, for example, is based, therefore, not on the amount of spent fuel that might actually be handled but on the original mass of uranium in the fuel when fresh.

TABLE 9.13. Once-through Path: World Annual Flows of Materials and Services at Reactors

Year	Natural Uranium (10^3tU/a)	Fuel Enrichment (10^6SWU/a)	Fabrication (10^3tU/a)	Spent Fuel (10^3tHM/a)
1970	3	2	1.9	0.4
1975	11	8	2.6	1.8
1980	22	16	5	4
1985	41	30	9	7
1990	54	40	12	10
1995	78	58	17	14
2000	130	95	30	25
2005	170	125	40	35
2010	225	165	50	45
2015	280	200	65	60
2020	350	255	85	75

The aggregated world flows of key materials and services in the once-through path, annual and cumulative, are given in tables 9.13 and 9.14, respectively. The values are all quoted at the time the fuel is loaded into the reactor. Obviously, the natural uranium must be mined at an earlier date, enriched at another time, and so on. The effects of estimated lead times will be discussed in section 6. Indeed, rather than discuss the significance of any of our once-through results at this point, we have chosen to present first the results of other fuel cycle scenarios and then to discuss all of them collectively in later sections. We need only draw attention here to what tables 9.13 and 9.14 imply to be the major obstacles on the once-through path: namely, natural uranium requirements and spent fuel accumulations.

Evolutionary Improvements in Uranium Utilization

The calculation of fuel requirements for the various reactors has been founded on the fixed set of performance characteristics described in Appendix B. These characteristics are firmly based, in turn, on the performance of reactors in commercial operation. It appears reasonable to expect that the reactor types currently in operation will not remain static in their performance, but will change in response to technical and economic developments. Since once-through operation is relatively demanding of natural uranium, and since the price of uranium is more likely to increase than to decrease, we expect these changes to be made in the direction of using less natural uranium.

There are a number of ways reductions in uranium requirements might occur. There might, for example, be a shift in capacity additions away from the LWR toward the HWR or high temperature reactor (HTR). HWRs could be operated on 1.2 percent enriched uranium — an expedient calculated to reduce their natural uranium consumption by almost 30 percent (table 9.A.3).

TABLE 9.14. Once-through Path: World Cumulative Flows of Materials at Reactors

Year	Natural Uranium (10^3tU)	Spent Fuel (10^3tHM)	Fissile Plutonium in Spent Fuel (tPu_f)
1970	6	0.4	2
1975	50	6	30
1980	135	20	105
1985	305	50	250
1990	540	90	490
1995	875	155	820
2000	1,420	255	1,350
2005	2,190	410	2,130
2010	3,200	610	3,200
2015	4,480	880	4,590
2020	6,090	1,230	6,350

Alternatively, LWRs might themselves be operated at higher fuel burnups. Several other design and operational changes have also been proposed. Finally, there is the option of lowering the enrichment plant tails assay, which is relevant to those reactors using enriched uranium. This option is discussed in greater detail in section 6.

It is not possible to predict the extent to which each of these improvements might be implemented. We have therefore chosen to represent them in a very general way. To illustrate their impact, we postulate that reactors completed in any year from 1986 onward will benefit from a 1 percent per annum improvement in natural uranium utilization. Thus, the reactors completed in year *t*, after 1985, require $0.99^{(t-1985)}$ times as much uranium as those described in tables 9.A.1 and 9.A.3. For example, the mix of converter reactors completed in 2000 requires only $(0.99)^{15}$, or 0.86, times as much natural uranium as the reference reactors operating once through. However, reactors in this particular group use uranium at this same relative rate throughout their life; our simple model does not allow for possible improvements during a reactor's lifetime, nor does it differentiate between first-core and reload requirements.

The effects of this evolutionary improvement on world natural uranium requirements are given in table 9.15. The annual flow and cumulative demand are compared with the base case, the "reference" once-through path in tables 9.13 and 9.14. Over the time period of the study, the cumulative demand shows a saving of 12 percent. In the last year, 2020, the annual demand is reduced by more than 20 percent, since all reactors by then in operation benefit from the improvements.

The effect that evolutionary improvement would have on other fuel cycle requirements, such as enrichment services, fabrication throughput, or spent fuel accumulation, has not been quantified. In fact, our simple approach of postulating a certain reduction in natural uranium demand without specifying the method precludes a detailed analysis of that nature. Since the main drive to improve the once-through cycle comes from a desire to save uranium, the

TABLE 9.15. Once-through Path: Effect of Postulated Evolutionary Improvement on Natural Uranium Demand at Reactors

Year	Annual Demand (10^3tU/a)		Cumulative Demand (10^3tU)	
	Reference	Improved	Reference	Improved
1975	11	11	50	50
1980	22	22	135	135
1985	41	41	305	305
1990	54	54	540	540
1995	78	75	875	865
2000	130	120	1,420	1,380
2005	170	150	2,190	2,080
2010	225	195	3,200	2,960
2015	280	230	4,480	4,040
2020	350	275	6,090	5,350

resulting changes in other areas may be regarded as of secondary importance. This is true as long as these other fuel cycle parameters do not impede evolutionary improvement, and this we do not expect to be the case.

5. CLOSED CYCLE PATHS

Closed Cycle Options

Spent fuel discharged from converter reactors contains fissile materials. In the case of converters operating in the once-through mode and fueled with UO_2 (UOX) (table 9.A.1), this fissile material is primarily plutonium, formed in the reaction of neutrons with fertile U238. In the case of the LWR, the spent fuel also contains U235 at a slightly higher concentration than occurs in natural uranium. The spent fuel can be chemically reprocessed, the highly radioactive fission products separated from the fissile and fertile constituents, and these latter constituents reused, in whole or in part, as fuel materials. This recycle of materials gives rise to the terms fuel cycle, closed cycle, and so on. As indicated in table 9.A.3, closed cycles tend to use uranium more efficiently than the once-through mode of operation.

The closed cycle on which most developmental work has been done involves recycling uranium or plutonium or both in LWRs. Data for that system are given in table 9.A.2. Since the expected source of plutonium to provide the initial loadings for FBRs is also the spent fuel from converter reactors, plutonium recycle in LWRs is potentially in competition with FBR fueling, should FBR growth ever become plutonium-limited. Although initial fueling of FBRs with enriched uranium is a technical possibility, it has not been considered economically attractive.

Almost from the inception of nuclear power, the closed cycle was viewed as a requisite for a civilian reactor strategy. The problems of reprocessing and handling plutonium were not regarded as serious, since military programs had demonstrated the technical feasibility of these operations. Of greater concern to utilities were the costs of reprocessing, but on this point they were reassured by very low contract prices in early demonstration facilities. It is even rumored that one reprocesser went out of business because the price for spent fuel reprocessing fell below \$20/kgHM (Siegel 1979).

Things have now changed: plans for reprocessing plants have been canceled or severely delayed, costs have increased by an order of magnitude, and opposition to reprocessing and what is termed the "plutonium economy" has become intense. At the same time, political pressures have developed to force the issue. Many countries have linked the granting of licenses for reactors to proof of safe disposal of nuclear waste. Reprocessing of spent fuel, with separation of waste, is seen by many to be an indispensable step along this route, and in some countries this requirement has been incorporated into the licensing process. Both at the Windscale inquiry in Britain and, more recently, at the Gorleben hearing in West Germany, construction of reprocessing plants was justified mainly on the grounds of waste disposal (Parker 1978).

With delays in reprocessing availability and soaring costs, plutonium recycle in thermal reactors, as a "stand-alone" option for fuel management, has lost much of its merit. But reprocessing is pursued in the interest of waste management and as the gateway to a fast breeder world. In the long run, the FBR offers independence of natural uranium resources, and this is frequently viewed as a prerequisite for nuclear power to make a major contribution to meeting world energy needs.

Our scenario paths to this nuclear energy future, which assume the same world nuclear capacity as in table 9.10, emphasize the key role of the FBR, with recycling in thermal reactors as an attendant operation only. In what follows, we sketch out two main paths:

1. FBR deployment in parallel with the use of thermal reactors in a once-through mode, with or without the recycle of uranium recovered as a by-product of reprocessing thermal reactor fuel to provide plutonium for FBRs (fast breeder path);
2. the expansion of reprocessing capacity beyond that needed for FBRs alone, to allow plutonium recycle also in thermal reactors in countries embarking on an FBR program (fast breeder/plutonium recycle path).

Construction of the Fast Breeder Path

Rule 1 A full-scale demonstration FBR begins operation five years after the installed nuclear capacity in a country reaches 25 GW.

The rationale behind this threshold value is that recovery of plutonium from converter reactor spent fuel is a necessary prologue to FBR deployment, and that economies of scale in spent fuel reprocessing argue for having the discharge from about 25 GW of converter reactor capacity. To bring the initial years of the scenario into closer conformity with particular national programs and plans, it was necessary, however, to make some adjustments in the application of this rule. The date for operation of the initial demonstration FBR was advanced or deferred in accordance with our judgment of these plans, and the following dates were thus assigned in different countries: France, 1985; Soviet Union, 1990; United States, 1995; West Germany, Japan, and Britain, 2000. Other countries were treated in accordance with the rule. Some predemonstration FBRs (e.g., the PFR reactor in Britain and the SNR reactor in West Germany) were listed in our results, but their operation alone was not considered to mark the beginning of demonstration as defined here. Except in cases where the rated power of a demonstration plant was already fixed (e.g., the Super-Phénix reactor in France, at 1,200 MW), a rating of 1,500 MW was used for all commercial-size FBR plants.

Rule 2 For nations beginning demonstration programs before 2000, operation of the first commercial FBR starts fifteen years after that of the first demonstration reactor. Two additional demonstration reactors start up at five-year intervals during that interim period. For nations beginning demonstration in 2000 or later, the interim period is shortened to ten years, during which one additional demonstration reactor is brought into operation.

Here we are using the term "commercial" to represent that situation in which a decision about adding to generating capacity, whether made by a nation or by a utility, can consider a breeder reactor as an option on an equal footing with an LWR, a coal-fired plant, or any other technique of electricity production. This means that the uncertainties affecting, for instance, cost estimates, construction times, and the licensing process are roughly equal. In the pre-2000 commercialization scenario here described, assuming a ten-year lead time, the first commercial order is placed when the first demonstration plant has been in operation for five years and the second has just come into operation. The inclusion of a third demonstration plant in the commercialization program is perhaps unnecessary; it is included here because it appeared to give reasonable rhythm and continuity to the engineering and construction work. The program is shortened in the case of countries that begin commercialization in 2000 and beyond, on the grounds that technology transfer should by then abbreviate the process.

Rule 3 For the first five postcommercialization years, one-third of the new reactors completed (net additions plus replacements) are FBRs. In the next five years this fraction is one-half. Thereafter it is two-thirds. The deployment rate is lower than this only if there is insufficient plutonium available to provide the initial inventory for additional FBRs.

Although in many reactor strategy studies it is the practice to allow FBR capacity to expand as rapidly as the plutonium supply allows, we did not find this a credible scenario. It seems unlikely that a nation would make such a total commitment of its nuclear capacity, especially in the first decades of

TABLE 9.16. Fast Breeder Path: FBR Deployment and Capacity (GW)

Country	1985	1990	1995	2000	2005	2010	2015	2020
France	1.5	3.0	4.5	6.0[a]	10.5	18.0	30.0	42.0
Soviet Union	1.0	1.0	2.5	4.0	5.5[a]	32.5	77.5	124.0
Britain	0.3	0.3	0.3	1.5	3.0	4.5[a]	9.0	16.5
West Germany	0.3	0.3	0.3	1.5	3.0	4.5[a]	10.5	25.5
Japan		0.4	0.4	1.5	3.0	4.5[a]	21.0	51.0
United States			1.5	3.0	4.5	6.0[a]	36.0	88.5
Canada					1.5	3.0	4.5[a]	12.0
Brazil						1.5	3.0	4.5[a]
China						1.5	3.0	4.5[a]
East Germany						1.5	3.0	4.5[a]
Italy						1.5	3.0	4.5[a]
Spain						1.5	3.0	4.5[a]
Czechoslovakia							1.5	3.0
India							1.5	3.0
Poland							1.5	3.0
Sweden							1.5	3.0
World FBR capacity	3.1	5.0	9.5	17.5	31.0	80.5	209.5	394.0
World nuclear capacity	280.0	385.0	550.0	895.0	1,220.0	1,610.0	2,010.0	2,550.0

[a] Commercial deployment starts according to rule 2.

FBR deployment. It is doubtful that the established converter reactor infrastructure would be abandoned so quickly. Furthermore, the deployment of the FBR at the rates we have indicated should result in lower uranium prices, thus making thermal reactors more economic. In other words, the interaction of FBR deployment and uranium prices could make some mix of breeder and converter reactors more economical than either type alone.

Application of those three rules leads to the country-by-country FBR deployment shown in table 9.16. According to this scenario, world nuclear capacity is 15 percent FBRs by 2020. The remaining converter reactors are operated in the once-through mode detailed in table 9.A.1.

Deployment of the breeder reactor on a commercial basis also requires the introduction of converter reactor spent fuel reprocessing, in order to supply the initial plutonium inventory for FBRs. (The fuel needed during the early part of the demonstration phase in a country is presumed to be provided by small-scale operations or imports.) In addition, the FBR system itself must include spent fuel reprocessing and fuel fabrication for its continued operation. Table 9.17 shows the amounts of converter reactor fuel of domestic origin that would have to be reprocessed to provide the initial inventories for the FBRs in table 9.16. This reprocessing schedule does not represent an attempt to track FBR plutonium needs year by year. Rather, it is intended to permit orderly growth of the reprocessing industry. In addition, no allowance is made for reprocessing foreign spent fuel. It should be noted, moreover, that the values in table 9.17 are those for the actual throughput of reprocessing plants. Finally, table 9.17 covers only the reprocessing of thermal reactor

TABLE 9.17. Fast Breeder Path: Converter Reactor Fuel Reprocessing Throughput Required (tHM/a)

Country	1981–1985	1986–1990	1991–1995	1996–2000	2001–2005	2006–2010	2011–2015	2016–2020
France	375	375	750	750	1,500	1,500	1,500	1,500
Soviet Union	375	375	750	1,500	3,000	4,500	6,000	6,000
Britain[a]		100	200	200	400	800	1,200	1,600
West Germany					750	750	1,500	2,250
Japan			375	375	750	1,500	3,000	6,000
United States				750	1,500	3,000	6,000	9,000
Canada[b]					600	1,200	2,400	3,600
Brazil						375	375	750
China						375	375	750
East Germany						375	375	750
Italy						375	375	750
Spain						375	375	750
Czechoslovakia							375	375
India[b]							750	750
Poland							375	375
Sweden							375	375
Total	750	850	2,075	3,575	8,500	15,125	25,350	35,575

Note: We assume average fissile plutonium content as follows: LWR: 6.29 kgPu$_f$/tHM (at 32 MWd/kg); HWR: 2.75 kgPu$_f$/tHM (at 7 MWd/kg); AGR: 4.78 kgPu$_f$/tHM (at 18 MWd/kg).

[a] AGR fuel.

[b] HWR fuel.

fuel and takes no account of the need to reprocess FBR fuel itself. The total reprocessing activity in a country would therefore be appreciably larger. In France, for instance, FBR reprocessing capacity in 2020 approaches 1,000 tHM/a. Details of the fuel cycle activities of the major countries for the FBR path are in Appendix C.

An important aspect of the construction of the FBR path is the decision process. Figure 9.7 shows the timing of affirmative decisions needed to implement it in the leading countries. Ten years is assumed as the lead time (from placing of an order to operation of the facility) for all facilities, with the exception of the initial LWR reprocessing plant, which is assigned a fifteen-year lead time. The "FBR fuel facility" referred to in figure 9.7 includes all such items as the reprocessing plant, fuel fabrication plant, transportation facilities, and storage facilities that are required to support breeder reactors on a commercial basis and that must operate in parallel with the reactors.

We regard the fast breeder path schedule in tables 9.16 and 9.17 as close to the upper limit of what can be expected for FBR deployment, given the electricity growth pattern chosen for this study. It assumes that the commercialization programs all lead successfully to commercial deployment, an outcome that implies a fair amount of good luck and widespread political acceptance. It would certainly be physically possible to have a more rapid global expansion of FBR capacity, particularly through multinational consortia. Some such cooperative efforts are already in operation. At present, however, it

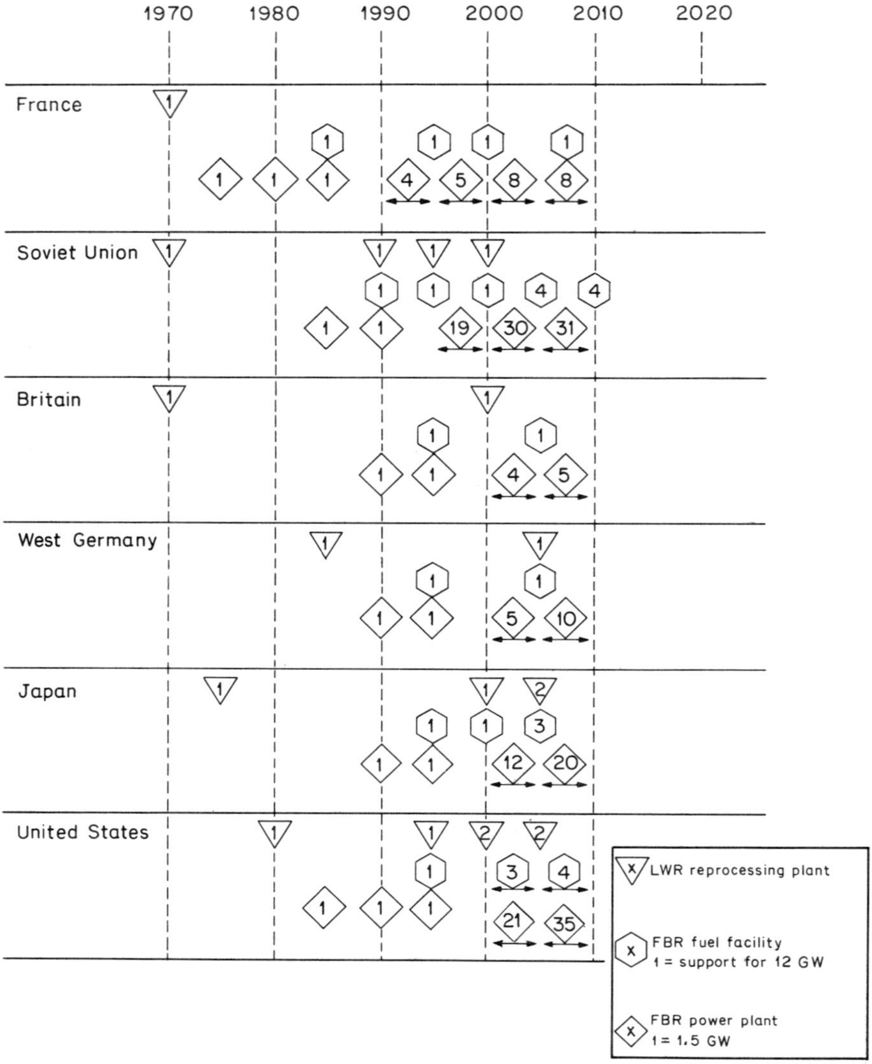

FIGURE 9.7. Fast breeder path: schedule of plant orders required.

seems unlikely that these efforts will bring about a more rapid FBR deployment than that given in table 9.16.

The Reference FBR

The characteristics of the FBR assumed in these calculations are described in Appendix B. A few comments about their effects on the calculations are in order. Table 9.A.2 shows, for example, that the first core of the FBR requires

3.5 t/GW of fissile plutonium. We have assumed that it will take two years to carry out the operations required to return to the reactor the fissile plutonium (Pu_f) contained in discharged fuel. Accordingly, we have provided from outside sources, whether a converter reactor or another FBR, a first core of 3.5 tPu_f plus two reloads of 1.3 tPu_f each: a total of 6.1 tPu_f for each GW of new FBR capacity (or 9.15 tPu_f for our nominal commercial-size FBR of 1,500 MW). The annual fuel discharge from the FBR contains 1.47 tPu_f/GW which, after 2 percent processing losses, yields 1.44 tPu_f. The annual net gain, therefore, is $1.44 - 1.30 = 0.14$ tPu_f/GW. This amounts to 2.3 percent per annum of the inventory. It thus requires 43 gigawatt years of operation to fuel another GW of FBR capacity. Expressed in another way, this FBR has an annually compounded doubling time of thirty years (since $1.023^{30} = 2.0$).

This means that these breeder reactors, by themselves, could sustain a growth rate of only 2.3 percent per annum per GW. In the scenario, however, most of the plutonium required for the initial fueling of FBRs comes from converter reactors. It so happens that the annual production of fissile plutonium in an LWR (table 9.A.1) is almost identical to the net production of an FBR: 0.14 tPu_f/GW. Consequently, in an LWR/FBR mix, FBR capacity can grow at 2.3 percent per annum of the sum of LWR and FBR capacity. To the extent that there is a backlog of spent fuel to reprocess, the rate can be temporarily higher. Likewise, importation of spent fuel can increase the FBR growth rate. In any case, low net plutonium yield does not prove to be a major constraint on any country in the fast breeder path scenario up to 2020. In the longer term, it probably would. There are, however, several technological paths by which the doubling time of FBRs might be shortened, given the incentive.

Results of the Fast Breeder Path

Some of the results of the fast breeder path are presented in tables 9.18, 9.19, and 9.20. They are somewhat more complicated than in the once-through case. For example, natural uranium savings can arise in two separate ways. The first is simply the displacement of converter reactors (CRs) by FBRs, which require no natural uranium. The second is the reuse of the uranium recovered from reprocessing LWR spent fuel. In the fast breeder path scenario, reprocessing is required to supply FBRs with plutonium, and recovered uranium is a by-product. In the case of LWR spent fuel, this uranium still contains 0.8 percent of fissile U235 (table 9.A.1). Having a higher U235 content than natural uranium, it is therefore worth somewhat more than natural uranium as a raw material from which to produce fresh fuel for converter reactors, even after this worth is discounted for the presence of the deleterious U236 isotope. If it were to be reused in that way, the recovered uranium would, of course, have to be made up to the enrichment level (if any) required by the relevant converter reactors, either by

TABLE 9.18. Fast Breeder Path: World Annual Flows of Materials and Services at Reactors

Year	Natural Uranium (10^3tU/a)		Enrichment (10^6SWU/a)	Fuel Fabrication (10^3tHM/a)		Spent Fuel (10^3tHM/a)	
	Without U Recycle	With U Recycle		CR	FBR	CR	FBR
1970	3	3	2	1.9	–	0.4	–
1975	11	11	8	2.6	–	1.8	–
1980	22	22	16	5	–	3.7	–
1985	41	41	30	9	–	7	–
1990	53	53	39	12	0.2	10	0.1
1995	77	75	56	17	0.8	14	0.2
2000	130	125	90	30	1.3	24	0.5
2005	165	155	120	38	1.8	33	0.8
2010	210	195	150	49	4	43	2
2015	250	230	175	58	9	52	6
2020	295	260	210	71	15	63	10

reenrichment or by blending with highly enriched uranium. At present, the latter route appears more likely (Appendix B).

The world annual natural uranium requirement in 2020 for the fast breeder path is 295,000 tU/a without uranium recycle and 260,000 tU/a if uranium is recycled. Compared with the reference once-through path, that is equivalent to a natural uranium saving of 16 and 26 percent respectively. However, the impact of the FBR can be much greater on the national level when a country has committed itself to a strong commercialization program. In France, for instance, where the FBR is introduced at an early date, the cumulative saving of natural uranium in the fast breeder path, with uranium recycle in thermal reactors, amounts to 33 percent by 2020, while the annual requirement has fallen to 45 percent of the once-through value.

TABLE 9.19. Fast Breeder Path: World Cumulative Flows of Materials at Reactors

Year	Natural Uranium (10^3tU)		Spent Fuel (10^3tHM)	
	Without U Recycle	With U Recycle	CR	FBR
1970	6	6	0.4	–
1975	50	50	6	–
1980	135	135	20	–
1985	305	300	50	–
1990	540	535	90	0.3
1995	870	855	155	0.8
2000	1,410	1,380	255	2
2005	2,160	2,080	405	5
2010	3,120	2,980	610	12
2015	4,280	4,030	860	32
2020	5,650	5,230	1,160	75

TABLE 9.20. Fast Breeder Path: World Fuel Cycle Balance

Year	Period	CR Capacity (GW) Capac	CR Close	CR Const	FBR Capacity (GW) Capac	FBR Close	FBR Const	Converter reactor Fuel (tHM) Fuel Charged	Fuel Disch	Spent F Import	Spent F Reprocessed UOX	Spent F Reprocessed MOX	Spent F Inventory	Fissile Plutonium (tPu_f) In Disch CR Fuel	From CR Reproc	Net from FBRs	To FBR Fuel	To CR MOX	Sep Pu Inventory
1975		71			0								0						0
	1976-80		0	70		0	0	19,520	13,300	0	0	0		72	0	0	0	0	
1980		141			0								13,300						0
	81-85		0	137		0	3	38,235	26,165	0	3,750	0		142	23	0	11	0	
1985		278			3								35,715						12
	86-90		0	104		0	2	52,130	43,030	0	4,250	0		236	25	1	15	0	
1990		382			5								74,495						23
	91-95		0	156		0	5	73,465	59,885	0	10,375	0		328	61	3	21	0	
1995		538			10								124,005						66
	96-00		0	340		0	8	123,025	92,720	0	17,875	0		500	104	5	40	0	
2000		878			18								198,850						135
	2001-05		71	381		0	13	172,965	145,410	0	42,500	0		779	236	11	76	0	
2005		1,188			31								301,760						306
	06-10		70	410		0	50	224,205	193,595	0	75,625	0		1,027	413	22	263	0	
2010		1,528			81								419,730						478
	11-15		137	408		3	132	271,495	247,030	0	126,750	0		1,299	669	63	742	0	
2015		1,799			210								540,010						468
	16-20		104	460		2	186	326,695	294,520	0	177,875	0		1,529	933	168	1,096	0	
2020		2,155			394								656,655						473

Year	Period	Fuel Cycle Activities (tHM/a) Converter Reactors Fabrication UOX	Converter Reactors Fabrication MOX	Converter Reactors Reprocessing UOX	Converter Reactors Reprocessing MOX	FBRs (Core + Blanket) Fabric	FBRs (Core + Blanket) Reproc
1975							
1980	1976-80	3,905	0	0	0	0	0
1985	81-85	7,650	0	750	0	60	0
1990	86-90	10,425	0	850	0	120	85
1995	91-95	14,695	0	2,075	0	220	130
2000	96-00	24,605	0	3,575	0	385	250
2005	2001-05	34,595	0	8,500	0	740	490
2010	06-10	44,840	0	15,125	0	2,040	1,130
2015	11-15	54,300	0	25,350	0	5,585	3,255
2020	16-20	65,340	0	35,575	0	10,335	7,010

Notes

KEY

CR	Converter reactor
FBR	Fast breeder reactor
Capac	Installed capacity
Close	Capacity withdrawn from service
Const	New capacity constructed
Disch	Discharged
Spent F	Spent fuel
UOX	Uranium oxide
MOX	Mixed (uranium/plutonium) oxide
Reproc	Reprocessing
Sep Pu	Separated plutonium
Fabric	Fuel fabrication

Table 9.20 gives an overview of the global development of the fast breeder path, and it clearly shows the strong dependence of FBR introduction on converter reactor plutonium. The corresponding national balances for the leading FBR countries are given in Appendix C. In all these tables, one quantity of interest that requires some explanation is the separated fissile plutonium inventory or stock. That quantity is seen to rise to a world value of more than 400 t by 2010 and to remain at about that level for the rest of the period. In these calculations, the magnitude of the plutonium inventory at any time is essentially the difference between the cumulative quantity separated in the plants that reprocess converter reactor fuel and the requirements for FBR first cores and first reloads. The emphasis in both FBR deployment (table 9.16) and reprocessing plant deployment (table 9.17) is on orderly growth. One result is the initial establishment of large plutonium inventories, since reprocessing plants produce plutonium at a higher rate than is immediately required by FBRs. In reality, there might be some international transfers that could help to reduce these stocks. By 2020, however, when the FBR is being deployed rapidly in some countries, this scenario shows plutonium supply constraints on FBR growth. International transfers of plutonium or converter reactor spent fuel could also serve to ease those constraints.

The quantities shown as separated plutonium in table 9.20 are primarily those separated from converter reactor fuel in reprocessing and not yet charged to or required in an FBR. There is no implication that this plutonium would necessarily be in some physically separate form, such as plutonyl nitrate or plutonium oxide. It could, of course, be admixed with uranium or fission products, depending on the particular safeguards programs developed. The separated plutonium inventory values do not include those quantities of plutonium within the FBR fuel cycle, except for the net production in FBRs. For every GW of FBR capacity, there are 2,600 kg of fissile plutonium in various parts of the out-of-pile cycle, including spent fuel, separated plutonium, and MOX fuel waiting to be loaded.

In the calculation of plutonium flows and stocks, there is one physical phenomenon that may appear to be a small detail but that could have significant effects. This concerns the fissile isotope plutonium 241. Pu241, which decays with a half-life of about fifteen years, constitutes some 20 percent of the fissile plutonium in once-through UOX fuel discharged from LWRs (but considerably less in fuel from natural uranium HWRs). In the fast breeder path scenario and, we believe, in reality, reuse of much of the LWR-produced plutonium will be delayed for times comparable to the Pu241 half-life. A delay of fifteen years would thus mean a loss of 10 percent of the total fissile plutonium. Although the computation of Pu241 decay in all spent fuel is, in principle, straightforward, it requires more detailed calculations than were performed here. Allowance for this phenomenon was limited to applying a percentage reduction of fissile plutonium quantity in cases where the scenario showed an extended delay in plutonium reuse (Appendix C).

Construction of the Fast Breeder/Plutonium Recycle Path

In the fast breeder path, all plutonium recovered from converter reactor spent fuel is dedicated to FBRs. An additional or alternative use for plutonium is as a component of LWR fuel. In this case it can be added to natural uranium to produce a fuel equivalent to enriched uranium. The "PWR Pu recycle" column in table 9.A.2 shows the composition of a mixed UO_2/PuO_2 (MOX) fuel that can substitute for 3.1 percent enriched uranium in the reference LWR. The plutonium contained in once-through UO_2 (UOX) spent fuel is 70 percent fissile. In the simulation, this plutonium was the only kind used in LWR MOX fuel. The plutonium contained in spent LWR MOX fuel has a lower fissile content of about 60 percent and was exclusively allocated to the FBR, on the ground that any penalty for isotopics should be less in the FBR. We have assumed that the plutonium separated from once-through UOX fuel is completely fungible, as far as the particular reactor to which it can be recycled is concerned. The reference recycle PWR uses 30 percent MOX and 70 percent UOX in its reloads. Alternatively, it would be possible to have a number of dedicated plutonium burners loaded only with MOX fuel.

There is a variety of situations in which a nation might find plutonium recycle in LWRs attractive. It might, for example, be an integral part of a fast breeder program. Much of the required infrastructure is in common. It will be noted that, in the fast breeder path (table 9.20), some large inventories of separated plutonium build up. These result from the fact that, in that scenario, the plutonium output from reprocessing plants and the inventory requirements of FBRs are not continuously matched. To the extent that this feature of the scenario coincides with reality, these plutonium stocks could be made available for LWR fuel. The net quantity of fissile plutonium available to FBRs would then be reduced by about 50 percent of the quantity used in MOX fuel for LWRs. This is easily seen from the figures in tables 9.A.1 and 9.A.2. For each kilogram of fissile plutonium used in MOX fuel for LWRs, there remains, after irradiation, 0.62 kg for FBRs. Had UOX been used in LWRs instead, 1 kg of fissile plutonium would have been saved and 0.2 kg produced: a total of 1.2 kg. Furthermore, the plutonium recycled to LWRs would remain unavailable for five to ten years from the time it was removed from inventory for use in thermal reactors, depending on the speed with which spent MOX fuel was reprocessed.

Plutonium recycle in LWRs may also be justified in its own right on economic grounds, depending on projected uranium prices and other fuel cycle costs. This subject is discussed in section 8. Finally, plutonium recycle in thermal converters may well have a role in a total nuclear energy plan that includes converter spent fuel reprocessing both for recovery of uranium and plutonium and as a step in the disposal of nuclear waste.

The fast breeder/plutonium recycle path has been constructed in order to display some of the features of plutonium recycle in LWRs. As the scenario

TABLE 9.21. Fast Breeder/Plutonium Recycle Path: Converter Reactor Fuel Reprocessing Throughput Required (tHM/a)

Country	1981–1985	1986–1990	1991–1995	1996–2000	2001–2005	2006–2010	2011–2015	2016–2020
France	375	375	1,500	1,500	2,250	2,250	3,000	3,000
Soviet Union	375	375	1,500	2,250	3,750	6,000	7,500	7,500
Britain[a]		100	1,200	1,200	1,400	1,800	3,200	3,200
West Germany					750	1,500	3,000	3,000
Japan			375	375	750	1,500	4,500	7,500
United States				750	1,500	4,500	7,500	10,500
Canada[b]					600	1,200	2,400	3,600
Brazil						375	375	750
China						375	375	750
East Germany						375	375	750
Italy						375	375	750
Spain						375	375	750
Czechoslovakia							375	375
India[b]							750	750
Poland							375	375
Sweden							375	375
Total	750	850	4,575	6,075	11,000	20,625	34,850	43,925

Note: We assume average fissile plutonium content as follows: LWR: 6.29 kgPu_f/tHM (at 32 MWd/kg); HWR: 2.75 kgPu_f/tHM (at 7 MWd/kg); AGR: 4.78 kgPu_f/tHM (at 18 MWd/kg).

[a] AGR fuel.

[b] HWR fuel.

name implies, plutonium recycle is made an integral part of FBR programs. The three rules for the commercialization and deployment of FBRs in the fast breeder path are still followed. Only those countries that meet the qualifications for a reprocessing plant under those rules are assumed to reprocess fuel for plutonium recycle, and first priority is given throughout to the plutonium requirements of FBRs. However, when the inventory of separated plutonium exceeds one year's output of a nation's converter reactor reprocessing plants plus the annual net output of its FBR fuel cycle, the production of MOX fuel for LWRs is permitted. Importation of spent fuel is also permitted, and the throughput of reprocessing plants is increased somewhat over the fast breeder path. The general rule followed for this expansion of reprocessing throughput is that, after the first ten years of reprocessing operation, a nation's throughput is increased by 750 tHM/a. This value rises to an increase (over the fast breeder path) of 1,500 tHM/a by 2020. Production of MOX fuel for use in HWRs is not considered in this scenario. The material balances are made up as if the MOX fuel were used domestically, but MOX fuel could certainly be exported. No attempt is made to identify countries from which spent fuel is imported or countries to which fresh MOX fuel might be exported. The resulting converter reactor spent fuel reprocessing throughputs are given in table 9.21. The cumulative increments of reprocessed LWR fuel over the fast breeder path amount to more than 150,000 tons by 2020.

TABLE 9.22. Fast Breeder/Plutonium Recycle Path: World Annual Flows of Materials and Services at Reactors

Year	Natural Uranium (10^3tU/a)	Enrichment (10^6SWU/a)	Fuel Fabrication (10^3tHM/a)			Spent Fuel (10^3tHM/a)	
			UOX	MOX	FBR	CR	FBR
1970	3	2	1.9	–	–	0.4	–
1975	11	8	2.6	–	–	1.8	–
1980	22	16	5	–	–	4	–
1985	41	30	9	–	0.2	7	–
1990	52	38	12	0.3	0.2	10	0.1
1995	73	53	16	0.7	0.7	14	0.2
2000	110	85	29	1	1.3	24	0.5
2005	145	110	36	1.7	1.8	33	0.8
2010	175	145	47	1.7	4	43	2
2015	205	170	56	1.5	9	52	6
2020	235	200	69	1.6	15	63	10

Results of the Fast Breeder/Plutonium Recycle Path

The world flows of materials and services at the time fuel is loaded or unloaded at reactors are given on an annual basis in table 9.22 and as cumulative figures in table 9.23. The activities within the various sectors of the fuel cycle for individual countries are displayed in Appendix C, from which the global balance in table 9.24 has been compiled.

The fast breeder path with uranium recycle alone (table 9.18) brought a 26 percent fall in annual natural uranium requirements in 2020, when compared with the reference once-through path (table 9.13). The introduction of plutonium recycle in thermal reactors, with the associated increase in reprocessing capacity, yields an additional saving of 7 percent, giving a total reduction of 33 percent in the annual uranium requirement (table 9.22). The additional saving of separative work in enrichment is smaller – some 3 percent in 2020 when compared with the fast breeder path without plutonium recycle (table 9.18) – since it reflects only the replacement of some enriched uranium by fissile plutonium. The recycled uranium recovered from reprocessing is assumed to be reenriched.

The cumulative reductions in natural uranium and separative work requirements resulting from the addition of plutonium recycle to the fast breeder path are modest, as a comparison of tables 9.19 and 9.23 reveals. By 2020, the cumulative uranium saving has risen to 19 percent, when compared with the reference once-through path, as against 7 percent for the basic fast breeder path and 14 percent for the fast breeder path with uranium recycle alone. The contributions of different measures to that total saving are shown in figure 9.8. A point to note is that, in the fast breeder path, it is the recycle of uranium in LWRs, rather than FBR operation, that accounts for the major share of uranium savings in arlier years and for half the total saving over the period to 2020. Moreover, it is the reuse in LWRs of more uranium recovered

TABLE 9.23. Fast Breeder/Plutonium Recycle Path: World Cumulative Flows of Materials at Reactors

Year	Natural Uranium (10^3tU)	Spent Fuel (10^3tHM)		
		CR(UOX)	CR(MOX)	FBR
1970	6	0.4	–	–
1975	50	6	–	–
1980	135	20	–	–
1985	300	50	–	–
1990	530	90	–	0.3
1995	830	155	0.5	0.8
2000	1,320	250	4	2
2005	1,990	395	8	5
2010	2,820	590	15	12
2015	3,780	830	25	32
2020	4,910	1,130	30	75

from reprocessing that again explains half the additional uranium saving in the fast breeder/plutonium recycle path.

As that indicates, a combined strategy allows better utilization of nuclear materials than FBR deployment alone, particularly during the early stages of a program. Taking the fast breeder/plutonium recycle path as a basis, the relative contribution of the FBR itself to cumulative uranium savings rises from less than 15 percent in 2000 to only about 37 percent in 2020. In absolute terms, measured against the reference once-through path, the figures are 1 percent and 7 percent respectively. Thus, given the FBR deployment program postulated in this study, the immediate benefit from breeder reactors alone is no more than modest, and the introduction of FBRs seems to be justified only in the context of a very long-term strategy. However, one must be careful not to generalize from the uranium savings results of the fast breeder/plutonium recycle path. How much uranium is ultimately conserved will depend on the rate of breeder reactor introduction relative to converter reactor deployment. It is necessary to have a very specific case in mind in order to know whether any gains due to plutonium recycle are eventually nullified by a limitation on FBR introduction.

It can be seen in table 9.24 that, from 2010, the plutonium requirements of FBRs exceed the quantity available for recycle to LWRs. The global inventory of separated plutonium is generally lower in this scenario (compare tables 9.20 and 9.24). As the period progresses, however, the global inventories in the two scenarios gradually converge, since the requirement to maintain a minimum working reserve represents an ultimate constraint in both cases. In fact, turning to the back end of the fuel cycle as a whole, we find that spent fuel of the MOX type accumulates only slowly in the fast breeder/plutonium recycle path, at the rate of some 2–3 percent of the converter reactor fuel discharged annually (table 9.23). Of the cumulative total of some 1.12 MtHM thus discharged during the study period, only 215,000 tHM have

TABLE 9.24. Fast Breeder/Plutonium Recycle Path: World Fuel Cycle Balance

Year	Period	CR Capacity (GW) Capac	CR Close	CR Const	FBR Capacity (GW) Capac	FBR Close	FBR Const	Converter reactor Fuel (tHM) Fuel Charged	Fuel Disch	Spent F Import	Spent F Reprocessed UOX	Spent F Reprocessed MOX	Spent F Inventory	Fissile Plutonium (tPu_f) In Disch CR Fuel	From CR Reproc	Net from FBRs	To FBR Fuel	To CR MOX	Sep Pu Inventory
1975		71			0								0						0
	1976-80		0	70		0	0	19,520	13,300	0	0	0		72	0	0	0	0	
1980		141			0								13,300						0
	81-85		0	137		0	3	38,235	26,165	0	3,750	0		142	23	0	11	0	
1985		278			3								35,715						12
	86-90		0	104		0	2	52,130	43,030	0	4,250	0		236	25	1	15	20	
1990		382			5								74,495						3
	91-95		0	156		0	5	73,465	59,885	(4,475)	22,630	0		336	129	3	21	92	
1995		538			10								111,750						22
	96-00		0	340		0	8	123,025	92,720	(6,850)	28,920	675		529	180	5	40	134	
2000		878			18								174,875						33
	2001-05		71	381		0	13	172,965	145,410	(11,370)	51,330	3,070		825	339	11	76	233	
2005		1,188			31								265,885						74
	06-10		70	410		0	50	224,205	193,595	(16,900)	97,715	4,430		1,112	615	22	263	310	
2010		1,528			81								357,335						138
	11-15		137	408		3	132	271,495	247,030	(27,435)	163,795	7,760		1,413	1,017	63	742	229	
2015		1,799			210								432,810						247
	16-20		104	460		2	186	326,695	294,520	(27,450)	207,045	10,315		1,604	1,281	168	1,096	276	
2020		2,155			394								509,985						324

Year	Period	Fuel Cycle Activities (tHM/a): Converter Reactors Fabrication UOX	Fabrication MOX	Reprocessing UOX	Reprocessing MOX	FBRs (Core + Blanket) Fabric	FBRs Reproc
1975							
1980	1976-80	3,905	0	0	0	0	0
1985	81-85	7,650	0	750	0	60	0
1990	86-90	10,290	135	850	0	120	85
1995	91-95	14,080	615	4,575	0	220	130
2000	96-00	23,715	890	5,940	135	385	250
2005	2001-05	33,045	1,550	10,335	665	740	490
2010	06-10	42,775	2,065	19,685	940	2,040	1,130
2015	11-15	52,775	1,525	33,230	1,620	5,585	3,255
2020	16-20	63,505	1,835	41,770	2,155	10,335	7,010

Notes

1. See key to Table 9.20.
2. 'Spent fuel import' in the world balance indicates total transfers between countries, which are here also included in the figures for fuel discharged.
3. Average annual figures for reprocessing in the fuel cycle activities section are adjusted to allow for reprocessing of imported LWR fuel, as well as indigenous AGR fuel, in British facilities (see notes to Table 9.A.12).

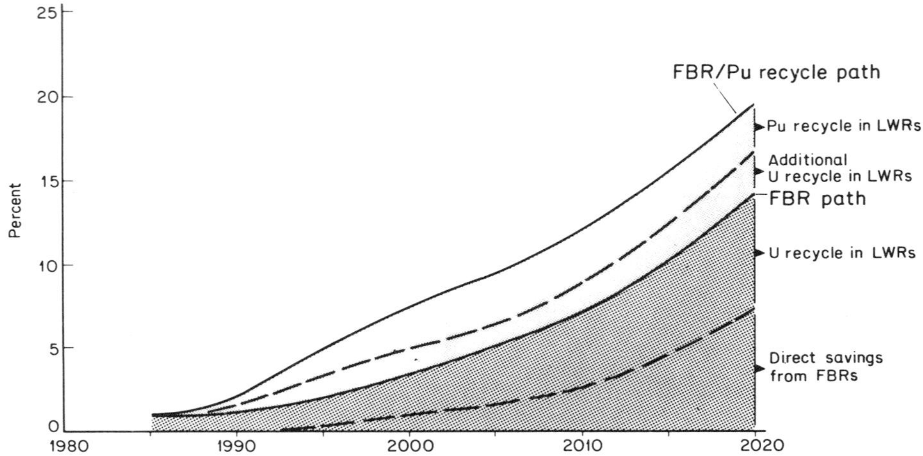

FIGURE 9.8. Fast breeder paths: natural uranium savings compared with the reference once-through path.

arisen in non-FBR (and nonreprocessing) countries. The remaining 900,000 tHM have originated in countries adopting a fast breeder path, where 605,000 tHM have accordingly been reprocessed. The figures for "spent fuel imports" in table 9.24 show that nearly 95,000 tHM of discharged fuel have meanwhile been shipped to countries in that group with large reprocessing capacities, such as France, Britain, and the Soviet Union, where it is assumed that most of the MOX fuel derived from reprocessing that material will also be used. However, if we also assume that this international trade in spent fuel, needed to even out reprocessing activity, takes place only between countries adopting FBRs, the corollary is still that, by 2020, the unreprocessed inventory those countries retain is limited to some 295,000 tHM.* Thus, compared with the world reprocessing capacity of 44,000 tHM/a in 2020, their inventory has already declined to less than seven years' throughput.

6. FUEL SUPPLY

Uranium

Although the nuclear future is not going to unfold according to any of the scenarios developed in previous sections, it will undoubtedly contain some elements of each. In this and following sections, we look at the principal issues on which our collection of scenarios seems to shed some light. These include questions of uranium supply and the supply of enrichment services, the distribution of various materials associated with the nuclear fuel cycle

* The sum of the individual inventories for FBR countries in Appendix C (390,000 tHM) exceeds the total implied in table 9.24 (295,000 tHM) because *exports* of spent fuel (95,000 tHM) have not been allocated to particular countries or subtracted from their inventories.

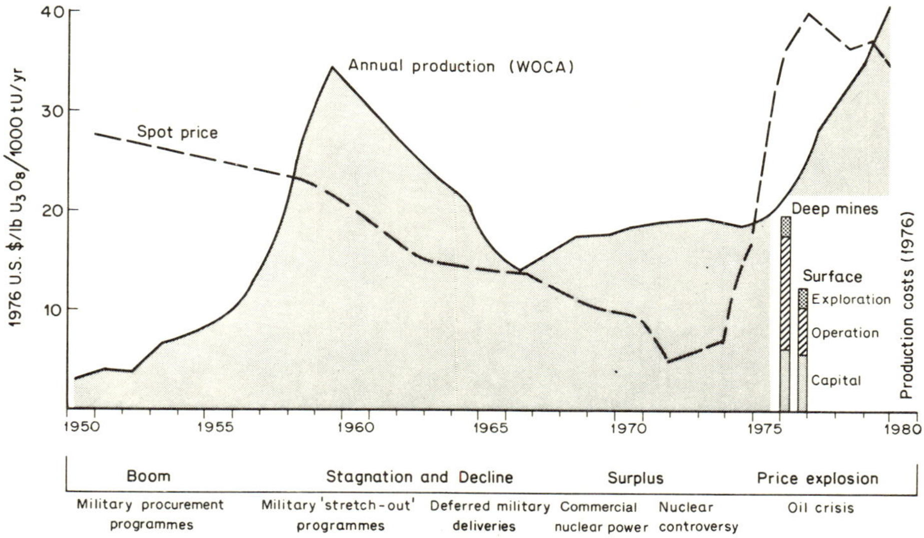

FIGURE 9.9. Natural uranium: production and prices, 1950–79.

(particularly at its back end), the economics of nuclear power, and the deployment of advanced fuel cycles, with special emphasis on the fast breeder reactor. We have chosen to highlight what we see as major points, rather than to attempt a complete and detailed analysis of a nuclear energy future.

Uranium – a limited, nonrenewable resource – is central to all discussions of the expansion of nuclear energy and the motivation for advanced reactor development. Information about the resource itself and about its timely conversion into a viable supply is a key component of any nuclear projection. Unfortunately, knowledge of the world's uranium resources is rather limited, and there are wide variations in what experts think could be made available and should serve as the basis for future planning. Uranium is not a rare mineral in the earth's crust; it occurs as frequently as tin or molybdenum. The question is rather one of the effort required to find it and the cost of producing it.

The history of uranium production and pricing, which goes back only over some thirty or forty years, has been dictated by changing military needs and by the ups and downs of civilian nuclear programs (fig. 9.9). It has hardly provided a climate allowing the mining industry to grow in an orderly manner. The outlook for the industry now depends uniquely on the prospects for nuclear power. Uranium is a single-purpose mineral with no alternative uses, whereas other mineral industries can rely on multiple uses for their products. In addition, substitution effects tend in those cases to even out the demand curve and to allow stable long-term operation of mining facilities.

Public awareness of nuclear power has also had its impact on the uranium mining industry. It has led to more stringent health, safety, and environmen-

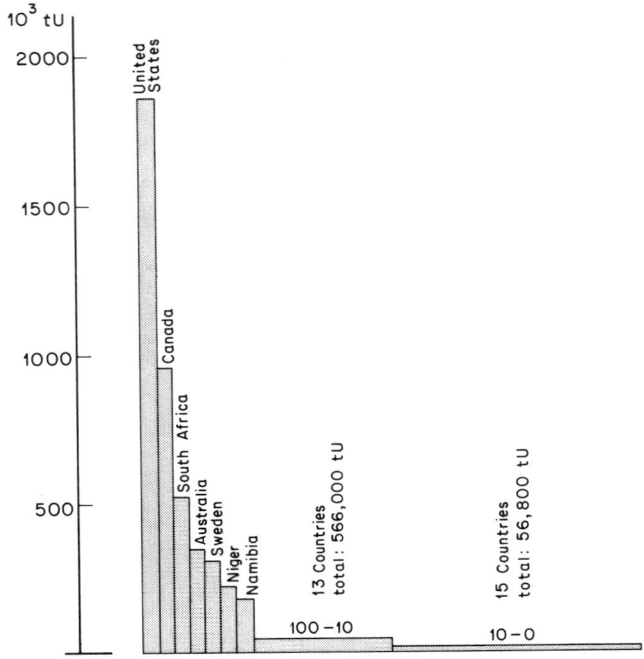

FIGURE 9.10. Natural uranium: distribution in WOCA of reasonably assured and estimated additional resources.

tal standards for mining operations. New mine openings have become increasingly difficult in some areas, owing to environmental and political opposition, as, for example, in Australia and Sweden.

Known Resources

Estimates of uranium reserves (i.e., well-identified and economically workable resources) are steadily increasing. In the period from 1965 to 1979, additions averaged 7 percent a year, and as of 1 January 1979 the joint NEA/IAEA study reported that 2.6 MtU within the WOCA area could be classified as "reasonably assured resources" (RAR) (NEA/IAEA 1979). The NEA/IAEA study also lists 2.4 MtU of "estimated additional resources" (EAR), which are extensions of well-known deposits or in related geological formations. For the present purpose, we equate RAR with reserves and define RAR plus EAR as "known resources." The total of known uranium resources in the WOCA area, obtainable with present mining technology and at costs lower than \$50/lbU_3O_8 or \$130/kgU, is thus in the neighborhood of 5 MtU. (These costs cover capital and operation but exclude exploration expenditure and profits.)

The uranium position of the centrally planned economies is more difficult to evaluate. However, although detailed estimates are not available, some data on uranium resources there have been published in West Germany (BGR 1976), and an overall estimate has been constructed by the NEA and IAEA (NEA/IAEA 1978). Assuming the ratio of RAR plus EAR to the total to be the same in the CPE countries as in the WOCA area, we obtain a total value of about 6 MtU for the world's known uranium resources.

As figure 9.10 shows, the known resources of uranium are distributed unevenly among the thirty-five countries surveyed by the NEA and IAEA (NEA/IAEA 1979). Almost 90 percent of the resources are concentrated in seven countries, and two-thirds of that amount are in the United States and Canada alone. This appears to reflect more extensive exploration in these last two countries rather than any special endowment of the North American continent. It is therefore reasonable to expect that uranium reserves will increase as activities of similar intensity are carried out worldwide. There has, in fact, been renewed interest in uranium prospecting since the oil crisis of 1973–74. This has followed a decade or more of low activity as the uranium industry waited for more favorable winds in the doldrums of terminating military programs. It should be kept in mind, however, that half the known resources are not yet sufficiently well delineated to allow mining, and that a first result of increased exploration is likely to be a substantiation of the reserve position, rather than an expansion of the "known" total.

Unknown Resources

Since this study involves scenarios extending over forty years, it would not be reasonable to base uranium supply projections only on today's known resources. Continuing exploration will undoubtedly produce additional known resources in the future. A quantitative estimate of such resources leads one into resource categories described variously in the literature as "possible," "potential," "speculative," and so on. We shall refer to these simply as "unknown resources."

Uranium of workable concentration has been found in deposits of numerous geological types. Over the years, the knowledge of uranium genesis has increased, and exploration has recently uncovered many new uraniferous formations. From this experience and their knowledge of the world's geography, mining experts have attempted to estimate the size of the unknown resources.

One approach is to perform an extrapolation from a well-known uranium area: in most cases, the United States. The choice is the more justified since estimates made by the United States Department of Energy (Voigt 1978) have also been extended to possible and speculative resources (fig. 9.11). Starting with these data, Harris has used three different models to estimate total world

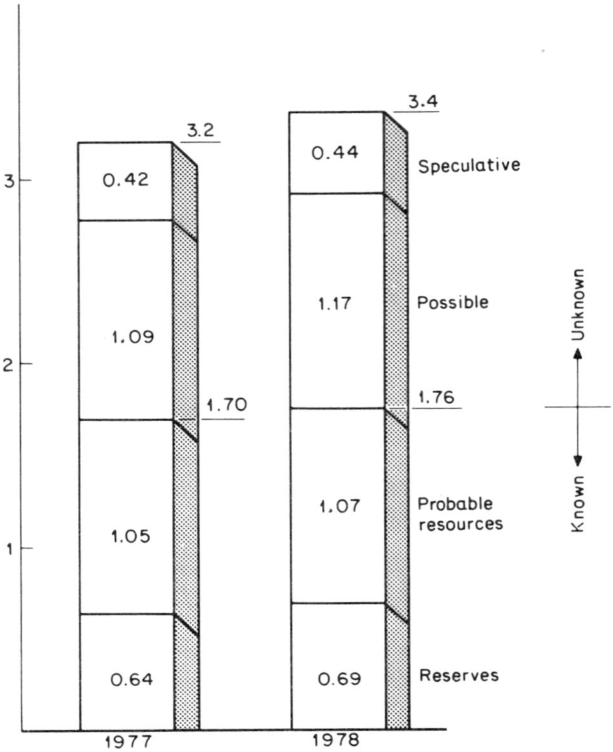

FIGURE 9.11. Natural uranium: known and unknown United States resources at up to $\$50/lbU_3O_8$.

resources of uranium in the cost range up to $\$50/lbU_3O_8$ or $130/kgU (Harris 1979):

1. by extrapolating United States figures for the ratio of total resources to reserves to the twenty-seven countries reported in the NEA/IAEA survey of 1975;
2. by geographic analogy, extending United States figures to the whole of the WOCA area using a mineral density measure;
3. by normalizing model 2 for exploration followed by reserve expansion and adjusting for geological and economic factors for the whole of the WOCA area.

Harris's results are given in table 9.25. It should be noted that the estimates depend heavily on the quantification of possible and speculative resources in the United States and on economic assumptions for the rest of the world. Another recent world survey by the NEA and IAEA, based on indirect geological evidence from 185 countries, estimates speculative resources of uranium minable at costs up to $\$50/lbU_3O_8$ to be in the range of 10–20 MtU

TABLE 9.25. Natural Uranium: Estimates of Total World Resources at up to $\$50/lbU_3O_8$

Method/Source	Result (MtU)
1. Extrapolation of United States resource/reserve ratio to 27 countries of 1975 NEA/IAEA survey (Harris 1979)	12
2. Geographic analogy of United States to WOCA area (Harris 1979)	22
3. Geographic analogy of United States to WOCA area adjusted for exploration and economic factors (Harris 1979)	15
4. Geological analysis of 176 WOCA area countries (NEA/IAEA 1978)[a]	11–19
5. Geological analysis of the world (185 countries) (NEA/IAEA 1978)[a]	14–26

[a] These figures represent the sum of "reasonably assured resources," "estimated additional resources," and speculative resources, although the expert group responsible warns explicitly against this simple approach to establishing the total.

(NEA/IAEA 1978). The sum of these speculative resources and RAR plus EAR is also given, for the WOCA area and the world, in table 9.25.

From these analyses, we conclude that the present state of information yields an estimate for the world's resource base of uranium extractable at costs up to $\$50/lbU_3O_8$ of 15–25 MtU, the larger part of which is currently unknown and not likely to be made available until sometime in the next century. Positive or negative confirmation of these estimates can be made only after extensive exploration throughout the world.

These estimates have so far addressed only "conventional" uranium resources. There are also "unconventional" low-grade resources of uranium, in shales, igneous rocks, and seawater, in addition to which uranium can be extracted as a by-product of other mining operations, for instance from phosphatic rocks. Although these resources are very large, uranium is mostly obtainable from them only at costs well above $\$50/lbU_3O_8$. However, some additions in the lower-cost category can be expected from these sources over the next fifty years. An indication of their possible scale is given by the United States Department of Energy's estimate that some 100,000 tU could thus be recovered by 2000 (NEA/IAEA 1978).

Natural Uranium and Separative Work Requirements

We have chosen to represent the natural uranium requirements of our different scenarios at three levels. First is the mine production rate needed to maintain current reactor operations; second is the addition to uranium reserves needed to justify mine openings; third is the cumulative lifetime requirement of all the reactors built or ordered by a given date. Each level reflects a different set of decision points in the development of a nuclear future.

TABLE 9.26. Natural Uranium: Annual Production at the Mine for Different Scenarios (10^3tU/a)

Year	Once-through Path		Fast Breeder Path		Fast Breeder/ Plutonium Recycle Path
	Reference	Improved	Without U Recycle	With U Recycle	
1975	15	15	15	15	15
1980	30	30	30	30	30
1985	45	45	45	45	45
1990	60	60	60	60	60
1995	100	95	100	95	90
2000	145	130	140	135	125
2005	190	170	185	170	160
2010	245	205	225	205	185
2015	305	250	265	240	220
2020	380	290	310	270	250

TABLE 9.27. Natural Uranium: Cumulative Production at the Mine for Different Scenarios (10^3tU)

Year	Once-through Path		Fast Breeder Path		Fast Breeder/ Plutonium Recycle Path
	Reference	Improved	Without U Recycle	With U Recycle	
1975	75	75	75	75	75
1980	190	190	190	190	190
1985	390	390	390	390	390
1990	660	660	660	660	660
1995	1,070	1,050	1,060	1,040	1,010
2000	1,710	1,650	1,690	1,650	1,550
2005	2,560	2,420	2,520	2,440	2,310
2010	3,680	3,380	3,570	3,410	3,210
2015	5,080	4,540	4,800	4,530	4,240
2020	6,830	5,930	6,260	5,830	5,420

The annual and cumulative mine production implied by the current uranium requirements in our scenarios are given in tables 9.26 and 9.27. Annual requirements for natural uranium and separative work are plotted in figures 9.12 and 9.13. In each case, it should be noted that figures for uranium production at mines or for separative work at enrichment plants differ from those tabulated in sections 4 and 5 as requirements measured at reactors, because of the delays between mining uranium, enriching it, and loading it as fuel into a reactor. We have assumed that uranium must be mined two years and enriched one year before it is loaded as fuel.

The range bounded in figures 9.12 and 9.13 by the once-through path and the fast breeder/plutonium recycle path represents the maximum spread of the various scenarios in this study. Included in the figures are recent estimates of the potential uranium mine and enrichment plant capacities in the WOCA area to 1985 (Voigt 1978). Since both those capacities exceed the projected world requirements for our once-through path, it seems that global uranium

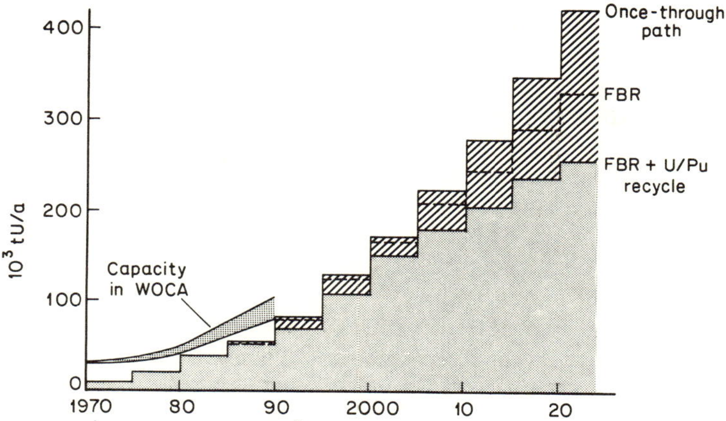

FIGURE 9.12. Natural uranium: annual production requirements at the mine.

mining and enrichment capacity will be high enough through the 1980s. There may, in fact, be a problem of oversupply.

The fast breeder paths obviously have lower uranium requirements than the once-through path, although the amount by which they reduce uranium requirements depends on the degree to which uranium and plutonium are recycled in converter reactors (primarily LWRs). For the scenarios of sections 4 and 5, the reductions are indicated in tables 9.26 and 9.27. The limiting case, with uranium and plutonium recycle, brings reductions of some 33 percent in the annual production rate and 19 percent in cumulative production by 2020. For comparison, the evolutionary developments we have postulated in the improved once-through scenario result in uranium savings of 21 percent in annual production and 12 percent in cumulative production by 2020. Clearly a combined strategy, involving the evolutionary improvement of converter reactors within a fast breeder path, would have a proportionately

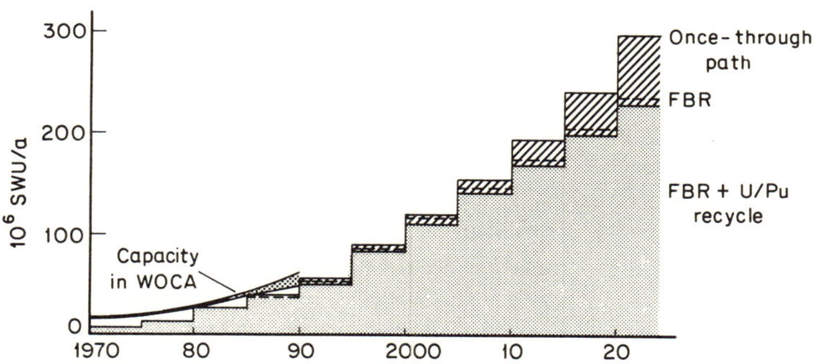

FIGURE 9.13. Enrichment: annual separative work requirements at the enrichment plant.

greater impact on uranium demand than either course alone. As between the two options, however, it must be emphasized that much larger reductions would be in prospect for the fast breeder path after 2020.

The average annual rate of growth in mine production from 1980 to 2020 implied by tables 9.26 and 9.27 ranges from 6.5 percent for the once-through path to 5.4 percent for the fast breeder/plutonium recycle path. Typically, the increase called for up to the end of the century is roughly 5,000 tU/a. In the most buoyant years of uranium mining, from 1955 to 1959, output was raised from 10,000 to 35,000 tU/a, so that future expansion at the required rate does not seem inconceivable.

To achieve any given trajectory of mine production, it is necessary to open new uranium mines, both to replace mines that have been exhausted and to meet net production increases. For a mine to be opened, it is necessary that the decision maker be satisfied that the uranium ore body is well defined and that uranium is present in adequate quantity and concentration. The uranium resource on which a mine-opening decision is based must be known well enough to fall within the "reserves" category of uranium resources. (Not all reserves, on the other hand, lead immediately to mine openings, because the controlling party may have reasons not to act, such as conservation policy or an expectation of higher prices in the future.) To calculate the reserves needed for mine openings, we have used a simple model. A mine is assumed to produce at its rated capacity for fifteen years; a mine opening, therefore, signifies the presence of reserves fifteen times as large as the rated annual capacity. We further assume that the uranium ore in question must reach reserve status five years before mine production actually begins. From these assumptions, it is a straightforward calculation to obtain the uranium reserves required for the mine openings to meet any given production projection. The results are given for the reference once-through path in table 9.28 and figure 9.14. These indicate that, to support mine openings up to 2020, cumulative reserves secured by that time would have to amount to some 10 MtU.

The third level at which uranium resource information is required is that used by nations or utilities in ordering a reactor power plant of a given type. The degree of certainty in this information can be much less than that required for mine openings, but the prospect of being able to obtain fuel material for a proposed reactor throughout its life must nevertheless be considered. Calculating the total amount of uranium implicitly "committed" on that basis, by reactor orders, is a simple matter: the commitment at any time is obtained by multiplying the number of reactors of each type that have been ordered by the lifetime natural uranium requirement of each type (table 9.A.3). All orders are assumed to be placed ten years ahead of operation. The results for the different scenarios are given in table 9.29. Uranium commitments in 2020 vary from 18.3 MtU in the once-through path to 12.6 MtU in the fast breeder path with uranium recycle. Plutonium recycle in converter reactors has not been included because this use of plutonium ultimately com-

TABLE 9.28. Natural Uranium: Cumulative Reserves Required for Mine Openings (Once-through Path)

		Mine Production (10^3tU/a)		Mine Opening Model (10^3tU/a)			
Year	Period	Level	Average	Average Net Expansion	Average Retirement	Average Total Openings	Cumulative Reserves Required[a] (10^3tU)
1975		15					
	1976–80		24	3	1	4	300 (1973)
1980		30					
	1981–85		39	3	1	4	600 (1978)
1985		45					
	1986–90		54	3	1	4	900 (1983)
1990		60					
	1991–95		84	8	4	12	1,800 (1988)
1995		100					
	1996–2000		127	9	4	13	2,800 (1993)
2000		145					
	2001–5		172	9	4	13	3,900 (1998)
2005		190					
	2006–10		223	11	12	23	5,500 (2003)
2010		245					
	2011–15		281	12	13	25	7,400 (2008)
2015		305					
	2016–20		350	15	13	28	9,500 (2013)
2020		380					

[a] Years in parentheses are the mean dates by which the resource must attain reserve status.

petes with FBR introduction, so that an estimate of its effect on long-term natural uranium commitments would require a very detailed post-2020 analysis. Our calculation of commitments up to 2020 in any case required some assumption concerning reactor deployment from 2020 to 2030, for which purpose a linear extrapolation of the period 2015–20 was used. The actual production at the mine of all the uranium in the commitments as of 2020 would therefore extend to 2058 (allowing a ten-year reactor lead time, a thirty-year reactor life, and a two-year delay between mine production and fuel loading).

The different levels of uranium resources required as time unfolds are shown for the reference once-through path in figure 9.14. The effects of different scenarios on each level of resource requirement are illustrated in figures 9.15, 9.16, and 9.17. It is of interest that the impact of advanced reactor systems is less on uranium produced by 2020 than on uranium commitments inherent in reactor ordering.

Besides giving a global view, our disaggregated model provides further insight into regional and national needs. The regional distribution of natural uranium requirements in 2000 and 2020 is depicted in figure 9.18, where the savings from introducing the FBR and recycling surplus fissile material in thermal reactors are shown as black areas in the regional squares. Not surprisingly, the largest effect is in the industrialized regions. Cumulative

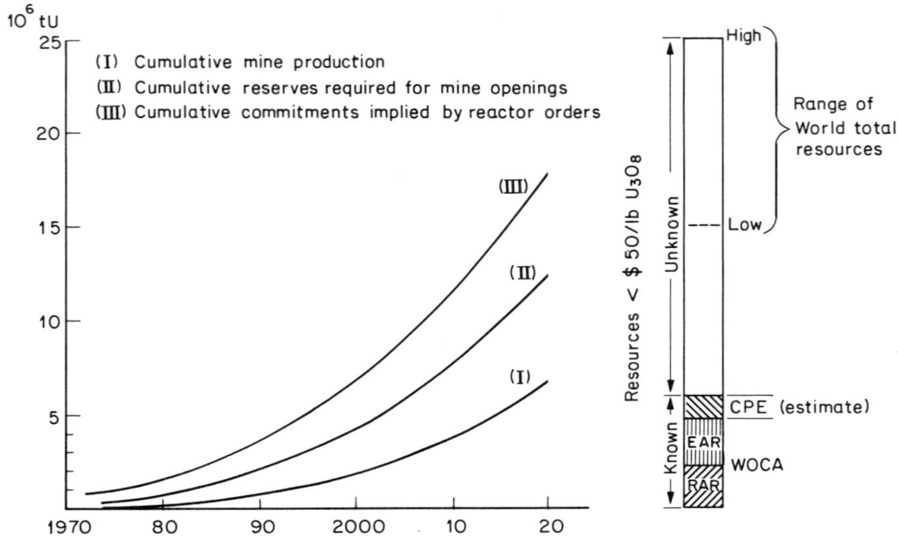

FIGURE 9.14. Natural uranium: cumulative requirements, levels I–III (once-through path).

natural uranium consumption for the different WEC regions is also given in table 9.30, together with NEA/IAEA estimates of uranium resources potential (NEA/IAEA 1978, 1979). The supply-demand comparison at this level of aggregation is of limited use, however, since there are wide disparities within regions (of which the difference between the uranium supply/demand ratios of Australia and Japan within region 3 is an obvious example). It is an individual nation, rather than a region, that normally strives for a supply/demand balance. In any case, traditional commercial ties exist between nations across regional boundaries.

TABLE 9.29. Natural Uranium: Cumulative Commitments Implied by Reactor Orders (MtU)

Year	Once-through Path: Reference	Once-through Path: Improved	Fast Breeder Path: Without U Recycle	Fast Breeder Path: With U Recycle
1975	1.1	1.1	1.1	0.9
1980	1.6	1.6	1.6	1.3
1985	2.2	2.1	2.2	1.8
1990	3.7	3.4	3.6	3.0
1995	5.2	4.7	5.1	4.2
2000	7.0	6.2	6.7	5.6
2005	9.2	7.8	8.4	6.9
2010	11.8	9.7	10.2	8.5
2015	14.6	11.7	12.3	10.2
2020	18.3	14.3	15.1	12.6

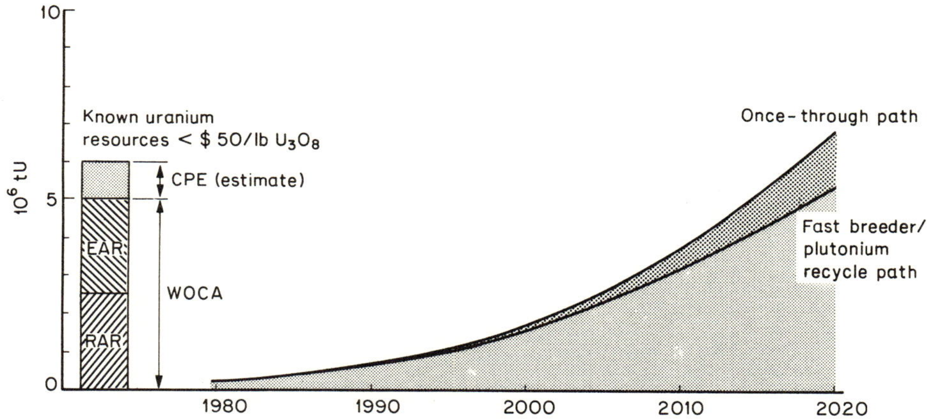

FIGURE 9.15. Natural uranium: cumulative mine production for different scenarios.

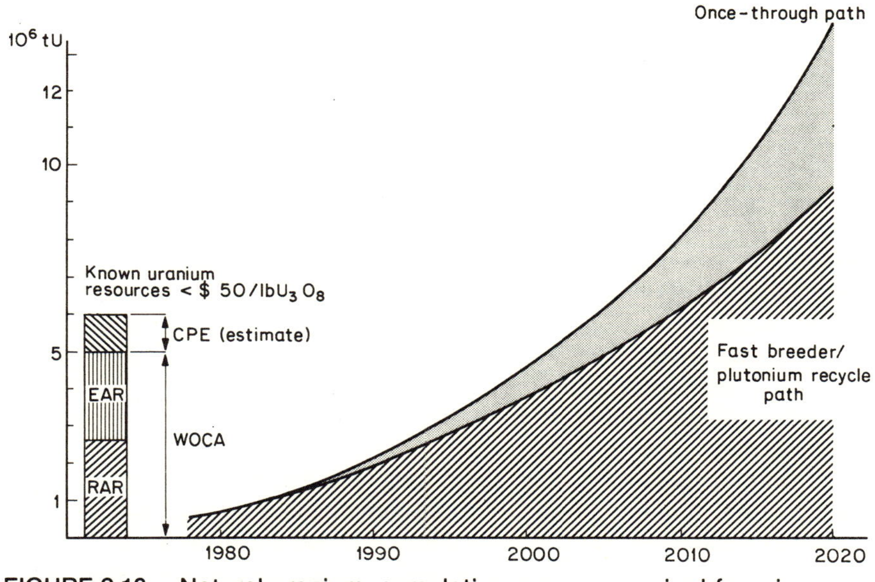

FIGURE 9.16. Natural uranium: cumulative reserves required for mine openings for different scenarios.

Post-2020 Uranium Requirements

A well-known characteristic of studies of energy futures, particularly nuclear energy futures, is the long time required for the differences between alternative policies to reach their potential numerical values. The system responds slowly on the time scale of human plans. Comparing, for example, differences in cumulative uranium mine production to 2020 as between the

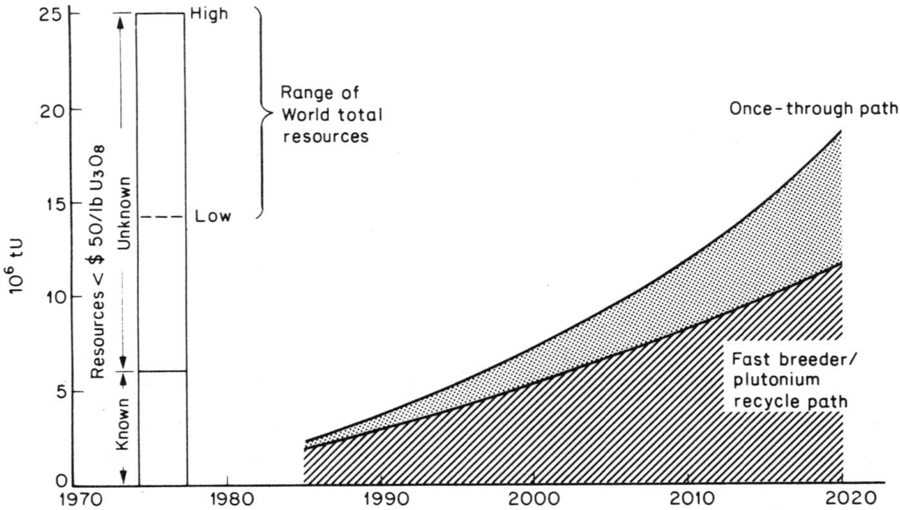

FIGURE 9.17. Natural uranium: cumulative commitments implied by reactor orders for different scenarios.

improved once-through path and the fast breeder path with uranium recycle (table 9.27), one finds only a modest advantage with the latter, considering the greater investments and uncertainties involved. However, the magnitude of the difference between the uranium requirements of the two paths is not properly assessed by the values for 2020; differences beyond 2020 can be much larger. Energy models will indicate the differences if the time horizon is moved to a later date. The difficulty of more distant time horizons, however, is that the sensitivity of results to assumptions increases rapidly, the effect of unforeseen developments becomes potentially dominant, and, in general, the credibility of any causal relationship suggested by the results is greatly diminished.

In spite of these misgivings about long-term projections, we have computed natural uranium consumption and commitments up to 2070 along three possible extensions of our scenarios. The results are shown in figures 9.19 and 9.20. In all cases, world nuclear capacity is assumed to increase linearly from 2020 at a rate of 100 GW/a, reaching 7,550 GW in 2070. The extended once-through path (A) assumes a continuing reduction in the lifetime uranium requirements of new converter reactors (fig. 9.19). The extended fast breeder paths (B, C) assume that the converter reactors in the mix include LWRs in which uranium from spent fuel is recycled, and again that new converter reactors have a progressively decreased lifetime uranium requirement. They also assume that there is no limitation on reprocessing capacity, to which all the world's reactors have unrestricted access. The principal distinguishing variable in the three extended scenarios is the rate of expansion of FBR

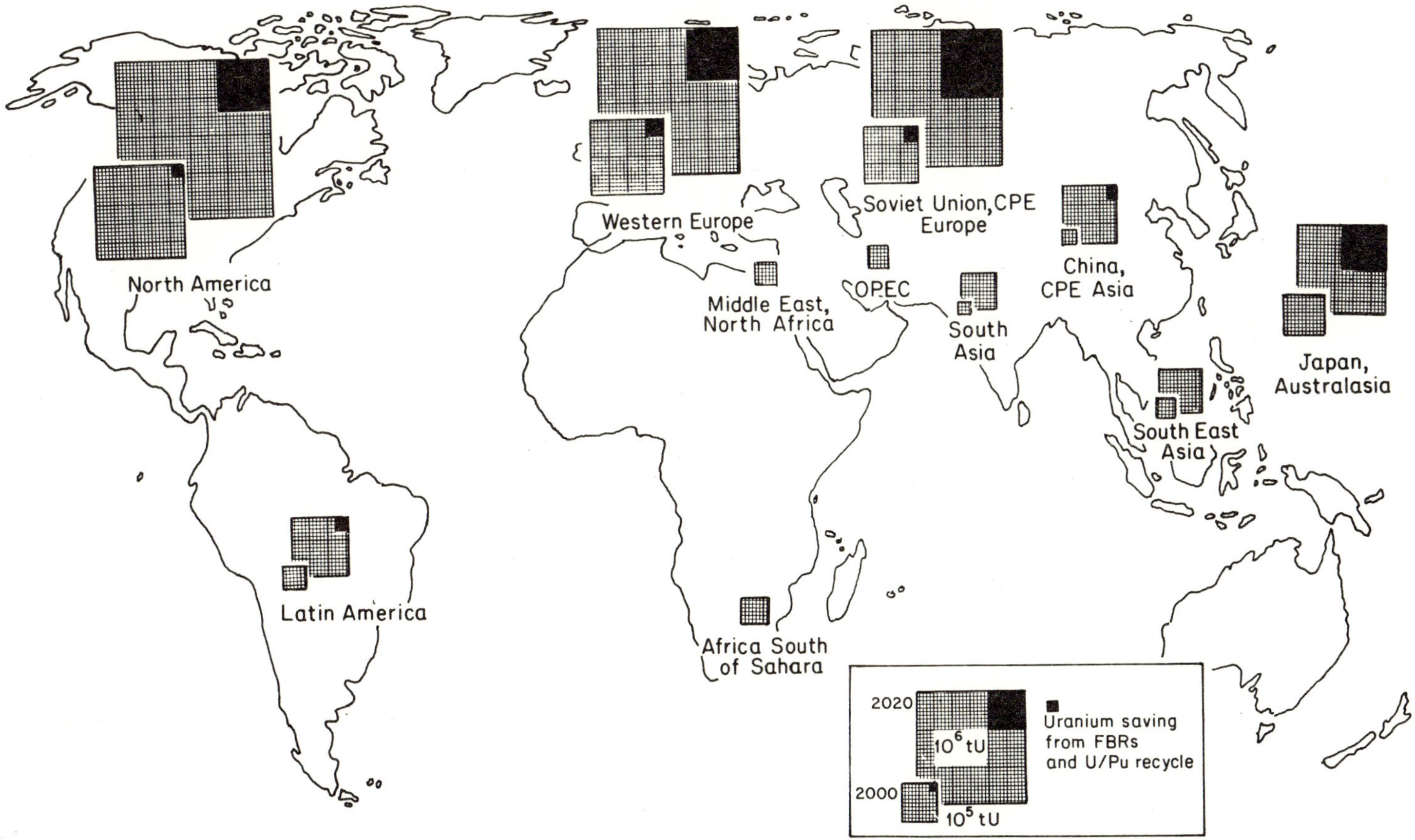

FIGURE 9.18. Natural uranium: cumulative requirements by region in 2000, 2020.

TABLE 9.30. Natural Uranium: Regional Estimates of Cumulative Consumption and Resource Potential (Once-through Path) (MtU)

WEC Region	Cumulative Consumption		Estimate Resource Potential		
	2000	2020	RAR	EAR	Unknown
1. North America	0.56	1.73	0.94	1.85	2.1–3.6
2. Western Europe	0.42	1.51	0.42	0.11	0.3–1.3
3. Japan, Australasia	0.11	0.59	0.31	0.05	2.0–3.0
4. Soviet Union, CPE Europe	0.22	1.32	?	?	3.3–7.3 (4 and 5 combined)
5. China, CPE Asia	0.02	0.24	?	?	
6–11. LDCs	0.10	0.66	0.92	0.45	2.2–6.9

capacity, which is set respectively at 0, 1.15, and 2.3 percent per annum of world installed nuclear capacity. At the beginning of 2021, for example, total installed capacity is 2,650 GW, so that, at the 2.3 percent rate, 61 GW of FBR capacity are added in that year. The highest expansion rate of 2.3 percent per annum reflects the rate of net plutonium production in the FBR and LWR (section 5), as well as FBR inventory requirements, and leads to an FBR share amounting to 82 percent of nuclear capacity by 2070 (as against 44 percent for the 1.15 percent annual expansion rate).

The FBR expansion rates used in calculating figure 9.19 are clearly sensitive to technological improvements. If the FBR is succeeding as well as the extended fast breeder paths imply, there is no reason to think its breeding ratio could not be increased and its doubling time decreased. Consequently, the values of 1.15 and 2.3 percent per annum should not be taken too seriously. However, there is a question as to the range of expansion rates or shares of total capacity that should be used in projecting a long-term fast breeder path, given the prospect that the FBR, even if an economic success in some countries, may not prove suitable for all. This subject is discussed more fully in section 8.

As figure 9.19 illustrates, the arbitrary cutoff at 2070 means that the last reactors notionally built in our extended scenarios would cease operating in 2100, resulting in a fictitious pattern of decline in the model. Nevertheless, the scenarios show that some 2.4×10^{15}kWh of electricity would be produced by nuclear plants between 1980 and 2100. To supply the same amount from conventional generating stations would require about 25,000 EJ of primary energy, which is equivalent to the whole of the world's reserves of fossil fuels (coal, oil, and gas) as classified in 1977 (table 9.6).

Figure 9.20 shows the implications of the three extended paths in figure 9.19 for three different aspects of uranium demand: cumulative commitments, indicating the resources ultimately needed, annual commitments, reflecting the rate at which the known resource base must be expanded, and annual consumption, representing the amount used to fuel reactors in any one year. Thus, cumulative commitments in a particular year include the lifetime requirements of all reactors commissioned by that date plus reactors

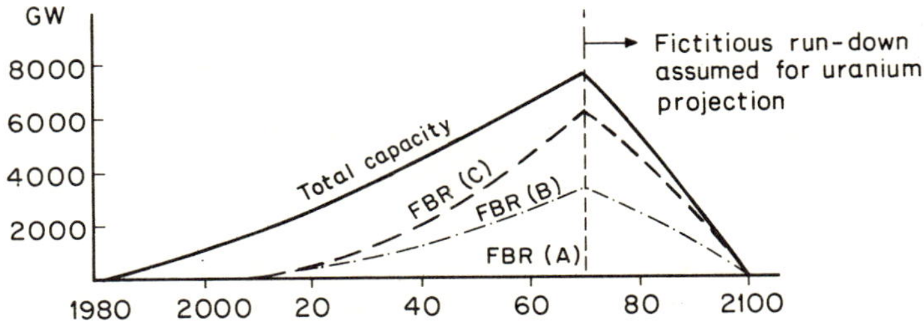

	FBR penetration		Converter reactors	Total	Converter reactor lifetime uranium requirements	
	Growth rate	Capacity 2070	Capacity 2070	Capacity 2070	2070	2070
	%/a	GW	GW	GW	tU/GW	tU/GW
(A) Improved once-through	0	0	7,550	7,550	2,800	1,700
(B) FBRs + converter reactors with uranium recycle	1.15	3,270	4,280	7,550	3,200	1,950
(C) As (B) with high FBR growth	2.30	6,150	1,400	7,550	3,200	1,950

FIGURE 9.19. Extended nuclear generating capacity projections to 2070.

being built or planned to enter service within the following ten years. By 2070 their total reaches a maximum of between 16.5 and 33 MtU, depending on the path followed: a range comparable with the estimate of 15–25 MtU for total world uranium resources discussed earlier. Annual commitments, indicating the pace at which uranium finds must be delineated to justify mine openings

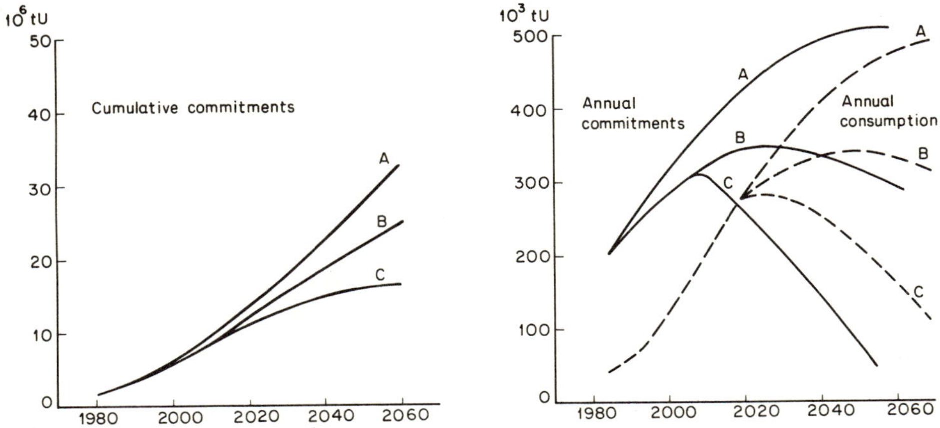

FIGURE 9.20. Extended uranium requirements projections to 2070.

in support of new orders for nuclear plants, follow a different trajectory. In the case of the fast breeder paths, assuming a ten-year lead time for attaining full mine production (as for building reactors), they reach a peak in about 2020. Meanwhile, annual consumption lags roughly twenty-five years behind commitments, reaching a maximum of 275,000–340,000 tU between 2030 and 2050.

The message of figures 9.19 and 9.20 seems to be that a successful FBR could have a profound effect on uranium demand in the twenty-first century, but that this effect is sensitive to the FBR expansion rate and to any limits on FBR penetration. We must, however, leave open the question of the values one should use in making such long-term projections, on either an aggregated or a country-by-country basis.

Uranium Costs and Prices

The price of uranium in the recent past has undergone rather large fluctuations (fig. 9.9). From 1973 to 1978 the price went from $6/lb to $40/lb$U_3O_8$, though it has since fallen back. Over the thirty-year period from 1950 to 1979, however, the price changes (in constant dollars) were less dramatic. In the early 1950s the United States government, in order to provide a sufficient inducement for the mining industry to enter the uranium business, had to set the price at $12.50/lb: equivalent in 1976 dollars to $26/lb. For comparison, production costs per pound of $10.42 for surface mining and $17.65 for deep mining were quoted in 1976 (Gärtner 1976). Production costs depend on ore grade, infrastructure, labor costs, and other variables. At present there is a considerable spread in production costs throughout the world, ranging from below $10/lb to as high as $30/lb for some new mines in North America.

Uranium at a price of $6–8/lb was a passing and short-lived phenomenon that can be explained only as the mining industry's attempt to maintain capacity for future and better days. Recent prices in the range of $30–40/lb certainly appear to allow a reasonable profit margin, and they have proved an incentive to increased exploration. Future uranium price formation is a question of concern to everyone in the nuclear business, as one of the key considerations in planning a nuclear strategy.

Although the success of some recent exploration efforts, combined with the deferral of reactor orders, suggests that there may be some real uranium price decline in the 1980s, we expect that the long-term prospect is for rising prices. As a base, we take a price of $35/lb, which represents a balance between the somewhat higher spot prices and the average price of about $20/lb actually paid for uranium delivered in 1979. The reference prices in 2020 that correspond to different average annual real price increases from that base are then shown in table 9.31.

In figure 9.21 we attempt to form some impression of the long-term development of uranium prices. A detailed explanation will permit the reader

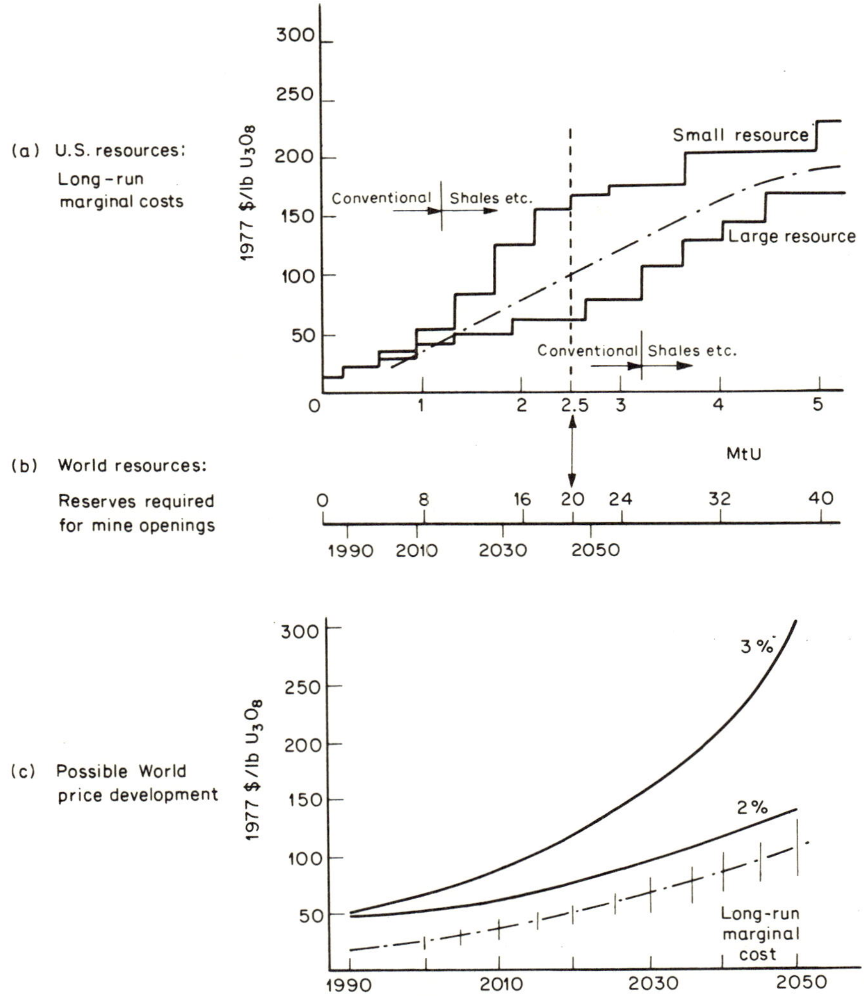

FIGURE 9.21. Natural uranium: model for long-run price formation.

to judge how uncertain the analysis is. We start, in the upper part of figure 9.21, with two estimates of uranium resources in the United States, termed "small" and "large," and the corresponding long-run marginal cost curves (ORNL 1978), between which we draw a single smooth curve. From those estimates we take an intermediate value of 2.5 MtU for "conventional" uranium resources in the United States, as defined earlier in this section, compared with the average value of 20 MtU for world resources extractable at up to $50/lb.* On that basis we assume that a long-run marginal cost curve for

* We are aware of the possible discrepancy between the estimates of United States uranium resources in figures 9.11 and 9.21 and can only draw attention to their different sources (Harris 1979; Voigt 1978; ORNL 1978).

TABLE 9.31. Natural Uranium: Reference Price Projections to 2020 (1979$/lb$U_3O_8$)

	Average Annual Rate of Increase from 1979 (%)					
	0	1	2	3	4	5
Price in 2020	35	55	80	120	175	260

the world has the same shape as that for the United States, but that there are eight times as much uranium at each cost point on the world curve. We next establish a global uranium demand, in the central part of figure 9.21, by equating demand to the cumulative uranium reserves required for mine openings on the once-through path (table 9.28) extrapolated to 2050. Commitments for mine openings exceed actual mine production. We have, however, taken them as a basis in order to reflect the reality that all producers throughout the world will not be operating at the same point on the supply curve, and that those working ores that are higher on the supply curve will tend to be the price setters. Finally, in the lower part of figure 9.21, we use these two previous diagrams to plot a curve of long-run marginal cost versus year for the world. For reference, prices corresponding to average annual increases of 2 and 3 percent from a base of $35/lb in 1979 are also shown.

It appears from this analysis that a real price increase of 2–3 percent per annum should offer a sufficient profit inducement for the necessary investment in exploration and mine openings. The model is subject to continuing revision as uranium resource information unfolds. In addition, it assumes a well-functioning world market, with fair price setting and a minimum of political interference; it would not have worked well if applied to the world petroleum prices of the past decade.

Enrichment Tails and Uranium Utilization

For all reactors requiring uranium enriched in U235, whether reprocessing is involved in the fuel cycle or not, the natural uranium needed, in tons of uranium per gigawatt per annum (tU/GW/a), is a function of the proportion of U235 extracted at the enrichment plant for use in fuel. That proportion varies inversely with the remaining percentage of U235 left unused in the "tails" from the plant, as measured by the "tails assay" at which the plant is operated. The latter value can be changed over time and can differ from one enrichment plant to another. As the tails assay is decreased, the natural uranium required to yield one kilogram of a specified enrichment (e.g., 3.1 percent U235) goes down, but the amount of separative work (i.e., energy) required to effect the separation goes up. Depending on the relative prices of separative work and natural uranium, therefore, the tails assay to achieve a minimum cost of enriched product will vary. However, because reactor and

TABLE 9.32. Enrichment: U235 Extraction and Natural Uranium Requirements

Tails Assay (% U235)	Proportion of U235 Extracted from Natural Uranium (%)				Natural Uranium Input per Ton Enriched Product (tU)			
	U235 Concentration in Enriched Product				U235 Concentration in Enriched Product			
	1.2%	3.0%	20%	93%	1.2%	3.0%	20%	93%
0.00	100.0	100.0	100.0	100.0	1.69	4.22	28.1	131
0.05	97.0	94.5	93.2	93.0	1.74	4.46	30.2	141
0.10	93.7	88.9	86.4	86.0	1.80	4.75	32.6	152
0.15	90.2	83.1	79.5	79.0	1.87	5.08	35.4	156
0.20	86.2	77.0	72.6	72.0	1.96	5.48	38.7	182
0.25	81.9	70.7	65.7	65.0	2.06	5.97	42.8	201
0.30	77.1	64.2	58.7	58.0	2.19	6.57	47.9	226
0.35	71.7	57.5	51.7	51.0	2.35	7.34	54.4	257

TABLE 9.33. Enrichment: Separative Work/Natural Uranium Ratio

	Tails Assay (% U235)						
	0.05	0.10	0.15	0.20	0.25	0.30	0.35
Ratio	10.59	4.17	2.19	1.29	0.80	0.51	0.33

enrichment plant operators may be looking at different sets of costs and prices, they will not necessarily agree on the same optimum tails assay.

The effect of the tails assay on natural uranium requirements is shown in table 9.32. It is there expressed in two complementary ways: as the percentage of the original U235 content of the natural uranium feedstock that is extracted and contained in the enriched product, and as the quantity of natural uranium needed to produce a unit of enriched product. From these values one can, as an example, compute that, if enriched uranium with 3 percent U235 is being produced, it will require 5.48/5.97, or 0.918, times as much natural uranium if the tails assay is 0.2 percent as if it is 0.25 percent.

Natural uranium can be saved by lowering the tails assay, but only at the expense of additional separative work in the enrichment plant. Whether such a saving is rational therefore depends, at any given point, on how much additional separative work must be expended to save how much uranium and on the relative prices of the two commodities. An illustration of all this is given in table 9.33, which expresses the ratio between separative work expended and natural uranium input (in kgU) at various tails assays. Given any two of the three quantities involved—the price (or marginal amount) of natural uranium, the price (or marginal amount) of separative work, and the tails assay—that ratio can be used to establish the rational level of the third. If we want, for example, to save one kilogram of natural uranium by lowering the tails assay from a particular level, the ratio tells us how many additional separative work units (SWU) must be expended. Conversely, for given unit prices of separative work and natural uranium, it tells us the optimum tails assay. In all cases, interestingly enough, the relationship is independent of the U235 concentration in the enriched product. Thus, table 9.33 indicates that, if the existing tails assay is 0.1 percent, an additional 4.17 SWU must be expended to reduce the natural uranium input by 1 kg (by also reducing the tails assay), regardless of the enrichment level required. However, it also indicates that it would only be rational to set the tails assay at 0.1 percent in the first place if the price of natural uranium per kilogram were 4.17 times that of separative work per SWU. Only if the uranium price rose above that level in relation to that of SWU would one want, therefore, to reduce the natural uranium input further. Similarly, if the natural uranium price were $130 kgU (equivalent to $50/lbU_3O_8$), the price of separative work would have to be as low as $31/SWU (= $130/4.17) to justify a tails assay of 0.1 percent. For prices of $130/kg for natural uranium and $100/SWU for enrichment, a tails assay of 0.2 percent is, in fact, about right.

We can look at the meaning of these numbers more specifically in the context of this study. We have used a tails assay of 0.2 percent in computing the natural uranium requirements of reactors requiring enriched uranium. Table 9.14 indicated that, between 1970 and 2000, the enriched uranium fuel charged to the world's reactors would have required some 1.42 MtU of natural uranium as feed for enrichment plants (ignoring the relatively small amount used directly in reactors fueled with natural uranium). Of that amount, by 2000, something like 1 MtU would be left as enrichment plant tails, containing 0.2 percent U235. The question arises whether the fissile U235 potentially still available from this stock of tails constitutes a hedge against rapidly rising natural uranium prices. Do the tails, in effect, increase the supply elasticity?

To obtain a crude answer, let us suppose that 1980 prices are $110/kgU for natural uranium and $85/SWU for separative work. These prices would justify a tails assay of 0.2 percent (table 9.33). Let us further assume that, in the same dollars, the corresponding prices in 2000 are $186/kgU and $85/SWU, which would justify reducing the tails assay to 0.15 percent. The production cost of 3 percent enriched uranium would then be $1,377/kg (5.08 kgU at $186/kg plus 5.08 SWU at $85/SWU). The cost of the same product obtained alternatively by reworking the 1 MtU of enrichment tails would depend somewhat on the cost of getting these tails back into the enrichment plant. If we allow $2/kgU for this, then the 1 MtU of accumulated tails could be stripped to the new tails assay of 0.12 percent to produce about 30,000 t of 3 percent enriched uranium at a total cost of about $1,377/kg. This would save about 150,000 t of natural uranium, or approximately a year's requirement in 2000. However, reworking the tails in this way would require three times more separative work than enriching the same amount of natural uranium. So it would hardly seem practicable to rework much of the tails inventory in a short space of time, given constraints on enrichment capacity.

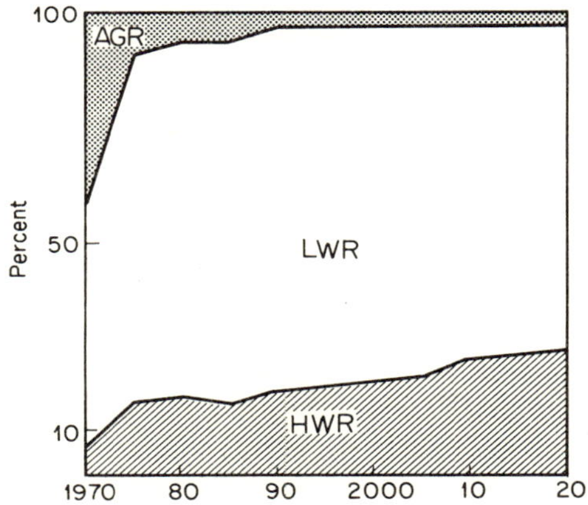

FIGURE 9.22. Fuel fabrication by reactor type (once-through path).

The more likely future seems to be a slow reduction in tails assay over the years, as the real average (not spot) price of natural uranium increases. There may be periods of surplus enrichment capacity when reworking a small portion of the total tails stock will prove economical. Global reworking of the tails does not, however, appear likely.

These remarks are, of course, limited to multistage isotope separation technologies, such as the diffusion and centrifuge plants in operation today. No dramatic decrease in the real costs of these is foreseen. If a laser separation process is successfully developed, it might have lower costs, but it hardly seems warranted to plan on such an outcome at this time. In any case, the increase in the ultimate yield of 3 percent enriched uranium achievable by reworking 0.2 percent tails has an upper limit of 30 percent, regardless of enrichment process development.

We judge the outlook for increasing uranium utilization by lowering enrichment tails assays to be one of slow progress over the years. We believe its cumulative effect up to 2020 will be small enough to be included within the 1 percent per annum reduction in the uranium requirements of new converter reactors that we have assumed for the improved once-through case in section 4.

Fuel Fabrication

The amount of fuel fabricated each year has been given in tables 9.13, 9.18, and 9.22, with reference to the time of loading it into the reactor. A breakdown of the requirement for various reactor types is given, on a percentage basis, in figures 9.22 and 9.23. With a lead time of one year between manufacturing and loading, the required fabrication capacity would have to be on line roughly one year ahead of the dates quoted in the tables referred to above. The timely availability of fuel fabrication capacity seems unlikely to

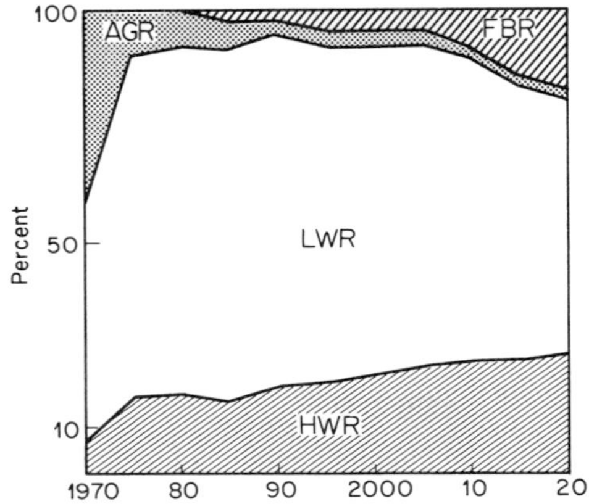

FIGURE 9.23. Fuel fabrication by reactor type (fast breeder paths).

be a problem, not least because the construction and commissioning times for fabrication facilities are shorter than for power plants.

It might be interesting to consider when a smaller nuclear country would plausibly decide to include fuel fabrication in its nuclear program. Fabrication of uranium-based fuel is technically the simplest and least capital-intensive part of the nuclear fuel cycle and is logically the first step toward fuel cycle independence. In this sense it may serve as one indicator of the speed with which a country approaches the capability to implement the more demanding nuclear technologies, such as reprocessing and the refabrication of MOX fuel. Other factors influencing a decision are obviously the size of the national nuclear program and the security of international supplies of fuel cycle services under bilateral or multilateral agreements. The economic scale for a fuel fabrication plant starts in the range of 200–300 tU/a. This would supply something like 10 GW of LWRs. In our scenario, fourteen countries are above this level by 2000 and thirty-seven countries by 2020.

7. SPENT FUEL AND WASTE ACCUMULATION

Waste and Radiation

Like any industrial process, the nuclear fuel cycle produces waste. Nuclear waste is seen by many, however, as a more severe problem than wastes from other industrial processes. It is interesting to note that the concern over nuclear waste has to date focused less on mine tailings than on treated and encapsulated wastes from reprocessing plants, though the former are larger in volume by orders of magnitude and produce a higher radiation dose. Could the reason be that the public is so accustomed to mines and mining operations that the excavation and processing of uranium fall into a category of accepted practice?

The accumulation of tailings at uranium mills is proportional to the quantity of uranium produced at the mines, which again depends on the natural uranium requirements of the various paths. The tailings are also dependent on the ore grade and type of deposit. But on the whole this is a relatively straightforward calculation and a problem we shall dismiss from further discussion.

A subject of great discussion is the storage of spent fuel and the disposal of nuclear waste. The underlying concern is the fear of radioactive exposure and the apprehension arising from the longevity of decaying nuclides. We think that the management of spent fuel and waste can be solved in an environmentally acceptable manner without undue health hazards, though we have no new arguments to support this view. Here our aim is to assess the magnitude of the operation by compiling the spent fuel and waste arisings from the different scenarios we have studied.

Spent Fuel

The cumulative quantities of unreprocessed spent fuel in the world and the fissile plutonium contained therein are given for the various scenarios in tables 9.14, 9.19, and 9.23. Some country-by-country values for cumulative spent fuel are given in Appendix C. The problems of spent fuel storage and disposal are obviously greater in the case of the once-through path. Their nature depends heavily on whether a country adopts the stowaway version of the once-through cycle, in which storage may only be a prelude to eventual reprocessing, or the throwaway version, in which spent fuel is to be disposed of irretrievably.

For the country that has not made the choice between stowaway and throwaway, there are circumstances that will force the issue. One may be political pressure to demonstrate and proceed with nuclear waste disposal. This will encourage disposal of waste in the form of spent fuel simply because it is the shortest route, obviating, as it does, the need for reprocessing. It seems that the disposal method would then have to be designed to make the spent fuel irretrievable. Retrievable storage would entail the absence of one or more barriers between the waste and the environment and would open up the possibility of subsequent diversion. In any case, "retrievable disposal" would only be a semantic substitute for storage.

Acting against the disposal of spent fuel would, of course, be its potential value in a more economic closed cycle, if and when one were developed. Estimates of future uranium prices would also have to be taken into account. It is also maintained that the better path to ultimate disposal of nuclear waste lies through reprocessing. In the meantime, the growing inventory of unreprocessed spent fuel will apply increasing pressure to reach a decision. The severity of this problem can vary widely from country to country; it depends not only on tonnage but on the particular storage procedures that are

TABLE 9.34. Spent Fuel: Distribution by Quantity and Age (Once-through Path)

	Number of Countries				Quantity (10^3tHM)			
	1990	2000	2010	2020	1990	2000	2010	2020
All spent fuel								
<1,000 tHM	18	29	24	20	7	10	10	8
1,000–5,000 tHM	5	16	24	24	10	35	60	62
>5,000 tHM	7	10	16	27	73	210	540	1,160
Total	30	55	64	71	90	255	610	1,230
Spent fuel over ten years old								
<1,000 tHM	20	18	29	24	5	7	10	10
1,000–5,000 tHM	4	5	16	24	7	10	35	60
>5,000 tHM	1	7	10	16	8	73	210	540
Total	25	30	55	64	20	90	255	610

developed. Spent fuel can be stored at the utility site, at a reprocessing plant, at a disposal site, or in some interim storage facility.

Assessment of how much storage would or should be provided at reactor sites has changed considerably in recent years. It used to be standard practice to provide storage space for two to three reloads and a full core loading. However, utility storage pools in the United States are now being retrofitted to raise their holding capacity to as much as ten years of spent fuel discharge. The capital cost of this capacity is reported to be as low as $10/kgU (*Nuclear Industry,* August 1979). It appears to be by far the lowest-cost form of spent fuel storage and is presumably the form that would be adopted in most countries until reprocessing or disposal facilities became available or until reactor-site capacity limits were reached.

Table 9.34 assesses the development of the spent fuel storage problem in the once-through path. It shows the total quantity of spent fuel over time. Spent fuel that is more than ten years old, and presumably straining utility pool capacity, is also shown. The spent fuel is further categorized into quantities of less than 1,000 tHM, 1,000–5,000 tHM, and more than 5,000 tHM. This categorization was chosen because studies have suggested that 1,000–5,000 tHM capacity represents the range in which economies of scale for away-from-reactor (AFR) storage are achieved (IAEA 1977). If this is taken as a criterion, the table indicates, for instance, that thirty countries hold a total of 90,000 t of fuel over ten years old in 2000, of which 83,000 t could economically be placed in AFR storage facilities. The need for AFR storage could, on the other hand, be postponed if spent fuel from the pools of older reactors were transferred to the unfilled pools of newer reactors.

The figures that have so far been reported indicate that AFR storage may be far more expensive than storage at the reactor. The IAEA study of regional fuel cycle centers contains a capital cost estimate of $35–70/kgHM

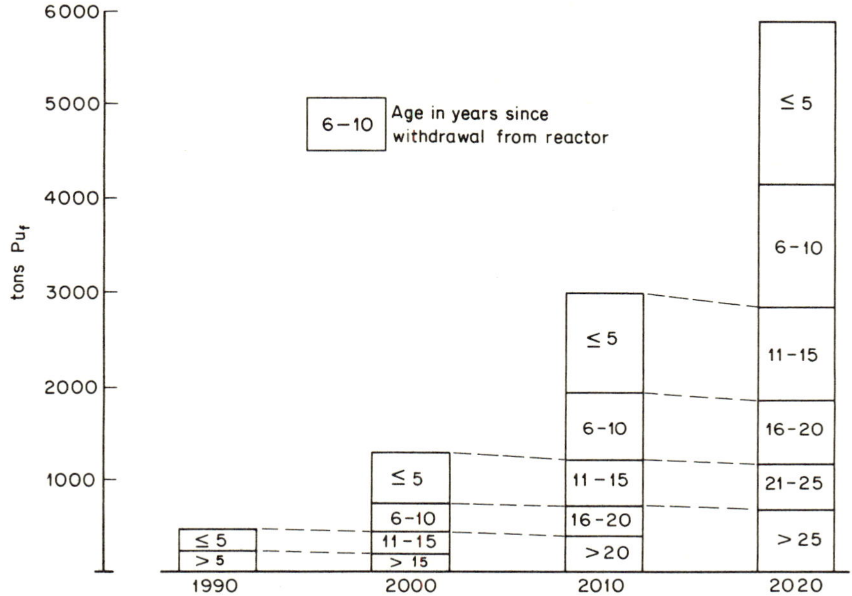

FIGURE 9.24. Spent fuel: accumulation of fissile plutonium by age group (once-through path).

for a 2,000 tHM facility. At the Windscale inquiry in Britain, an even higher capital cost estimate of £150/kgHM (about $300/kgHM) was presented (Parker 1978). Undoubtedly, these figures represent facilities built to different specifications. Nevertheless, they all suggest that storage of spent fuel away from the reactor may be relatively expensive. In addition, there would be costs of transporting spent fuel, in the range of $25–50/kgHM for domestic and $80–120/kgHM for intercontinental transport (IAEA 1977). For a smaller country without plans for reprocessing or a disposal facility, therefore, it may prove to be the economical decision to hold spent fuel indefinitely, stretching reactor-site fuel storage as necessary, rather than to pay for any combination of storage, reprocessing, and disposal in another country. It is possible that, for a long time to come, the annual cost of indefinite reactor-site storage will be less than the interest charges on the bill for any alternative.

This focus on the economics of spent fuel storage should not obscure the fact that other aspects may be attributed overriding importance. The licensing of alternative storage schemes, for instance, or political pressure to demonstrate and execute spent fuel disposal, may lead to different routes in individual countries. It is unlikely, therefore, that the world will see a universal and uniform approach to the spent fuel problem.

Another point that may influence the question of storage, disposal, or reprocessing is the plutonium contained in the spent fuel. On the one hand,

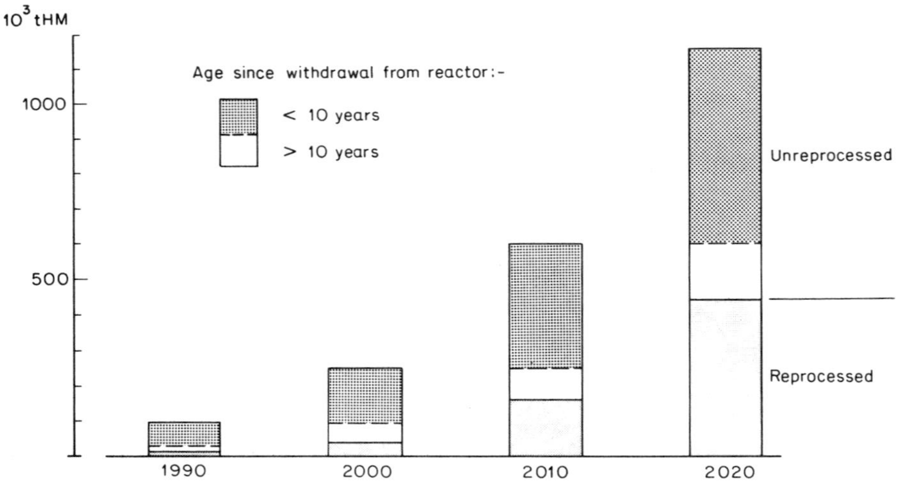

FIGURE 9.25. Spent fuel: accumulation of converter reactor fuel (fast breeder paths).

plutonium is a valuable fissile material on which the fast breeder option depends. On the other hand, it features largely in discussions of environmental and proliferation aspects of the nuclear fuel cycle. The content of fissile plutonium in discharged fuel is shown in figure 9.24 for the once-through path. Values are given as a function of age for several reasons. The age of spent fuel, together with burnup, determines the magnitude and character of radioactivity and its heat generation. These in turn affect transportability, reprocessing, disposal of waste, either as spent fuel itself or as processed waste, and the degree of "protection" to be afforded to contained plutonium.

The numbers in figure 9.24 take into account the decay of Pu241, discussed in section 5; over the whole of the period to 2020, some 500 t of fissile material is thus lost in the radioactive process. However, Pu241 also has another effect on fuel cycle logistics that has not been considered here. Its decay produces an equivalent amount of americium 241 (Am241), a gamma emitter that increases the difficulty of handling separated plutonium. Am241 may have to be chemically separated from plutonium shortly before the latter is incorporated into fuel. This will in turn influence decisions, for instance, on where and in what chemical form separated plutonium is to be stored.

Interim Storage and Reprocessing Waste

In a future with reprocessing of spent fuel, there still exists a need for interim storage. The fast breeder paths we have explored assume that reprocessing facilities are introduced toward the end of the century and then expanded gradually as confidence and needs increase. In the meantime, discharged fuel is accumulating in the world, as shown in figure 9.25. A distinction is again made there between unreprocessed fuel older and younger than ten years. For the former, AFR storage facilities will most probably be required. The circumstances depicted in figure 9.25 suggest that AFR storage in the range of 100,000 tHM of converter reactor fuel will be needed during our study period. Spent fuel at reactor sites grows to several hundred thousand tons, assuming that extended storage space is incorporated into the layout of plants.

The geographic distribution of spent fuel between the various WEC regions in 2000 and 2020 is shown in figure 9.26. The areas of the squares in the figure indicate the relative amounts of spent fuel generated by converter reactors in the once-through path, while the contained black squares show the reduction in unreprocessed inventories achieved by closing the fuel cycle in the fast breeder/plutonium recycle path. Tables 9.20 and 9.24 similarly showed that, in the fast breeder paths, the inventories of unreprocessed spent fuel and their contained plutonium are reduced by 2020 to less than half of their once-through values. This effect is accompanied by the accumulation or disposal of large quantities of reprocessing waste. The volume and distribution of that waste are shown in tables 9.35 and 9.36 and in figure 9.27. Waste is classified in table 9.35 according to radiation intensity (high, medium, and low) and the physical form in which it is fixed for final disposal (e.g., glass or concrete). The classifications and quantities in that table are taken from a study conducted in West Germany and based on LWR spent fuel (DGWK 1977). It is assumed that, in the case of HWRs, AGRs, and FBRs, the waste produced per ton of spent fuel would be proportional to the burnup or fission product concentration. The cumulative quantities in each waste category are shown in table 9.35 and the total volume in cubic meters (m^3) is shown in figure 9.27; the distribution by country is given in table 9.36.

8. THE ECONOMICS OF NUCLEAR POWER

Relative Electricity Costs

It is generally assumed that a technology penetrates the market when an economic advantage arises. The data basis for the cost evaluation involved is, however, very difficult to establish. It differs from country to country and varies with time. Even for such demonstrated technologies as LWRs or coal-fired power plants with flue gas scrubbers, there exists an uncertainty band in cost data. Only after considerable construction and operation experience can a final economic conclusion be reached. For new nuclear systems like the

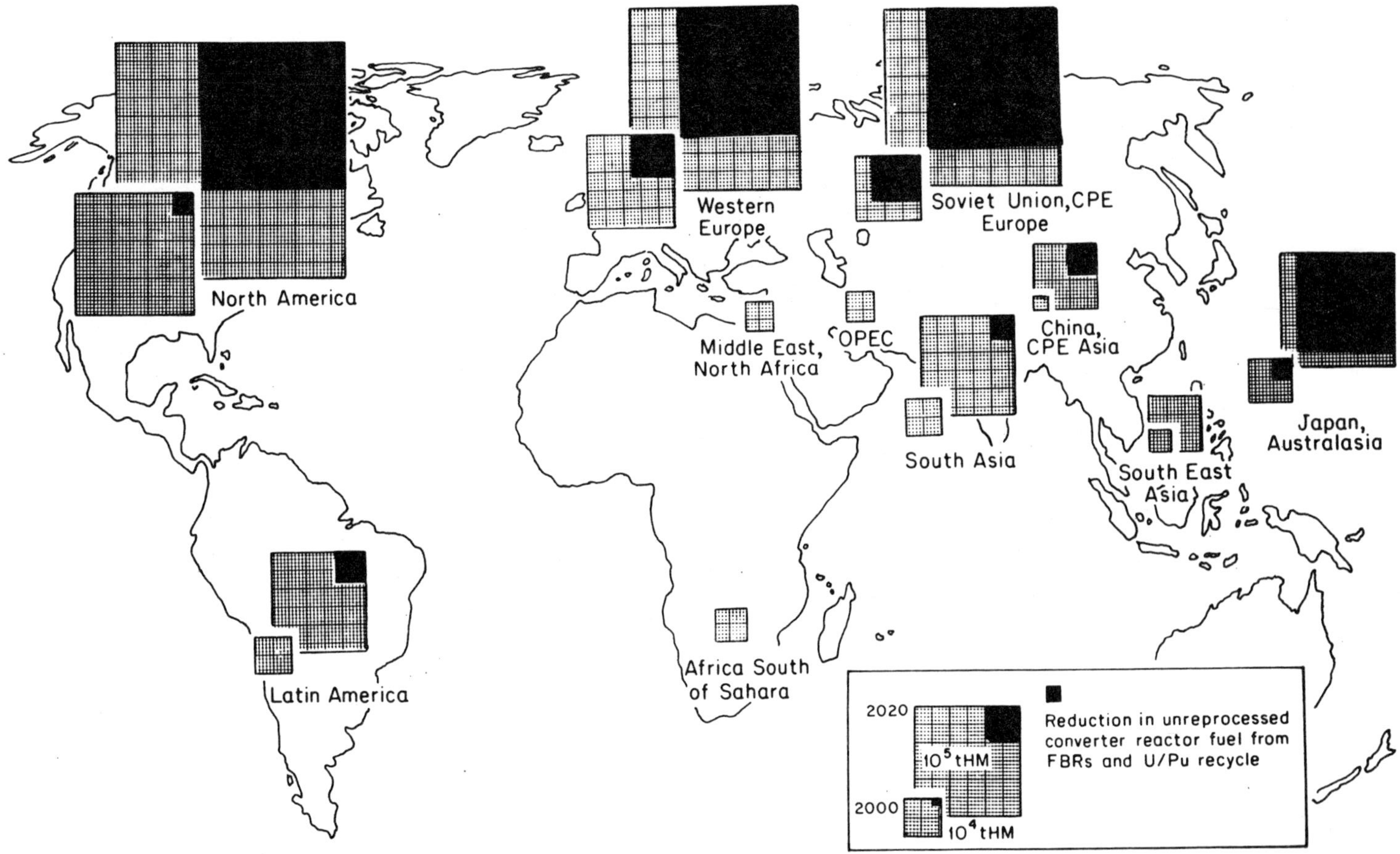

FIGURE 9.26. Spent fuel: accumulation by region in 2000, 2020.

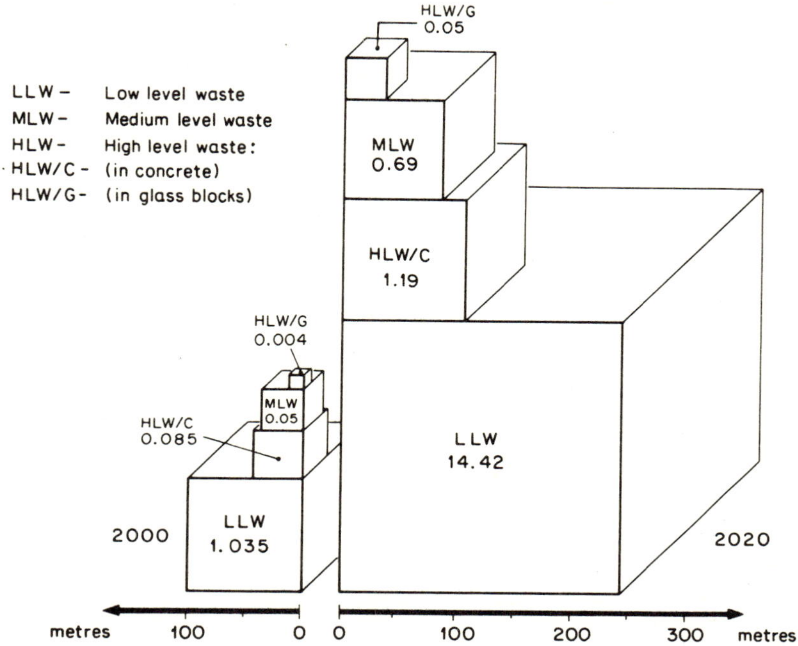

FIGURE 9.27. Radioactive waste: cumulative volume for disposal (10^6m^3).

FBR or advanced converter reactors, the economic merits elude a rigorous treatment. We must resort to more general economic arguments.

Today, nuclear reactors operating in the once-through mode are competitive with fossil-fired power stations. The high capital investment of a nuclear station is viewed as a disincentive. But, in a world of continuous inflation, the high proportion of the generating costs attributable to fixed charges may prove a hedge against future price increases. Figure 9.28 illustrates this point. By comparing an LWR to a coal-electric station, it shows that, if all operating costs, including coal and uranium, increase at the same rate, the economic advantage shifts increasingly to the nuclear station with the passage of time. In short, nuclear power has in most cases proved a sound investment and is likely to continue to be so in the future.

In the long run, there are few options open to electricity generating utilities, and the competitive test for nuclear energy is against coal. The magnitude of the coal industry and the time scale of coal development mean that it is not necessarily an easy option. Nor is it open to all countries. At present one can only say that, within the time frame of this study, coal and nuclear energy will probably share the base-load electric market in most countries. The exact division will depend as much on political as on economic considerations.

TABLE 9.35. Radioactive Waste: Cumulative Quantities for Disposal (Fast Breeder Paths)

Year	Radiation Category[a]	Packing Unit[b]	Number of Packing Units (10^3)				
			LWR	HWR	AGR	FBR	Total
1990	LLLW	400 l drums	160	–	6	15	181
	LLLW(α)	200 l drums	25	–	1	2	28
	LLSW	400 l drums	95	–	3	9	107
	MLLW	200 l drums	25	–	1	2	28
	MLOW	200 l drums	5	–	0.2	0.5	6
	HLLW	70 l blocks	10	–	0.3	0.8	11
	HLSW	2 m³ containers	2	–	0.1	0.2	2
2000	LLLW	400 l drums	725	–	30	85	840
	LLLW(α)	200 l drums	105	–	4	10	119
	LLSW	400 l drums	430	–	20	50	500
	MLLW	200 l drums	120	–	5	15	140
	MLOW	200 l drums	20	–	1	2	23
	HLLW	70 l blocks	40	–	2	4	46
	HLSW	2 m³ containers	10	–	0.5	1	12
2010	LLLW	400 l drums	2,935	45	100	485	3,565
	LLLW(α)	200 l drums	420	6	15	70	511
	LLSW	400 l drums	1,750	25	60	290	2,125
	MLLW	200 l drums	490	7	15	80	592
	MLOW	200 l drums	90	1	3	15	109
	HLLW	70 l blocks	155	2	5	25	187
	HLSW	2 m³ containers	45	1	1	7	54
2020	LLLW	400 l drums	8,360	210	270	2,850	11,690
	LLLW(α)	200 l drums	1,200	30	40	410	1,680
	LLSW	400 l drums	4,985	175	160	1,700	7,020
	MLLW	200 l drums	1,390	35	45	475	1,945
	MLOW	200 l drums	250	7	8	85	350
	HLLW	70 l blocks	445	10	15	150	620
	HLSW	2 m³ containers	125	3	4	45	177

Source: DGWK 1977.

[a] LLLW, LLLW(α): low-level liquid waste (α: alpha-emitting) in concrete
LLSW : low-level solid waste in concrete
MLLW : medium-level liquid waste in bitumen
MLOW : medium-level organic waste in PVC
HLLW : high-level liquid waste in glass blocks
HLSW : high-level solid waste in concrete.

[b] l: liter; m³: cubic meter.

Economics of the Fast Breeder Reactor

The fast breeder reactor introduces another dimension into the problem because, unlike converter reactors, it is not a commercially developed system. The path to a breeder reactor future is envisaged in section 5 as a two-stage process of demonstration/commercialization followed by deployment. It is our judgment that the global deployment rate used here (table 9.16) is as rapid as we can anticipate from a technical standpoint. Although the effects of breeder reactor penetration on global uranium requirements to 2020 are not dramatic, this deployment rate, if continued beyond 2020, would reduce

TABLE 9.36. Radioactive Waste: Distribution from Reprocessing (Fast Breeder Paths)

Country	Share of Total (%)			
	1990	2000	2010	2020
France	49	33	18	11
Soviet Union	49	43	36	27
Britain	2	2	1	1
West Germany	–	–	5	6
Japan	–	11	10	14
United States	–	11	18	24
Canada	–	–	4	5
Brazil, China, East Germany, Italy, Spain	–	–	8	9
Czechoslovakia, India, Poland, Sweden	–	–	–	3
Total	100	100	100	100

uranium requirements to levels that today appear reasonable in terms of uranium resource information. Moreover, the effect on uranium requirements for the nation that successfully deploys the breeder reactor is much more pronounced than the global effect.

There are serious questions about the economics of the breeder reactor, its required scale of operation, and the details of introducing it into any country. It is implicit in our fast breeder reactor deployment scenarios that the FBR becomes economically competitive in each nation concerned on the basis of marginal economics, including allowances made for security of supply, future uranium prices, and so on. It is by no means a foregone conclusion that these conditions will be satisfied. Cohen, for example, has discussed the problem of high capital cost as an obstacle to market penetration by the FBR (Cohen 1978).

Reduced to its simplest terms, the breeder reactor offers the promise of eliminating the natural uranium costs of nuclear energy. In solely microeconomic terms, this saving has to be matched against any additional capital cost of a breeder as against a converter reactor, as well as against FBR development and commercialization costs.

The large range of current estimates of fuel cycle unit costs makes comparisons of capital and fuel costs highly uncertain. However, we believe that the fuel cycle cost advantage of the FBR over the LWR at equilibrium is not likely to exceed the LWR cost of natural uranium. This assumes that any amount by which fuel fabrication and reprocessing costs for the FBR exceed those for the LWR will not be greater than LWR fuel enrichment costs. The

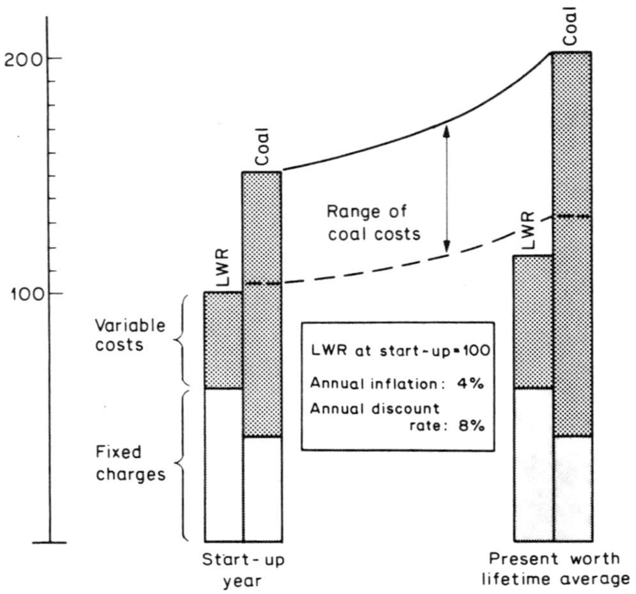

FIGURE 9.28. Effect of inflation on electricity generating cost in LWR and coal-fired plants.

difference in fuel cycle costs comes down, therefore, to the cost of the natural uranium that the LWR needs but the FBR does not. On the basis of this simplification, we can say something about the relationship between converter reactor fuel costs and higher FBR capital costs. We assume a converter reactor that requires 3,000 t of natural uranium for thirty gigawatt years at a 65 percent capacity factor, or an average of 100 t of natural uranium a year. (This represents a 25 percent reduction from the once-through PWR in table 9.A.3, to allow for improved efficiency.) If we take the price of uranium at some future time to be \$50/lbU_3O_8 (\$130/kgU), the "present worth" of that uranium over thirty years is given in table 9.37.

Although the capital cost of an LWR has its uncertainties, a reasonable working number in the time frame of interest is \$1,000 (in 1979 dollars) for each kilowatt (kW) of capacity: that is, \$1 billion for a 1 GW reactor. The question is therefore whether the capital cost of building an FBR will eventually exceed that level by an amount that falls inside or outside the limits of the lifetime LWR uranium costs indicated by table 9.37 (or by any other expected combination of discount rate and uranium price development). That question is, of course, the subject of much discussion and disagreement. On this limited basis alone, however, table 9.37 suggests that, if the real annual discount rate were, for instance, 5 percent and the real uranium price were to rise at 2 percent a year, the capital cost of an FBR would have to be not more than \$1,265/kW to compete with an LWR at \$1,000/kW.

TABLE 9.37. Present Worth of Uranium at \$130/kgU over Thirty Years (\$/kW)

Real Uranium Price Escalation	Real Discount Rate	
	0%/a	5%/a
0%/a	390	210
2%/a	527	265
4%/a	729	341

Note: We assume reactor uranium needs of 0.1 kgU/kW/a.

In addition to considerations of incremental capital cost, there is the question of development and commercialization costs. In the construction of the fast breeder paths in section 5, we postulated an extended period during which full-scale demonstration plants were built and operated. Presumably, the net cost of development should be seen as the amount by which the expenses associated with demonstration reactors and their associated fuel cycle facilities exceed the cost of alternative commercial electricity generating capacity. From an inspection of table 9.16, we judge that, for the leading FBR nations, an investment with a present value at the start of the program of \$5 billion is a reasonable estimate of that cost, excluding development costs already sunk. Let us then define a successful program as one that results in the deployment of 100 GW of FBR capacity in the period from twenty to fifty years after the start of demonstration, beginning with 1.5 GW in year 21 and reaching 100 GW in year 50 (an average 15 percent per annum growth). We must therefore ask what levelized capital cost (in constant dollars) has to be assessed against these 100 GW to match the \$5 billion present worth of the investment in development. The answer is \$50/kW and \$390/kW for a 0 percent and 5 percent annual discount rate respectively. These values do not include any allowance for the risk that the development program will not lead to deployment. They are inversely proportional, moreover, to the amount of FBR capacity included in the assessment, which points to the importance of the scale of the future market, as determined by factors including international licensing. In general, however, it is fair to conclude that FBR commercialization costs represent a significant addition to future capital costs when assessed in this way.

One item of the initial cost of the FBR that deserves special mention because of its size is the cost attributable to reprocessing the converter reactor fuel necessary to provide an FBR with its initial plutonium inventory (6,100 $kgPu_f$/GW in our calculations). This quantity requires the reprocessing of 1,000 t of LWR spent fuel. At a unit reprocessing cost (including waste disposal) of \$600/kgHM, the total is \$600 million. If one allows a \$200/kgHM credit for disposal of the reprocessing waste, which must be done in any case, and a \$100/kgHM credit for the uranium recovered, there is still a cost of \$300/kW to be charged against the startup of an FBR. We do not attach high credence to these unit costs (which are discussed in a little more detail below), but we do note that they represent a cash outlay before reactor operation that is significant compared with the capital cost estimates for an FBR power plant itself.

Our rather simple analysis suggests that a breeder reactor development program cannot conclusively be justified on the basis of microeconomics and marginal cost comparisons alone. Not only must macroeconomic advantages be foreseen, but other incentives or values must also play a role. Classical economic considerations represent only one level in a hierarchy of criteria upon which a country bases decisions. Other motives of a general political nature include national political independence, weapons proliferation, stability in the supply of energy and other vital commodities, and environmental issues. Some of these considerations may well be given more weight than purely microeconomic evaluations. They can lead to decisions both for and against nuclear power, but some of them must operate positively to bring about a vigorous, sustained breeder reactor development program. Whether developments based on noneconomic incentives can push a technology over the commercialization threshold, or can only create a temporary distortion of market forces, is an open question.

Certainly the FBR represents a commercialization problem quantitatively different from the converter reactor. It is not just a nuclear steam supply system; it is an entire fuel cycle system. Unlike the situation with the AGR, HWR, or LWR, it is not feasible to start up fast breeder reactors and then decide whether the country will or will not reprocess the spent fuel or recycle the recovered products. These steps are a concurrent necessity if the breeder reactor is to reach a significant scale of deployment. There is thus a certain "all-or-nothing" character to FBR success. Expressed in another way, the FBR has a high and a steep commercialization threshold. Because of economies of scale, economic operation is unlikely to be achieved before a substantial number of breeder reactors are deployed.

Once commercial deployment had started, there would be strong pressure to move as rapidly as possible. Deployment would almost certainly have to be a nationally orchestrated operation. For example, if the FBR commercialization program went well from a national standpoint, breeder reactor deployment would have to be mandated nationally, rather than left to the decisions of individual utilities or states. Otherwise, the economically necessary penetration rate and scale of operations could probably not be assured. For smaller countries, therefore, the breeder reactor may never prove a suitable system, for reasons having nothing to do with technological development. For developing countries, the diffusion of breeder reactor technology has its obvious difficulties.

In summary, the development and deployment of the breeder reactor on the part of any nation will require a high level of political determination and perseverance over an extended period of time. The only country in the Western world that gives convincing evidence of possessing that determination at present is France.

TABLE 9.38. Costs of Fresh UOX and MOX Fuel for LWRs

Material/Service	Quantity (kg)	Unit Cost ($)	Cost Range ($)
1 kg UOX (3.1% U235)			
Natural uranium	5.68	50–300	284–1,704
Enrichment (SWU)	4.53	50–200	226–906
Fuel fabrication	1.00	100–300	100–300
Subtotal			601–2,910
Storage/disposal of spent UOX	4.95	50–400	248–1,980
Total			858–4,890
1 kg MOX (3.0% Pu_f)			
UOX reprocessing and waste disposal	4.95	200–800	990–3,960
Fuel fabrication	1.00	200–800	200–800
(Credit for recovered uranium)	3.34	0–300	(1,002–0)
Total			188–4,760

Recycle of Plutonium in LWRs

In recent years there have been large increases in estimates of fuel cycle unit costs, most notably for spent fuel reprocessing. There are, moreover, some rather large variations in cost estimates from different sources (Siegel 1979). These changes and variations have raised questions about the economics of plutonium recycle in LWRs. To illustrate the point, we have shown in table 9.38 a range of the unit costs with which an operator contemplating the use of fresh UOX fuel in an LWR might be faced, and a comparison with the cost of MOX fuel derived from spent UOX fuel and designed to give the same burnup (MWd/kgHM) in the same reactor (table 9.A.2). To obtain the fissile plutonium to produce 1 kg of MOX fuel, 4.95 kg of UOX spent fuel must be reprocessed. (No allowance is made for the decay of Pu241, discussed in section 5.) At the same time, the equivalent of 4.29 kg of natural uranium is recovered, of which 0.95 kg is used in the MOX fuel, while the balance of 3.34 kg is credited as natural uranium, assuming it will be used elsewhere as fuel material. As to the alternative of using fresh UOX, without reprocessing, the spent fuel must presumably be stored indefinitely or disposed of, so that it seems reasonable, in this comparison, to assess the associated cost against the UOX fuel.

The ranges of unit costs in table 9.38 represent a judgment of the bounds within which a current worldwide set of estimates might fall. The grouping of all the minima and all the maxima is not meant to imply that they would in fact occur simultaneously. Also, a proper evaluation should be made over reactor and fuel cycle facility lifetimes, with judgments being made for changes in the various unit costs over time, correct discounting rules being applied, and so on. Although the values in table 9.38 show that a cost comparison of UOX and MOX fuel for LWRs could favor either one, more recent estimates from Western industrialized nations are not favorable to MOX. An IAEA study of regional fuel cycle centers gives an estimate of $420/kgHM for reprocessing, including waste disposal, and $440/kgHM for MOX fabrication, based on a 1,500 tHM/a reprocessing plant and a 25 percent per annum fixed charge rate (IAEA 1977). The report of the 1977 Windscale inquiry gives £260/kgHM (about $520/kgHM) as the cost of reprocessing spent UOX fuel (Parker 1978). Press reports in 1979 have quoted charges of over $700/kgHM for spent fuel reprocessing at Cap de la Hague in France, not including waste disposal (Siegel 1979).

That reprocessing contracts have been made and executed in the $25–50/kgHM range, and that such prices were commonly quoted up to the early 1970s, thus gives a good measure of the dramatic cost increases in this area. The reasons most obviously relate to more stringent standards on plant effluents, waste disposal, materials safeguards, and such. Some of the increase may also represent a large contingency allowance against the many uncertainties that lie in the path of final execution of all the operations involved.

At present there are large differences between the fuel cycle unit costs estimated in different countries. Whether these differences are due to different standards, different financing methods, different contingency allowances, or other factors, we cannot say. However, they suggest that the economics of plutonium recycle might be viewed quite differently from country to country.

There are, of course, other considerations that influence a plutonium recycle decision. For instance, some countries have decided that spent UOX must be reprocessed in order to solve the waste disposal problem. In that case the marginal economics of plutonium recycle are clearly favorable. A second circumstance influencing the economics of plutonium recycle in LWRs is one in which a nation decides as a matter of policy to develop the fast breeder system. In that case much of the infrastructure required for plutonium recycle in LWRs, most importantly LWR spent fuel reprocessing and MOX fuel fabrication, must be developed. The introduction of plutonium recycle in thermal reactors (either in the reprocessing country or elsewhere), as an adjunct activity or an interim step in the FBR program, could then become attractive.

The Case for the Advanced Converter Reactor

The relatively low uranium utilization of once-through converter reactors and the potential difficulties of the FBR lead naturally to the question of alternative possibilities. Among these is the improvement of HWR or LWR systems operating in the once-through mode. Here, recent studies seem to indicate that a reduction of between 15 and 40 percent in uranium requirements, relative to current LWRs operated in a once-through mode, would be the best that could be achieved. The operation of HWRs on low-enriched (rather than natural) uranium appears to be the most direct path to this goal (table 9.A.3). For some nations such an improvement may be entirely sufficient, in the light of their individual uranium supply and requirements prospects. For others, however, something intermediate between the breeder reactor and once-through thermal reactors appears to be a real need. Meeting this need would require a closed cycle, which is to say reprocessing and the recycle of recovered fuel materials.

If a closed cycle is accepted as necessary, then the most obvious candidate is the LWR operating in the plutonium recycle mode. We say "obvious" because of the widespread deployment of LWRs and the extensive commitment to further deployment during the rest of this century. Furthermore, the LWR fueled with MOX fuel differs slightly, if at all, from the LWR using only UOX. This means that the reactor system itself can be regarded as developed. The single difficulty is that the LWR of current design, operating with uranium and plutonium recycle, offers rather little in the way of uranium utilization improvement. It appears unlikely that it can achieve better than a 30 to 50 percent reduction from the present once-through LWR. Nor does the introduction of a thorium-based fuel cycle in this reactor type achieve any significant improvement (Matzie and Rec 1976). Our evaluation of the potential uranium efficiency obtainable in LWR systems has not, however, been exhaustive. The realistic extent to which improvements can be made in LWRs operating in a closed cycle is a question that merits serious analysis.

The next question is, What are the alternatives? Let us first enumerate the desirable features of an alternative system, recalling the difficulties with the LWR and FBR and presuming that a closed fuel cycle is acceptable.

1. The reactor in question should be currently developed and licensable. Conditions do not appear favorable at this time for embarking upon an entirely new development.
2. The reactor fuel cycle system should have the potential to achieve a substantial reduction in natural uranium requirements from those of the current LWR. Perhaps one can be no more definite than to require that a factor of ten reduction, to 400 t of natural uranium for a lifetime of thirty gigawatt years (table 9.A.3), should be possible.
3. Although appropriate to a closed cycle, the reactor should be able to operate economically in the once-through mode with today's uranium prices and other fuel cycle costs (albeit with a higher uranium requirement). This flexibility is desirable so that the system can take advantage of relatively low-cost uranium as long as it lasts and so that it can operate without excessive economic penalty should reprocessing services not always be available.

4. The reactor should have sufficient flexibility in its fueling to permit changing the fuel composition (e.g., from enriched uranium to Pu/U or Pu/Th), preferably during a given reactor's lifetime but certainly from one reactor to the next. Such flexibility would afford protection against breakdowns in the reprocessing system as well as against rapid increases in uranium prices.
5. The reactor should be able to use the plutonium contained in the spent fuel of HWRs and LWRs efficiently. The importance of this feature would depend on the extent to which FBRs could otherwise consume the plutonium.

Present LWR concepts with Pu/U recycle fail on the point of uranium utilization. The thorium cycle in combination with other types of thermal reactors gives more promise in this respect. Reference to table 9.A.3 shows that the high temperature reactor (HTR) has been calculated to achieve a net uranium consumption of 690 tU over thirty gigawatt years when operated in a closed cycle. This represents a factor of six improvement over current LWRs. The HWR has also been shown to achieve similar uranium consumption values when operated on a closed thorium cycle (Foster and Critoph 1975). Moreover, both the HTR and the HWR systems appear to have the potential to reduce specific uranium consumption still further, were future uranium prices to advance more rapidly than other fuel cycle costs.

The use of thorium in a thermal neutron spectrum reactor has the advantage of a good neutron economy, which leads to low uranium consumption. Thorium's abundance in the earth's crust is twice that of uranium. It has not, however, been the object of as much exploration. The broadening of the nuclear resource base to include thorium alleviates the uranium supply problem, but the potential resource limitation would still be natural uranium, as the thorium requirements of a converter reactor amount to only fractions of its uranium requirements.

This study has not included an advanced converter reactor scenario, partly for reasons of time limitation, but also because the most rapid deployment that can reasonably be assigned at present to an HTR or HWR with a closed thorium cycle is far too slow to have significant effects on global uranium requirements before 2020, and probably until much beyond. Even at the national level, the short-term effects of operating such a closed fuel cycle would hardly be noticeable, though the operation of an HTR or HWR in the once-through mode could affect some uranium savings. In the post-2020 period, however, national dedication to a thorium strategy could bring uranium savings in the range of those achievable by FBR deployment (Engelmann et al. 1979).

In most of the industrialized countries, the FBR is the advanced system to which the heaviest commitment is made. In some countries, any large increase in advanced converter reactor development could probably be seen as detracting from FBR development. From the global standpoint, as from that of some individual countries, there can be no doubt that successful breeder reactor deployment is the path with long-term minimum uranium resource requirements. For other countries, in their individual circumstances, the position is less clear. There, the advanced converter reactor may be more accessible and flexible and offer a quicker return. In an ideal world, one would wish, of course, to see some division of effort in order to make complementary options available. Cohen has pointed out that advanced converter reactors may offer no advantage in the trade-off between capital costs and uranium savings (Cohen 1978). The possibly telling difference, however, is that development of the HTR or HWR does not have quite the same "all-or-nothing" character as does that of the FBR. The movement toward more efficient fuel systems can take place in stages or smaller increments, and incremental development can proceed in parallel with deployment of earlier systems. The commercialization threshold may be just as high, but it may not be quite as steep, as in the case of the FBR. It may be that these hypothetical advantages of an advanced converter reactor system are more illusory than real. We suggest, however, that they deserve more careful analysis, with due attention to the long-term implications of introducing new techniques (e.g., thorium/uranium reprocessing) and to the problems of individual nations.

Systems have also been proposed in which fast breeder reactors and efficient converter reactors operate in a cooperative or symbiotic union (Häfele and Manne 1974; Melese-d'Hospital and Simnad 1977). These proposals envisage fast breeder reactors fissioning plutonium, the fuel they use most efficiently, while producing uranium 233 from thorium, which is transferred to those thermal converter reactors in which U233 is the most efficient fuel. Different countries might operate different parts of the total system, depending on individual needs and circumstances. In the perspective of this study, however, the world is a long way from being able to implement such a complex system.

9. CLOSING REMARKS

In the course of this study, a number of ideas and tentative judgments have germinated, without fully developing into carefully formulated conclusions. We nevertheless take this opportunity to present them. In so doing, we start from the position that nuclear fission technology has the potential to provide a good solution for some portion of the future energy supply problem in many countries. Questions of what schedule, what energy share, which energy forms, which countries, and so on, we regard as open. We recognize that many people hold the view that nuclear energy should either be phased out with all deliberate speed or sharply curtailed, whether quantitatively or qualitatively. To them we leave the task of analyzing non-nuclear futures, which we suspect will not be free from very difficult problems. For us the outstanding questions converge on a single issue: Is nuclear energy to develop in an orderly manner to maturity, or is it to pass away in early childhood or, perhaps, go through life as a malformed, poorly functioning adolescent, neither fully accepted nor fully rejected by society?

The disaggregated study presented here offers information and insights not revealed in reactor strategy studies for the aggregated world or large subregions of it. It serves to make explicit the fact that different countries face different situations, or similar situations at different times. It nevertheless suggests that, globally speaking, nuclear fission, in spite of present difficulties and delays, is technically and economically capable of being expanded by 2020 to supply the primary energy equivalent of twice today's output of coal. However, it is unlikely to contribute the share of electric energy projected for it in the WEC L4 scenario. In the years after 2000, the absolute magnitude of nuclear investment could be larger if economic and electricity growth is higher than in the L4 scenario or if there is an aggressive program to obtain non-electric energy from fission reactors. In any case, some national programs may develop well in excess of the international or regional average. However, the introduction of the breeder reactor or of any advanced, uranium-efficient fuel cycle will take place only slowly, and its global effects up to 2020 will at best be modest. Consequently, the medium-term prospect for such components of the nuclear industry as uranium mines and enrichment plants will be much more affected by the levels of nuclear generation reached than by the choice between fuel cycle strategies.

A very serious obstacle in the path of any nuclear future is the prospect that the number of reactor orders in the coming decade may be too low to keep the infrastructure intact, particularly in the Western industrialized nations. By infrastructure, we mean the entire fabric of people, skills, and equipment that permits the nuclear industry to respond to a request to supply a power reactor. If that infrastructure cannot be maintained, then the resulting decrease from the nuclear energy supply in our reference scenario could be much greater than that represented by a simple shift of the capacity values five or ten years back in time. One can even ask seriously whether the political will to rebuild a nuclear industry could then be developed.

Our assumption that to be economically viable nuclear power requires a large electrical grid and a high level of technical skill results in nuclear reactors being concentrated in our scenarios mainly in the industrialized countries. In fact, of the fission energy deployment in 1990, almost 95 percent is situated in WEC regions 1–4—that is, in the OECD and Comecon countries—and even at the end of study period this figure is still 80 percent. We regard this finding, though not new in itself, as a significant aspect of a world nuclear future, bearing on almost all the questions that arise in that context. Apart from the obvious observation that the energy economy of developing countries will depend to a greater extent on the future availability and cost of fossil fuels, their nuclear sectors will also be strongly dependent on developments in the more experienced nuclear countries. Improved reactor types as well as more resource-efficient fuel cycles will be needed to limit the world's future demand for uranium. The required technologies will inevitably be even more demanding of technical and managerial skills than those employed today. At present these skills are more likely to be found in the industrialized countries, and, consequently, advanced reactor systems will probably first be introduced in those countries. Moreover, because the major part of nuclear capacity is concentrated in that area, introducing improved fuel cycles there will have a higher payoff, in terms of restricting global uranium consumption, than if similar measures were required in a large number of small countries.

We have passed through a period in which some projections of fuel cycle unit costs appear to have been too low. One can argue that there were good reasons for these low estimates (e.g., unforeseen changes in environmental regulations). Nevertheless, they did lead to unrealized expectations. At present the opposite problem may exist: namely, that wealthier countries, by imposing unreasonable standards of performance on various fuel cycle operations such as spent fuel storage or reprocessing, will force up costs, thus discouraging poorer countries from taking part in international programs and persuading them either to abandon nuclear energy or to build and operate less elaborate facilities on their own. One likely result of the higher costs we foresee in this study is that many small countries will find it economically advantageous to store spent fuel at their own reactor sites almost indefinitely, rather than to ship it to another nation or a multinational facility for storage, disposal, or reprocessing.

The particular scenarios described here all require that a major uranium mining industry be developed steadily over the next forty years. If uranium is used at the efficiency represented by thermal converter reactors operating once through, it follows that, as a global energy resource, uranium is of the same order of magnitude as petroleum. There are, of course, large uncertainties in our knowledge of these resources, particularly in the case of uranium. From a global perspective, however, it seems reasonable to continue the petroleum analogy. It is perceived that the world will have to reduce its reliance on petroleum over the next fifty years by developing substitutes. So will it have to reduce its reliance on inefficient uranium converters. Fission technology offers the possibility of using uranium much more efficiently through technical developments. Just as the move toward petroleum substitution must get under way now, so also must the move toward more efficient nuclear technologies.

From a national planning point of view, uranium price movements over the coming decades are a more important factor than the magnitude of uranium ore bodies. In this context, one should perhaps resist the petroleum analogy at present, but it is also difficult to ignore it. However, it is important to note the differences in producer-consumer relations as between the petroleum and uranium industries, the latter being marked by long-term contracts and frequent joint ventures by utility and mining companies. Whether this will be an effective protection against excessive prices in a market of tight supplies is, of course, difficult to foresee. As to parallels, petroleum demand has proved very price-inelastic in the short run, and uranium demand should prove even more so, since the operator of a particular nuclear power station can do little to decrease its uranium requirement. We have also seen, in the petroleum case, that substitute technologies (e.g., coal gasification or liquefaction), although technically feasible, have not offered a rapid remedy. The lesson seems to be that, if a country defers deployment of a more uranium-efficient system until it can be justified on the basis of marginal costs, then it may never be in the position to deploy it.

In making simple microeconomic comparisons between, say, the cost of power from converter and breeder reactors, it is hard to avoid the conclusion that comparisons of marginal cost tend to favor the status quo. The capital cost of a new system must carry a high contingency allowance to cover uncertainties, while the countervailing rise in the operating cost of the present system may never appear great enough to justify the development of the new system. Thus a major decision, such as the development of a breeder reactor system, can probably only be made on the basis of an integrated view of the future that attributes significant secondary benefits to the new system.

As one attempts to analyze the economics of the breeder reactor, it is difficult to escape the paradox that a successful FBR program would immediately reduce the apparent need for the FBR. The reason is that breeder reactor deployment on a significant scale can be expected to depress world uranium prices by reducing demand. Yet high uranium prices are required for the microeconomic justification of the FBR. In an ideal world, one would like to see the breeder reactor penetrate the market far enough to keep uranium prices in check and thus make other reactor systems accessible to countries and situations to which the breeder reactor is not well suited. In this context, we think a broader view of the economics of the FBR, or for that matter of any advanced uranium-efficient reactor, is in order. Rather than comparing individual reactor types, the economics of the nuclear option, with its various subsystems, should be evaluated as a whole against those of competing systems based on other energy resources such as coal.

In trying to analyze longer-term nuclear futures, we note that judgments about the market penetration rate of a successful FBR, with any national or global limits on its penetration, are important characteristics of any nuclear strategy. There is a need for more careful analysis to test projections that allow the FBR, once established commercially, to penetrate the market at a rate and to an extent limited only by plutonium availability.

For those who are understandably wary of placing major reliance on nuclear fission as a future energy source, it is tempting to propose the hedging strategy of "keeping the option open." Some difficulties arise, however, if one examines carefully the question of when an option is indeed open. The fast breeder paths in section 5 illustrate this point. If we assume that, for each country, the breeder reactor option is open at the point where commercial orders can be placed, we see that, to reach that point, an aggressive and expensive program must first be conducted over ten or fifteen years. Moreover, if commercial orders cannot be placed economically by that time, a substantial continuing investment may be required to keep the infrastructure intact and the option truly open. The amount of political support and economic determination needed to open such an option and to keep it open may hardly be less, therefore, than that needed for a wholehearted commitment to nuclear power.

From a resource point of view, there seems little reason to doubt that nuclear power, as represented by present reactor types, could grow to become an important supplier of energy in the coming decades. Thereafter, our long-term projections for the post-2020 world indicate that, with efficient utilization, potential world uranium resources of some 15–25 MtU could be converted into an amount of electricity whose generation in conventional power plants would otherwise consume all currently proved reserves of coal, oil, and gas. This reassuring prospect should not, however, create false complacency. One problem of nuclear energy development is that, while there is a great deal of time in which to develop safe and economical systems, it requires a great deal of time to do so. For that reason, studies that reach the conclusion that the breeder reactor is "not needed" until 2000 or 2020 offer little basis for complacency. Even the aggressive development of an advanced system can take that long to produce results in the form of significant deployment. In the negotiations that will inevitably attend nuclear decisions, the great temptation will be to treat time as the most expendable of commodities and the cheapest of bargaining chips. The 1976–77 United States decisions on spent fuel reprocessing and FBR development are cases in point. If too much time is traded away initially, the danger is that, by the time a strong political consensus to move forward with nuclear energy has been forged, it may be too late to deploy the nuclear systems that can satisfy the requirement underlying that consensus.

With current projections of future nuclear capacity far below the levels of a few years ago, it is to be expected that individual nations will find it ever more difficult to justify the large expenditures involved in developing a new reactor type or fuel cycle. Even the continued development of such established systems as the heavy water reactor and the high temperature helium-cooled reactor is in jeopardy. It will be a substantial loss if the currently accumulated investment in reactor system development cannot be saved. One of the more serious long-term problems of the nuclear energy future that emerges from these considerations is that of global versus national interests. Will the collective result of national and multinational development programs lead to systems that can serve the world of the twenty-first century? The difficulty of constructing world reactor strategies is the absence of a world reactor strategist, or at least of one with authority to implement. Although the global incentive to deploy a breeder reactor or an advanced converter reactor is high, the incentive for any single nation may not be great enough to surmount the commercialization threshold. It follows that the need to share development costs internationally and to transfer technology equitably is greater than ever.

REFERENCES

Bundesanstalt für Geowissenschaften und Rohstoffe. 1976. *Die Künftige Entwicklung der Energienachfrage und deren Deckung: Perspektiven bis zum Jahr 2000.* Abschnitt 3. *Das Angebot von Energie-Rohstoffen.* Hanover: Bundesanstalt für Geowissenschaften und Rohstoffe.

Cohen, K. 1978. High capital cost: A deterrent to technological change. International Scientific Forum for an Acceptable World Energy Future, Miami Beach, Fla. Mimeographed.

Deutsche Gesellschaft für Wiederaufarbeitung von Kernbrennstoffen. 1977. *Entsorgungszentrum für Ausgediente Brennelemente aus Kernkraftwerken.* Hanover: Deutsche Gesellschaft für Wiederaufarbeitung von Kernbrennstoffen.

Engelmann, P.; Hansen, U.; Kolb, G.; Leushacke, D.; Teuchert, E.; and Werner, H. 1979. *Potential der Thorium-Nutzung im Hochtemperaturreaktor: Uranverbrauch und Kosten Langfristiger HTR-Strategien im Vergleich zu LWR und SBR.* Jül-1612. Jülich: Kernforschungsanlage Jülich.

Foster, J. S., and Critoph, E. 1975. The status of the Canadian nuclear power program and possible future strategies. *Annals of Nuclear Energy,* vol. 2.

Gärtner, E. 1976. Paper presented to Kolloquium der Geologischen Universität Köln. Cologne. Mimeographed.

Häfele, W., and Manne, A. S. 1974. *Strategies for a transition from fossil to nuclear fuels.* RR-74-7. Laxenburg: International Institute for Applied Systems Analysis.

Harris, DeV. P. 1979. World uranium resources. *Annual Review of Energy,* vol. 4.

International Atomic Energy Agency. 1977. *Regional nuclear fuel cycle centres.* 2 vols. Vienna: International Atomic Energy Agency.

Kempken, M. 1978. Verzeichnis der Kernkraftwerke der Welt. *Atomwirtschaft,* vol. 23, no. 10 (October).

______. 1979*a*. Verzeichnis der Kernkraftwerke der Welt. *Atomwirtschaft,* vol. 24, no. 4 (April).

______. 1979*b*. Verzeichnis der Kernkraftwerke der Welt. *Atomwirtschaft,* vol. 24, no. 11 (November).

Matzie, R. A., and Rec, J. A. 1976. Assessment of thorium fuel cycles in pressurized water reactors. Washington, D.C.: International Conference on World Nuclear Power. Mimeographed.

Melese-d'Hospital, G., and Simnad, M. 1977. Status of helium-cooled nuclear power systems. *Energy,* vol. 2.

Nuclear Energy Agency. 1978. *Nuclear fuel cycle requirements and supply considerations through the long-term: A report by an expert group of the OECD Nuclear Energy Agency.* Paris: Organization for Economic Co-operation and Development.

Nuclear Energy Agency/International Atomic Energy Agency. 1978. *World uranium potential: An international evaluation: A joint report by the OECD Nuclear Energy Agency and the International Atomic Energy Agency.* Paris: Organization for Economic Co-operation and Development.

______. 1979. *Uranium resources, production and demand: A joint report by the OECD Nuclear Energy Agency and the International Atomic Energy Agency.* Paris: Organization for Economic Co-operation and Development.

Oak Ridge National Laboratory. 1978. *The economics and utilization of thorium in nuclear reactors.* ORNL/TM-6332. Oak Ridge, Tenn.: Oak Ridge National Laboratory.

Parent, J. D., and Linden, H. R. 1977. *A survey of U.S. and total world production, proved reserves and remaining recoverable resources of fossil fuels and uranium as of December 31, 1975.* Chicago: Institute of Gas Technology.

Parker, Justice. 1978. *The Windscale inquiry: Report by the Hon. Mr Justice Parker.* 3 vols. London: Her Majesty's Stationery Office.

Siegel, J. R. 1979. Issues impacting the reprocessing decision. Hamburg: European Nuclear Conference. Mimeographed.

United Nations. 1977. *World energy supplies 1971–75.* Statistical papers, ser. J, no. 20. New York: United Nations.

Voigt, W. R., Jr. 1978. Natural uranium supply: The U.S. view. AIF/BNF International Conference on the Nuclear Fuel Cycle, London. Mimeographed.

Workshop on Alternative Energy Strategies. 1977. *Energy: Global prospects 1985–2000. Report of the Workshop on Alternative Energy Strategies.* New York: McGraw-Hill.

World Energy Conference. 1978*a*. *Coal resources: The full reports to the Conservation Commission of the World Energy Conference.* Guildford/New York: IPC Science and Technology Press.

______. 1978*b*. *World energy demand: The full report to the Conservation Commission of the World Energy Conference.* Guildford/New York: IPC Science and Technology Press.

______. 1978*c*. *World energy resources 1985–2020: Executive summaries of reports on resources, conservation and demand to the Conservation Commission of the World Energy Conference.* Guildford/New York: IPC Science and Technology Press.

______. 1978*d*. *Nuclear resources: The full reports to the Conservation Commission of the World Energy Conference.* Guildford/New York: IPC Science and Technology Press.

APPENDIX A
WORLD ENERGY CONFERENCE: REGIONAL GROUPINGS

Region 1: North America *

United States	Canada

Region 2: Western Europe

West Germany	Finland
Italy	Norway
Britain	Ireland
France	Cyprus
Spain	Luxembourg
Netherlands	Malta
Belgium	Iceland
Portugal	Channel Islands
Greece	Isle of Man
Sweden	Greenland
Austria	Faeroe Islands
Switzerland	Gibraltar
Denmark	

Region 3: Japan, Australasia

Japan	New Zealand
Australia	

Region 4: Soviet Union, CPE Europe

Soviet Union	Czechoslovakia
Poland	Hungary
Yugoslavia	Bulgaria
Romania	Albania
East Germany	

Region 5: China, CPE Asia

China	Kampuchea
Vietnam	Laos
North Korea	Mongolia

Region 6: OPEC Members

Indonesia	Ecuador
Nigeria	Libya
Iran	Kuwait
Algeria	United Arab Emirates
Venezuela	Gabon

* Within regions, countries are listed in order of their reported population (1977).

Iraq
Saudi Arabia
Qatar

Region 7: Latin America

Brazil
Mexico
Argentina
Colombia
Peru
Chile
Cuba
Guatemala
Bolivia
Dominican Republic
Haiti
El Salvador
Puerto Rico
Honduras
Uruguay
Paraguay
Nicaragua
Jamaica
Costa Rica
Panama
Trinidad and Tobago
Guyana
Surinam
Martinique
Guadeloupe
Barbados
Netherlands Antilles
Bahamas
Belize
Saint Lucia
Grenada
Saint Vincent
Virgin Islands
Dominica
Antigua
French Guiana
Bermuda
Saint Kitts-Nevis

Region 8: Middle East, North Africa

Turkey
Egypt
Morocco
Syria
Tunisia
North Yemen
Israel
Lebanon
Jordan
South Yemen
Oman
Bahrain

Region 9: Africa South of Sahara

Ethiopia
South Africa
Zaire
Sudan
Tanzania
Kenya
Uganda
Ghana
Mozambique
Madagascar
Cameroon
Ivory Coast
Burundi
Somalia
Benin
Sierra Leone
Togo
Central African Republic
Liberia
Mauritania
Congo
Lesotho
Namibia
Mauritius

Zimbabwe	Guinea-Bissau
Angola	Botswana
Mali	Gambia
Malawi	Swaziland
Upper Volta	Réunion
Senegal	Comoros
Zambia	Equatorial Guinea
Guinea	Cape Verde
Niger	Djibouti
Rwanda	São Tomé and Príncipe
Chad	Seychelles

Region 10: Southeast Asia

Philippines	Western Samoa
Thailand	Brunei
South Korea	French Polynesia
Taiwan	New Caledonia
Malaysia	Pacific Islands Trust Territory
Hong Kong	New Hebrides
Papua New Guinea	Guam
Singapore	Tonga
Fiji	Kiribati
Macao	American Samoa
Solomon Islands	Tuvalu

Region 11: South Asia

India	Sri Lanka
Bangladesh	Nepal
Pakistan	Bhutan
Burma	Maldives
Afghanistan	

APPENDIX B
REACTOR DATA

Mass Flows

The reactor types and fuel cycles chosen for our strategy evaluations are all based on present technologies or on extensions of them already under development. We have put our main emphasis on the LWR and its various operating modes. As the most likely advanced reactor system and complement to the LWR, we have focused on the liquid metal fast breeder reactor (LMFBR),

TABLE 9.A.1. Once-through Cycle Mass Flows over Thirty Years (65% Load Factor)

			PWR	HWR	AGR
Thermal efficiency		%	32.6	29.3	41.0
Burnup (average)		MWd/kg	32	7.3	18
Initial core	U	kg/GW	83,927	130,000	155,000
of which	U235	kg/GW	2,014	924	2,682
Initial period		years	1	0	0
Running-in					
Loading	U	kg/GW/a	19,018	88,065	–
of which	U235	kg/GW/a	590	626	–
Discharge	U	kg/GW/a	18,609	87,550	–
of which	U235	kg/GW/a	153	239	–
	Pu	kg/GW/a	121	309	–
of which	Pu_f	kg/GW/a	96	231	–
Running-in period		years	3	2	0
Equilibrium					
Loading	U	kg/GW/a	22,613	111,150	29,900
of which	U235	kg/GW/a	701	790	684
Discharge	U	kg/GW/a	21,618	110,500	29,250
of which	U235	kg/GW/a	181	254	222
	Pu	kg/GW/a	209	423	195[a]
of which	Pu_f	kg/GW/a	145	306	143
Equilibrium period		years	26	28	30
Final core	U	kg/GW	81,409	129,000	153,450[a]
of which	U235	kg/GW	1,160	503	2,340[a]
	Pu	kg/GW	594	340	800[a]
of which	Pu_f	kg/GW	447	270	600[a]

[a] Authors' estimates.

whereas alternative systems like the HWR and the HTR, the latter in combination with the thorium cycle, are treated more parenthetically. This treatment does not represent a negative judgment on the great potential of these HWR and HTR systems, but merely reflects the low de facto level of industrial commitment to them worldwide. It must also be borne in mind that the time horizon of 2020 leaves little opportunity to develop new concepts or deploy advanced technologies on a broad front.

Data bearing directly on the role of nuclear power as a future energy option are annual mass flows. These are given in tables 9.A.1 and 9.A.2. A distinction is made between initial core, running-in phase, equilibrium cycle operation, and final core inventory. This breakdown is necessary in order to analyze correctly variations during the introduction and expansion of a nuclear system and to evaluate the gross demand and net consumption of nuclear materials. The difference between gross demand and net consumption is the system inventory released when the reactor is shut down.

TABLE 9.A.2. Closed Cycle Mass Flows over Thirty Years (65% Load Factor)

			PWR		FBR	HTR	
Fuel cycle			Pu recycle		Oxide	Th recycle	
Thermal efficiency		%	32.6		38.0	40.0	
Burnup (average)		MWd/kg	32.0		55.0[a]	37.0	
Initial core	U	kg/GW	83,927		89,000	966	
of which	U235	kg/GW	2,014		180	898	
	Pu/Th[b]	kg/GW	–		5,000	29,560	
of which	Pu_f/U233[c]	kg/GW	–		3,500	–	
Initial period		years	1		1	0	
Running-in							
Loading	U	kg/GW/a	19,018		–	700	
of which	U235	kg/GW/a	590		–	651	
	Pu/Th	kg/GW/a	–		–	17,900	
Discharge	U	kg/GW/a	18,609		–	–	
of which	U235	kg/GW/a	153		–	–	
	Pu/Th	kg/GW/a	121		–	17,300	
of which	Pu_f/U233	kg/GW/a	96		–	345	
Running-in period		years	3		0	3	
Equilibrium							
Fuel proportion		%	70	30	–	60	40
Loading	U	kg/GW/a	15,830	6,490	25,000	142	752
of which	U235	kg/GW/a	491	46	50	132	–
	Pu/Th	kg/GW/a	–	293	1,850	9,400	5,940
of which	Pu_f/U233	kg/GW/a	–	205	1,300	–	445
Discharge	U	kg/GW/a	15,133	6,295	24,150	787	
of which	U235	kg/GW/a	127	23	40	–	
	Pu/Th	kg/GW/a	146	216	2,100	14,800	
of which	Pu_f/U233	kg/GW/a	101	129	1,470	452	
Equilibrium period		years	26		29	27	
Final core	U	kg/GW	56,987	23,245	86,500	2,400	
of which	U235	kg/GW	812	113	160	–	
	Pu/Th	kg/GW	416	894	5,700	41,030	
of which	Pu_f/U233	kg/GW	313	572	4,000	1,405	

[a] Figure is for core burnup; average burnup of core and radial blanket combined is 25 MWd/kg.
[b] Total plutonium (PWR/FBR) or thorium (HTR).
[c] Fissile plutonium (PWR/FBR) or bred U233 (including U235) (HTR).

Light Water Reactor (LWR)

The PWR used in our calculations is only one of the two different types of LWR commercially deployed at present (the other type being the BWR). The choice of the PWR as reference was due partly to its larger and expanding market share and partly to the more readily available data set. The data refer to the standard West German (KWU) 1,300 MW reactor.

In once-through operation, the reactor requires reload fuel of 3.1 percent enrichment to achieve a burnup of 32,000 MWd/t. At the design capacity (load) factor of 80 percent, a third of the core is replaced annually, accompanied by radial shuffling of the remaining fuel elements. With the reduced load factor of 65 percent used in the study, the annual replacement rate is closer to a quarter of the core.

In the recycle mode, the discharged fuel is reprocessed and the uranium and plutonium recovered, in either mixed or separated form, for fabrication into new fuel elements. As with other thermal reactors, a problem arises over the further use of the recovered uranium, which contains 0.8 percent U235 but which also contains a fraction of U236 and other uranium isotopes. At present it seems unlikely that enrichment plants would accept it as feed for reenrichment owing to this contamination. Blending is, of course, possible, but high-enriched uranium, on the order of 65 percent U235, would be needed to yield the same volume of LWR replacement fuel at the required enrichment. From detailed calculation of various ways of utilizing the degraded uranium, we have arrived at an average equivalence value of 0.94 compared with virgin uranium with the same U235 content.

For plutonium recycle, we have made no special assumptions concerning PWR design changes. It is thought that, without alterations to the core, a quarter to a third of the fuel elements in a batch could be of mixed oxide (MOX). This loading corresponds to designs of the "self-generated plutonium recycle" type, implicit in which is the assumption that discharged fuel is reprocessed without undue delay, and the plutonium returned to the reactor in which it was produced. On that basis, plutonium changes in composition as it progresses through generations of recycle. Such a model concept permits an exact assessment of the fissile value of LWR plutonium. However, the "self-generated plutonium recycle" concept is no longer compatible with the current development of reprocessing capacity. Owing to delays in introducing commercial reprocessing, large backlogs of spent fuel are building up, and it is assumed that, when reprocessing starts, the oldest spent fuel will be reprocessed first. The composition of most of the plutonium available for recycling during the time frame of this study, up to 2020, is thus likely to correspond to that of such "first generation" plutonium. Accordingly, in our model of plutonium recycle in the PWR, the composition of feed plutonium (70 percent fissile) does not change over time, and MOX fuel constitutes a constant 30 percent of the elements in a reload batch. The plutonium equivalence value is taken as 0.8: that is, 1 g of fissile plutonium (Pu_f) in MOX replaces 0.8 g of U235 in 3.1 percent enriched uranium. The make-up material used in our MOX fuel is natural uranium, although recovered degraded uranium at 0.8 percent U235 can be assumed to have a similar reactivity effect. Further use of reprocessed MOX fuel would have to take into account the fact that the equivalence value of fissile plutonium is reduced to 0.7 after one pass through the reactor.

For reasons of simplification, we have not assumed MOX fuel in the initial core of the "closed cycle" PWR, but have rather assumed that any operating PWR should be able to accept MOX in reload fuel.

Heavy Water Reactor (HWR)

The most widely deployed HWR is the CANDU type, and the data used here were taken from a recent evaluation of a 1,000 MW version of this system. The fuel is natural uranium and the operating mode once-through. After an initial period of one hundred full-power days, refueling is performed continuously, thus allowing improved neutron economy and operating flexibility.

The burnup with natural uranium is relatively low at 7,300 MWd/t. Improving fuel utilization in a once-through mode calls for higher burnups, and this could be achieved with slightly enriched uranium of 1.2 percent U235. Improvements in this direction are treated in a somewhat global manner in our evolutionary improvement strategy.

Heavy water reactors can also be operated with improved fuel cycle efficiency using thorium fuel. The necessary initial and make-up fissile material can be either high-enriched uranium or plutonium. To utilize the U233 bred from thorium calls for a closed cycle. In this operating mode, the HWR is very similar to the HTR.

Advanced Gas-Cooled Reactor (AGR)

The AGR is at the moment built and operated only in Britain. It is taken here to represent all past and future thermal reactors in that country. The data base available to us covered initial core and equilibrium operation only. The final core composition was constructed from the equilibrium data, based on the fact that the AGR is fueled continuously "on load."

Only once-through operation was considered, since British policy is to save discharged and recovered plutonium for the future FBR program.

High Temperature Reactor (HTR)

There exist two basic core designs for the HTR, with either prismatic or spherical fuel elements. The latter concept, with continuous refueling of the pebble-bed core, was chosen as our design basis. An average thermal efficiency of 40 percent was taken to represent a typical value for either steam or direct cycle gas turbine plants.

The HTR can operate in the once-through or closed cycle mode, the former being more likely during the introduction phase. To highlight the potential of thorium fuel, we selected an efficient converter reactor design, with an average conversion ratio of 0.82 and a burnup of 37,000 MWd/t. All spent fuel elements are reprocessed together, and the recovered uranium, in which the main fissile component is U233, is mixed with thorium and recycled. The fraction of MOX elements in a batch is 40 percent. The recovered thorium is stored for twenty years and may then be recycled using cold fabrication technology. Owing to the buildup of U236 during recycle, the fissile inventory increases from 900 kg in the just-critical initial core to 1,400 kg in the final core.

Fast Breeder Reactor (FBR)

A fast neutron spectrum reactor is the most efficient utilizer of recovered plutonium, and consequently of uranium, with the possibility of a net breeding gain. Thermal reactors may achieve self-sufficiency in fissile material, but hardly any net fissile material production. From the various FBR concepts, we chose the liquid metal cooled reactor for our calculations, because it enjoys the backing of most development and commercialization efforts.

Oxide fuel has reached an advanced stage of demonstration and constitutes the design base used here. More advanced oxide and, in particular, carbide fuel is under development and may be introduced commercially. However, it is our view that, during the first decades of FBR deployment, fuel will be of the "early oxide" type now planned for the French and West German demonstration stations.

It seems important to point out that these reactor designs, with a breeding ratio of some 1.2, vary rather widely in their net breeding gain, from 120 to 250 $kgPu_f/GW$ in each full-power year, allowing for a 2 percent loss in recycle operations. Even the more efficient rely to a large extent on plutonium from thermal reactors during their introduction phase. The net breeding gain depends largely on the radial blanket of U238, which has a longer residence time in the reactor than the core elements. The effect is that the FBR does not fulfill the role of the breeder until blanket management has reached its equilibrium state: that is, after some five years.

The FBR design used in this study (table 9.A.2) was chosen from a variety of concepts reported in the literature, somewhat modified to typify current development trends. Our "synthetic" breeder reactor has a net plutonium production of 215 $kgPu_f/GW$ in each full-power year, which is toward the higher end of the range mentioned above, albeit with a relatively large fissile inventory. The peak core burnup, estimated to be 25 percent higher than the average of 55,000 MWd/t, should be well within today's expected limit of 70,000 MWd/t. Owing to the rather crude manner in which the layout was selected, no attempt was made to simulate a proper running-in phase. Assuming a two-year hold-up time in the out-of-pile cycle, the necessary system inventory to start-up a 1 GW reactor amounts to 6.1 t of fissile plutonium.

TABLE 9.A.3. Lifetime Reactor Requirements over Thirty Years (65% Load Factor)

Type	Fuel	Cycle	Natural Uranium (tU/GW)		Separative Work (10^3SWU/GW)	
			Gross	Net	Gross	Net
PWR	3.1% U235	Once-through	4,062	–	3,205	–
HWR	Unat	Once-through	3,453	–	0	–
HWR	1.2% U235	Once-through	2,485	–	886	–
AGR	2.3% U235	Once-through	4,175	–	2,786	–
PWR	3.1% U235	U recycle	3,409	3,181	3,130	3,043
PWR	{3.1% U235 / MOX (UO_2/PuO_2)}	U/Pu recycle	2,727	2,387	2,328	2,125
HTR	{93% U235 / MOX (Th/U233)}	U233 recycle	1,266	690	1,642	895
FBR	MOX (UO_2/PuO_2)	Pu recycle	337	(–659)	269	(–526)

Enrichment tails assay: 0.2%
Fabrication losses: 1%
Recycle losses: 2%
1 gPu_f = 0.8 gU235
1 gU233 = 1.25 gU235

Natural Uranium and Separative Work Demand

For each of the reactor types used in the study, the requirements for natural uranium and separative work are listed in table 9.A.3. The gross demand is the amount the operator will have to provide during the lifetime of the reactor. After shutdown, the final core and the out-of-pile inventory of the fuel cycle – that is, the fuel discharged during the last two years and not yet returned to the reactor for recycle – become available for use in subsequent plants. Subtracting these inventories from the gross demand yields the net consumption of the reactor, equivalent to the "hole" its operation has left in the natural uranium resource. The FBR has been treated in an analogous manner, reflecting the fact that it operates in a market dominated by thermal reactors and that its plutonium input and output are measured in terms of the natural uranium required to produce that plutonium, or rather the saving of natural uranium if the plutonium were used in thermal reactors.

In a real deployment strategy, the annual requirement will lie between the gross demand and net consumption figures, because a mix of reactors of different ages will have to be fueled.

APPENDIX C
FAST BREEDER PATHS: MATERIAL FLOWS IN MAJOR COUNTRIES

The following worksheets (tables 9.A.4–9.A.22) are reproduced to illustrate the way material flow calculations were made for individual countries. The background to the worksheets, which constitute the foundation for tables 9.20 and 9.24 in the main text, deserves some explanation.

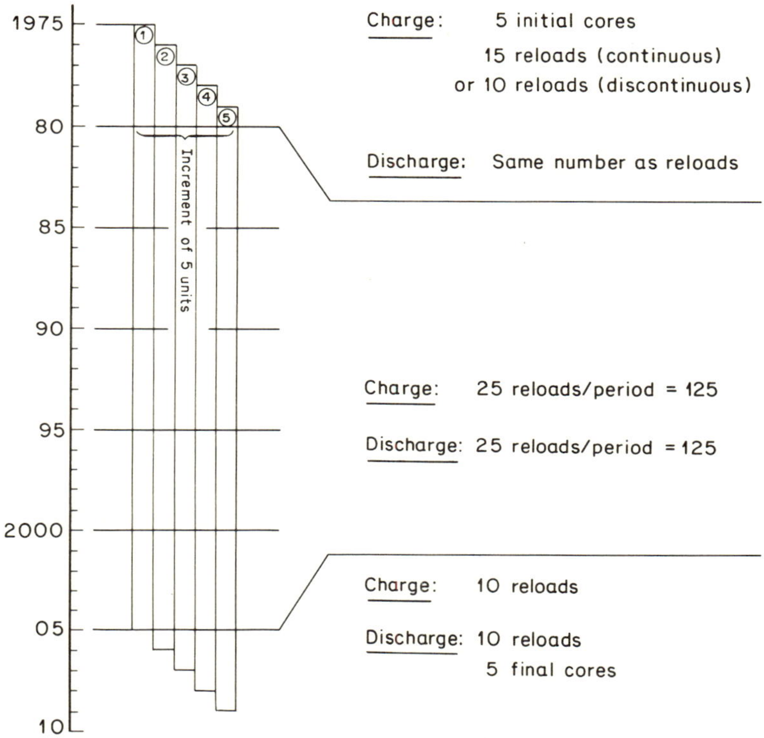

FIGURE 9.A.1. Model treatment of a five-unit capacity increment in the period 1976–80.

In parallel with our computer calculations, a simple model was developed of fuel cycle mass flows in individual countries. For fuel cycle paths involving recycling, this has the merit of making transparent the assumption on which the plutonium balance for FBRs and the use of MOX fuel in converter reactors are based. The model is not identical in all respects with the computer program; certain numerical differences between the results are therefore inevitable. However, the model does provide some additional insight, in addition to which it enables others who may be interested to retrace our methods (and errors).

For calculational simplicity, the model works in periods of five years. As throughout our study, installed generating capacities refer to 1 January of the last year in each period. Increments of capacity during a period are distributed equally between the five years. The main assumptions that then yield the mass flows in the model are illustrated in figure 9.A.1.

A linear growth during any period equivalent to introducing, say, five 1 GW units can be represented by a stepwise increment in capacity and an appropriately tailored mass flow to take account, in particular, of the introduction phase. In the calculation of fuel loaded into and discharged from a reactor, a distinction must, for example, be made between continuous (on-load) and discontinuous (off-load) refueling. In the latter case, which applies to the PWR and other LWRs, there is an initial delay of one year before refueling begins, and an average of two reloads instead of three must therefore be allowed for each unit starting operation in a given period.

The calculational algorithms used in constructing tables 9.A.4–9.A.22 are set out below. The numerical mass flow data were based on tables 9.A.1 and 9.A.2 in Appendix B.

CAP(I)	capacity at end of period I (GW)
CØN(I)	capacity constructed in period I (GW)
	= CAP(I) − CAP(I-1) + CLS(I)
CLS(I)	capacity closed in period I (GW)
	= CØN(I-6)
IC	initial core loading (t/GW)
RL	annual reload during running-in (t/GW)
EL	annual reload during equilibrium (t/GW)
$\overline{EL}$	average annual equilibrium reload for PWR (t/GW)
	= (1RL + 26EL)/27
FC	final core at discharge (t/GW)
	= IC (i.e., final core taken as equivalent to initial core, because all discharges are measured in "new" tons loaded).

AGR: Continuous Refueling

Material loaded in period I

CØN(I) * (IC + 3EL)	= CØN(I) * 244.7 t/GW
+ (CAP(I-1) − 3/5 * CLS(I)) * 5EL	= ( ) * 149.5 t/GW

Material discharged in period I

CØN(I) * 3EL	= CØN(I) * 87.7 t/GW
+ (CAP(I-1) − 3/5 * CLS(I)) * 5EL	= ( ) * 149.5 t/GW
+ CLS(I) * FC	= CLS(I) * 155.0 t/GW

HWR: Continuous Refueling

Material loaded in period I

CØN(I) * (IC + 2RL + 1EL)	= CØN(I) * 417.3 t/GW
+ (CAP(I-1) − 3/5 * CLS(I)) * 5EL	= ( ) * 555.8 t/GW

Material discharged in period I

CØN(I) * (2RL + 1EL)	= CØN(I) * 417.3 t/GW
+ (CAP(I-1) − 3/5 * CLS(I)) * 5EL	= ( ) * 555.8 t/GW
+ CLS(I) * FC	= CLS(I) * 130.0 t/GW

PWR: Discontinuous (Annual) Refueling

Material loaded in period I

CØN(I) * (IC + 2RL)	= CØN(I) * 122.0 t/GW
+ (CAP(I-1) − 3/5 * CLS(I)) * 5$\overline{EL}$	= ( ) * 112.4 t/GW

Material discharged in period I

CØN(I) * 2RL	= CØN(I) * 38.0 t/GW
+ (CAP(I-1) − 3/5 * CLS(I)) * 5$\overline{EL}$	= ( ) * 112.4 t/GW
+ CLS(I) * FC	= CLS(I) * 84.0 t/GW

The concentration of fissile plutonium in discharged fuel is taken as the weighted average over batches discharged during running-in and equilibrium operation. Since fuel from retired reactors contributes very little to the discharged tonnage, its effect on the concentration has been disregarded. The final core is assumed to have the same concentration as reload fuel; any consequential error has only minor significance by the end of the study period.

The average concentration of fissile plutonium in discharged fuel is taken as:

AGR: 4.783 $kgPu_f/tHM$
HWR: 2.746 $kgPu_f/tHM$
PWR: 6.292 $kgPu_f/tHM$.

During reprocessing, inevitable losses occur that reduce the fissile material available for recycling. In the materials balance, 2 percent reprocessing losses are assumed. In addition, the plutonium 241 isotope decays with a half-life of about fifteen years, thus progressively lowering the concentration of fissile isotopes in the total composition. For fuel recycled within a five-year period—that is, with an average out-of-pile time of 2.5 years—Pu241 has decayed to 89 percent of its value at discharge. Taking Pu241 typically as 20 percent of all fissile plutonium in PWR and AGR fuel (or 10 percent in HWR fuel), the original total is thus reduced by 2 percent (or 1 percent for HWRs). In the introduction phase of closed fuel cycles, the hold-up of discharged fuel in store may reach ten to twenty years. For countries that start reprocessing around the turn of the century or later, an average out-of-pile time of fifteen years is assumed, leading to a 50 percent loss of Pu241, equivalent to 10 percent of all fissile plutonium (or 5 percent for HWRs).

Plutonium balancing in FBRs is not as straightforward as counting loaded and discharged batches of converter reactor fuel. During the introduction and expansion phase of an FBR program, plutonium from reprocessed converter reactor fuel is essential. One aim of our calculation must therefore be to identify the amount required from such "external" sources. An out-of-pile time of two years is assumed for FBR fuel, so that, in addition to the initial core, the plutonium for the first two reloads must come either from converter reactors or from net breeding gain in another FBR. The net breeding gain (BG) is the difference between the fissile plutonium in a reprocessed discharge batch (assuming 2 percent reprocessing losses but disregarding Pu241 decay) and that in a reload batch. Only that difference is recorded in the worksheets as "net from FBRs." The final core is treated separately but included in the balance. The plutonium from FBRs enters the relevant column in the worksheets after a delay of two years. (Because there is always a sufficient reserve of discharged converter reactor fuel, no such delay is assumed for plutonium from converter reactor reprocessing.)

During the demonstration phase of an FBR program, the introduction date for each unit was specified individually. During the commercial phase, FBR additions are given for the period only, and a linear growth model, as for converter reactor capacity expansion, is used. An analysis of the number

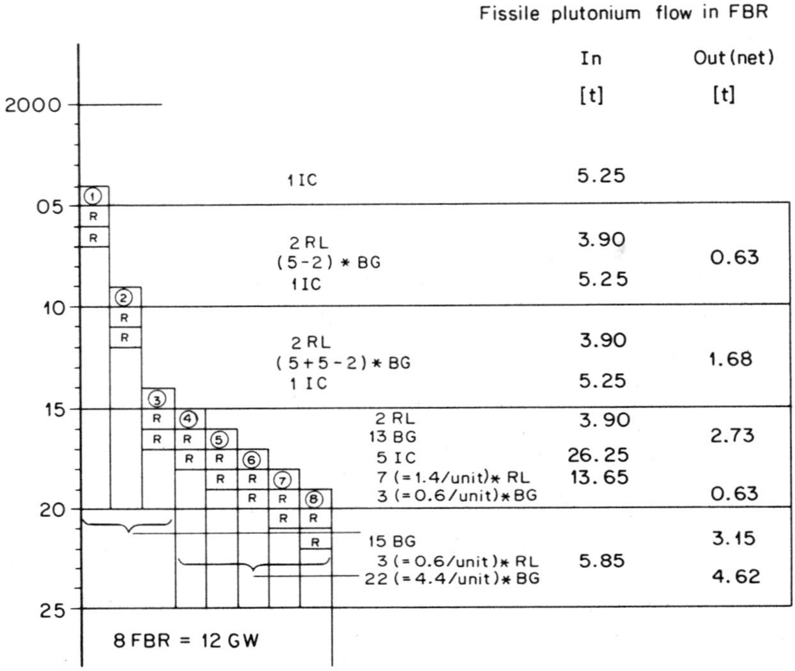

IC = 5.25 t Pu_f /1.5 GW
RL = 1.95 "
BG = 0.21 "
FC = 6.00 "

FIGURE 9.A.2. Calculation of net plutonium inputs and outputs for the Canadian FBR program.

of "external" reloads required, in addition to material derived from breeding gain, yields 1.4 such reloads for each unit during the period in which it is introduced. A further 0.6 "external" reloads are needed in the next period. In other words, an average of 0.6 reload batches becomes available from breeding gain for each new unit during the period of its introduction, and an average of 4.4 such batches becomes available in the following period. To illustrate the calculation of FBR plutonium flow, the inputs and outputs for the Canadian FBR program in the model are shown in figure 9.A.2.

Fuel cycle activities with respect to FBR fuel fabrication and reprocessing, as listed in the worksheets, are booked at the times the fuel is respectively loaded into and discharged from the reactor. In reality, lead and lag times of the order of one year exist, and a corresponding shift in time should therefore be applied. For simplicity, this has not been done. However, the other results are not affected, since the appropriate delays have been accounted for in the plutonium balance matrix.

In the fast breeder/plutonium recycle path, MOX fuel is used only in PWRs. From tables 9.A.1 and 9.A.2, we find that, for 1 tPu_f recycled, about 33 tMOX are required. In discharged equilibrium reload fuel, the fissile plutonium content is taken as:

UOX: 6.4 $kgPu_f/t$

MOX: 19.0 $kgPu_f/t$.

Figures in the column for fissile plutonium in discharged converter reactor fuel are adjusted accordingly.

The schedule for MOX flow in the fuel cycle is made as simple as possible. Fuel is loaded in one period, discharged in the next, and reprocessed in the one following. On average, therefore, MOX remains in the fuel cycle for ten years. Since its residence time in the PWR is almost four years, Pu241 decay is calculated by taking the remaining six years as the out-of-pile time. With a higher Pu241 proportion of about 30 percent of fissile plutonium in discharged MOX fuel, as compared with 20 percent in UOX, the losses in that time amount to 8 percent of total fissile plutonium. Correcting for all recycle losses, the output per tMOX reprocessed is therefore taken as 17.1 $kgPu_f$, with no distinction being made between countries that start reprocessing early or late in the study period.

Plutonium recovered from reprocessed MOX is assumed to be destined for use in FBRs, and no attempt is made to construct a second- or third-generation MOX recycle strategy for converter reactors.

Finally, it should be noted that, for the sake of numerical consistency, more digits are carried in the worksheets than we regard as significant. There are discrepancies, therefore, between certain of the values indicated in tables 9.A.4–9.A.22 and those in some other tables in the main text, where values have been rounded to avoid the illusion of extreme precision.

TABLE 9.A.4. Canada: Fast Breeder Path

Year	Period	CR Capacity (GW) Capac	CR Close	CR Const	FBR Capacity (GW) Capac	FBR Close	FBR Const	Converter reactor Fuel (tHM) Fuel Charged	Fuel Disch	Spent F Import	Spent F Reprocessed UOX	Spent F Reprocessed MOX	Spent F Inventory	Fissile Plutonium (tPu_f) In Disch CR Fuel	From CR Reproc	Net from FBRs	To FBR Fuel	To CR MOX	Sep Pu Inventory
1975		2.5			0								0						0
1980	1976-80	5.0	0	2.5	0	0	0	2,435	2,110	0	0	0	2,110	5.8	0	0	0	0	0
1985	81-85	9.0	0	4.0	0	0	0	4,450	3,930	0	0	0	6,040	10.8	0	0	0	0	0
1990	86-90	12.0	0	3.0	0	0	0	6,255	5,865	0	0	0	11,905	16.1	0	0	0	0	0
1995	91-95	15.0	0	3.0	0	0	0	7,920	7,530	0	0	0	19,435	20.7	0	0	0	0	0
2000	96-00	29.0	0	14.0	0	0	0	14,180	12,360	0	0	0	31,795	34.0	0	0	0	0	0
2005	2001-05	39.5	2.5	13.0	1.5	0	1.5	20,710	19,345	0	3,000	0	48,180	53.2	7.7	0	5.3	0	2.4
2010	06-10	53.0	2.5	16.0	3.0	0	1.5	27,795	26,040	0	6,000	0	68,180	71.6	15.3	0.6	9.2	0	9.1
2015	11-15	65.5	4.0	16.5	4.5	0	1.5	35,010	33,385		12,000	0	89,565	91.8	30.6	1.7	9.2	0	32.2
2020	16-20	78.0	3.0	15.5	12.0	0	7.5	41,875	40,250	0	18,000	0	111,815	110.7	45.9	3.4	43.8	0	37.7

Year	Period	Fuel Cycle Activities (tHM/a): Converter Reactors Fabrication UOX	Fabrication MOX	Reprocessing UOX	Reprocessing MOX	FBRs (Core + Blanket) Fabric	FBRs Reproc
1975							
1980	1976-80	485	0	0	0	0	0
1985	81-85	890	0	0	0	0	0
1990	86-90	1.250	0	0	0	0	0
1995	91-95	1,585	0	0	0	0	0
2000	96-00	2,835	0	0	0	0	0
2005	2001-05	4,140	0	600	0	30	0
2010	06-10	5,560	0	1,200	0	65	40
2015	11-15	7,000	0	2,400	0	105	75
2020	16-20	8,375	0	3,600	0	325	190

Notes

1. **See key to Table 9.20.**
2. Converter reactor: HWR
   - 2.75 $kgPu_f/tHM$ discharged
   - losses: 2% in reprocessing + 5% from Pu241 decay
   - 2.55 $kgPu_f/tHM$ reprocessed.

TABLE 9.A.5. France: Fast Breeder Path

Year	Period	CR Capacity (GW) Capac	Close	Const	FBR Capacity (GW) Capac	Close	Const	Converter reactor Fuel (tHM) Fuel Charged	Fuel Disch	Spent F Import	Spent F Reprocessed UOX	Spent F Reprocessed MOX	Spent F Inventory	Fissile Plutonium (tPu_f) In Disch CR Fuel	From CR Reproc	Net from FBRs	To FBR Fuel	To CR MOX	Sep Pu Inventory
1975		3.0			0								0						0
	1976-80		0	8.5		0	0	1,375	660	0	0	0		4.2	0	0	0	0	
1980		11.5			0								660						0
	81-85		0	22.0		0	1.5	3,975	2,130	0	1,875	0		13.4	11.4	0	5.3	0	
1985		33.5			1.5								915						6.1
	86-90		0	5.5		0	1.5	4,435	3,975	0	1,875	0		25.0	11.4	0.6	9.2	0	
1990		39.0			3.0								3,015						8.9
	91-95		0	6.5		0	1.5	5,175	4,630	0	3,750	0		29.1	22.7	1.7	9.2	0	
1995		45.5			4.5								3,895						24.1
	96-00		0	10.5		0	1.5	6,395	5,515	0	3,750	0		34.7	22.7	2.7	9.2	0	
2000		56.0			6.0								5,660						40.3
	2001-05		3.0	10.5		0	4.5	7,375	6,745	0	7,500	0		42.4	45.4	4.2	27.8	0	
2005		63.5			10.5								4,905						62.1
	06-10		8.5	8.0		0	7.5	7,540	7,580	0	7,500	0		47.7	45.4	7.6	43.4	0	
2010		63.0			18.0								4,985						71.7
	11-15		22.0	12.0		1.5	13.5	7,060	7,900	0	7,500	0		49.7	45.4	15.1	77.7	0	
2015		53.0			30.0								5,385						54.5
	16-20		5.5	6.5		1.5	13.5	6,380	6,295	0	7,500	0		39.6	45.4	32.0	82.4	0	
2020		54.0			42.0								4,180						49.5

Year	Period	Fuel Cycle Activities (tHM/a) Converter Reactors Fabrication UOX	Fabrication MOX	Reprocessing UOX	Reprocessing MOX	FBRs (Core + Blanket) Fabric	FBRs (Core + Blanket) Reproc
1975							
	1976-80	275	0	0	0	0	0
1980							
	81-85	795	0	375	0	30	0
1985							
	86-90	885	0	375	0	65	40
1990							
	91-95	1,035	0	750	0	105	75
1995							
	96-00	1,280	0	750	0	140	115
2000							
	2001-05	1,475	0	1,500	0	280	195
2005							
	06-10	1,510	0	1,500	0	475	340
2010							
	11-15	1,410	0	1,500	0	830	615
2015							
	16-20	1,275	0	1,500	0	1,130	910
2020							

Notes

1. See key to Table 9.20.
2. Converter reactor: PWR
   - 6.29 kg Pu_f/tHM discharged
   - losses: 2% in reprocessing + 2% from Pu241 decay
   - 6.05 kg Pu_f/tHM reprocessed.

TABLE 9.A.6. France: Fast Breeder/Plutonium Recycle Path

Year	Period	CR Capacity (GW) Capac	Close	Const	FBR Capacity (GW) Capac	Close	Const	Converter reactor Fuel (tHM) Fuel Charged	Fuel Disch	Spent F Import	Spent F Reprocessed UOX	Spent F Reprocessed MOX	Spent F Inventory	Fissile Plutonium (tPu_f) In Disch CR Fuel	From CR Reproc	Net from FBRs	To FBR Fuel	To CR MOX	Sep Pu Inventory
1975		3.0			0								0						0
	1976-80		0	8.5		0	0	1,375	660	0	0	0		4.2	0	0	0	0	
1980		11.5			0								660						0
	81-85		0	22.0		0	1.5	3,975	2,130	0	1,875	0		13.4	11.4	0	5.3	0	
1985		33.5			1.5								915						6.1
	86-90		0	5.5		0	1.5	4,435	3,975	0	1,875	0		25.0	11.4	0.6	9.2	5.9	
1990		39.0			3.0								3,015						3.0
	91-95		0	6.5		0	1.5	5,175	4,630	1,355	7,500	0		31.6	45.4	1.7	9.2	30.9	
1995		45.5			4.5								1,500						10.0
	96-00		0	10.5		0	1.5	6,395	5,515	1,985	7,305	195		47.6	47.5	2.7	9.2	41.0	
2000		56.0			6.0								1,500						10.0
	2001-05		3.0	10.5		0	4.5	7,375	6,745	5,255	10,225	1,025		59.6	79.4	4.2	27.8	48.8	
2005		63.5			10.5								2,250						17.0
	06-10		8.5	8.0		0	7.5	7,540	7,580	3,670	9,885	1,365		68.2	83.1	7.6	43.4	45.3	
2010		63.0			18.0								2,250						19.0
	11-15		22.0	12.0		1.5	13.5	7,060	7,900	7,850	13,375	1,625		68.7	108.7	15.1	77.7	40.1	
2015		53.0			30.0								3,000						25.0
	16-20		5.5	6.5		1.5	13.5	6,380	6,295	8,705	13,490	1,510		56.4	107.4	32.0	82.4	54.0	
2020		54.0			42.0								3,000						28.0

Year	Period	Fuel Cycle Activities (tHM/a) Converter Reactors Fabrication UOX	Fabrication MOX	Reprocessing UOX	Reprocessing MOX	FBRs (Core + Blanket) Fabric	FBRs (Core + Blanket) Reproc
1975							
	1976-80	275	0	0	0	0	0
1980							
	81-85	795	0	375	0	30	0
1985							
	86-90	845	40	375	0	65	40
1990							
	91-95	830	205	1,500	0	105	75
1995							
	96-00	1,005	275	1,460	40	140	115
2000							
	2001-05	1,150	325	2,045	205	280	195
2005							
	06-10	1,210	300	1,975	275	475	340
2010							
	11-15	1,145	265	2,675	325	830	615
2015							
	16-20	915	360	2,700	300	1,130	910
2020							

Notes

1. See key to Table 9.20.
2. See notes to Table 9.A.5.
3. Pu recycle in PWR
   - 19.0 kg Pu_f/tMOX discharged
   - losses: 2% in reprocessing + 8% from Pu241 decay
   - 17.1 kg Pu_f/tMOX reprocessed.

TABLE 9.A.7. West Germany: Fast Breeder Path

Year	Period	CR Capacity (GW) Capac	Close	Const	FBR Capacity (GW) Capac	Close	Const	Converter reactor Fuel (tHM) Fuel Charged	Fuel Disch	Spent F Import	Spent F Reprocessed UOX	MOX	Spent F Inventory	Fissile Plutonium (tPu_f) In Disch CR Fuel	From CR Reproc	Net from FBRs	To FBR Fuel	To CR MOX	Sep Pu Inventory
1975		3.5			0								0						0
	1976-80		0	6.0		0	0	1,125	620	0	0	0		3.9	0	0	0	0	
1980		9.5			0								620						0
	81-85		0	9.0		0	0.3	2,165	1,410	0	0	0		8.9	0	0	1.0	0	
1985		18.5			0.3								2,030						(-1.0)
	86-90		0	10.0		0	0	3,300	2,460	0	0	0		15.5	0	0.2	0.8	0	
1990		28.5			0.3								4,490						(-1.6)
	91-95		0	7.0		0	0	4,055	3,470	0	0	0		21.8	0	0.2	0	0	
1995		35.5			0.3								7,960						(-1.4)
	96-00		0	15.0		0	1.2	5,820	4,560	0	0	0		28.7	0	0.2	4.2	0	
2000		50.5			1.5								12,520						(-5.4)
	2001-05		3.5	16.0		0	1.5	7,390	6,340	0	3,750	0		39.9	20.8	0.9	8.4	0	
2005		63.0			3.0								15,110						7.9
	06-10		6.0	18.5		0	1.5	8,935	7,885	0	3,750	0		49.6	20.8	1.9	9.2	0	
2010		75.5			4.5								19,245						21.4
	11-15		9.0	13.0		0.3	6.3	9,465	9,130	0	7,500	0		57.4	41.6	3.6	37.4	0	
2015		79.5			10.5								20,875						29.2
	16-20		10.0	15.0		0	15.0	10,090	9,670	0	11,250	0		60.8	62.4	9.7	84.7	0	
2020		84.5			25.5								19,295						16.6

Year	Period	Fuel Cycle Activities (tHM/a) Converter Reactors Fabrication UOX	MOX	Reprocessing UOX	MOX	FBRs (Core + Blanket) Fabric	Reproc
1975							
	1976-80	225	0	0	0	0	0
1980							
	81-85	435	0	0	0	5	0
1985							
	86-90	660	0	0	0	10	10
1990							
	91-95	810	0	0	0	10	10
1995							
	96-00	1,165	0	0	0	30	10
2000							
	2001-05	1,480	0	750	0	65	40
2005							
	06-10	1,785	0	750	0	105	75
2010							
	11-15	1,895	0	1,500	0	290	180
2015							
	16-20	2,020	0	2,250	0	685	415
2020							

Notes

1. See key to Table 9.20.
2. Converter reactor: PWR
   - 6.29 kg Pu_f/tHM discharged
   - losses: 2% in reprocessing + 10% from Pu241 decay
   - 5.55 kg Pu_f/tHM reprocessed.

TABLE 9.A.8. West Germany: Fast Breeder/Plutonium Recycle Path

Year	Period	CR Capacity (GW) Capac	Close	Const	FBR Capacity (GW) Capac	Close	Const	Converter reactor Fuel (tHM) Fuel Charged	Fuel Disch	Spent F Import	Spent F Reprocessed UOX	MOX	Spent F Inventory	Fissile Plutonium (tPu_f) In Disch CR Fuel	From CR Reproc	Net from FBRs	To FBR Fuel	To CR MOX	Sep Pu Inventory
1975		3.5			0								0						0
1980	1976-80	9.5	0	6.0	0	0	0	1,125	620	0	0	0	620	3.9	0	0	0	0	0
1985	81-85	18.5	0	9.0	0.3	0	0.3	2,165	1,410	0	0	0	2,030	8.9	0	0	1.0	0	(-1.0)
1990	86-90	28.5	0	10.0	0.3	0	0	3,300	2,460	0	0	0	4,490	15.5	0	0.2	0.8	0	(-1.6)
1995	91-95	35.5	0	7.0	0.3	0	0	4,055	3,470	0	0	0	7,960	21.8	0	0.2	0	0	(-1.4)
2000	96-00	50.5	0	15.0	1.5	0	1.2	5,820	4,560	0	0	0	12,520	28.7	0	0.2	4.2	0	(-5.4)
2005	2001-05	63.0	3.5	16.0	3.0	0	1.5	7,390	6,340	0	3,750	0	15,110	39.9	20.8	0.9	8.4	2.9	5.0
2010	06-10	75.5	6.0	18.5	4.5	0	1.5	8,935	7,885	0	7,500	0	15,495	50.9	41.6	1.9	9.2	30.3	9.0
2015	11-15	79.5	9.0	13.0	10.5	0.3	6.3	9,465	9,130	0	14,900	100	9,625	70.0	84.4	3.6	37.4	41.6	18.0
2020	16-20	84.5	10.0	15.0	25.5	0	15.0	10,090	9,670	0	14,000	1,000	4,295	78.3	94.8	9.7	84.7	16.8	21.0

Year	Period	Fuel Cycle Activities (tHM/a): Converter Reactors, Fabrication UOX	MOX	Reprocessing UOX	MOX	FBRs (Core + Blanket) Fabric	Reproc
1975							
1980	1976-80	225	0	0	0	0	0
1985	81-85	435	0	0	0	5	0
1990	86-90	660	0	0	0	10	10
1995	91-95	810	0	0	0	10	10
2000	96-00	1,165	0	0	0	30	10
2005	2001-05	1,460	20	750	0	65	40
2010	06-10	1,585	200	1,500	0	105	75
2015	11-15	1,620	275	2,980	20	290	180
2020	16-20	1,910	110	2,800	200	685	415

Notes

1. See key to Table 9.20.
2. See notes to Table 9.A.7.
3. Pu recycle in PWR
   - 19.0 kg Pu_f/tMOX discharged
   - losses : 2% in reprocessing + 8% from Pu241 decay
   - 17.1 kg Pu_f/tMOX reprocessed.

TABLE 9.A.9. Japan: Fast Breeder Path

Year	Period	CR Capacity (GW) Capac	Close	Const	FBR Capacity (GW) Capac	Close	Const	Converter reactor Fuel (tHM) Fuel Charged	Fuel Disch	Spent F Import	Spent F Reprocessed UOX	Spent F Reprocessed MOX	Spent F Inventory	Fissile Plutonium (tPu_f) In Disch CR Fuel	From CR Reproc	Net from FBRs	To FBR Fuel	To CR MOX	Sep Pu Inventory
1975		5.0			0								0						0
	1976-80		0	8.0		0	0	1,540	865	0	0	0		5.4	0	0	0	0	
1980		13.0			0								865						0
	81-85		0	4.0		0	0	1,950	1,615	0	0	0		10.1	0	0	0	0	
1985		17.0			0								2,480						0
	86-90		0	9.5		0	0.4	3,070	2,270	0	0	0		14.3	0	0	1.4	0	
1990		26.5			0.4								4,750						(-1.4)
	91-95		0	10.0		0	0	4,200	3,360	0	1,875	0		21.1	10.4	0.2	1.0	0	
1995		36.5			0.4								6,235						8.2
	96-00		0	38.0		0	1.1	8,740	5,545	0	1,875	0		34.9	10.4	0.3	3.9	0	
2000		74.5			1.5								9,905						15.0
	2001-05		5.0	37.5		0	1.5	12,610	9,880	0	3,750	0		62.2	20.8	0.7	8.1	0	
2005		107.0			3.0								16,035						28.4
	06-10		8.0	46.5		0	1.5	17,160	13,925	0	7,500	0		87.6	41.6	1.7	9.2	0	
2010		145.5			4.5								22,460						62.5
	11-15		4.0	27.5		0	16.5	19,440	17,465	0	15,000	0		109.9	83.3	4.1	91.7	0	
2015		169.0			21.0								24,925						58.2
	16-20		9.5	29.5		0.4	30.0	21,955	20,275	0	30,000	0		127.5	166.5	16.4	172.5	0	
2020		189.0			50.6								15,200						68.6

Year	Period	Fuel Cycle Activities (tHM/a) Converter Reactors Fabrication UOX	Fabrication MOX	Reprocessing UOX	Reprocessing MOX	FBRs (Core + Blanket) Fabric	Reproc
1975							
	1976-80	310	0	0	0	0	0
1980							
	81-85	390	0	0	0	0	0
1985							
	86-90	615	0	0	0	10	0
1990							
	91-95	840	0	375	0	10	10
1995							
	96-00	1,750	0	375	0	30	10
2000							
	2001-05	2,520	0	750	0	65	40
2005							
	06-10	3,430	0	1,500	0	105	75
2010							
	11-15	3,890	0	3,000	0	575	280
2015							
	16-20	4,390	0	6,000	0	1,365	835
2020							

Notes

1. See key to Table 9.20.
2. Converter reactor : PWR
   - 6.29 kg Pu_f/tHM discharged
   - losses : 2% in reprocessing + 10% from Pu241 decay
   - 5.55 kg Pu_f/tHM reprocessed.

TABLE 9.A.10. Japan: Fast Breeder/Plutonium Recycle Path

Year	Period	CR Capacity (GW) Capac	Close	Const	FBR Capacity (GW) Capac	Close	Const	Converter reactor Fuel (tHM) Fuel Charged	Fuel Disch	Spent F Import	Spent F Reprocessed UOX	Spent F Reprocessed MOX	Spent F Inventory	Fissile Plutonium (tPu_f) In Disch CR Fuel	From CR Reproc	Net from FBRs	To FBR Fuel	To CR MOX	Sep Pu Inventory
1975		5.0			0								0						0
	1976-80		0	8.0		0	0	1,540	865	0	0	0		5.4	0	0	0	0	
1980		13.0			0								865						0
	81-85		0	4.0		0	0	1,950	1,615	0	0	0		10.1	0	0	0	0	
1985		17.0			0								2,480						0
	86-90		0	9.5		0	0.4	3,070	2,270	0	0	0		14.3	0	0	1.4	0	
1990		26.5			0.4								4,750						(-1.4)
	91-95		0	10.0		0	0	4,200	3,360	0	1,875	0		21.1	10.4	0.2	1.0	5.2	
1995		36.5			0.4								6,235						3.0
	96-00		0	38.0		0	1.1	8,740	5,545	0	1,875	0		37.1	10.4	0.3	3.9	6.8	
2000		74.5			1.5								9,905						3.0
	2001-05		5.0	37.5		0	1.5	12,610	9,880	0	3,575	175		65.0	22.8	0.7	8.1	13.4	
2005		107.0			3.0								16,035						5.0
	06-10		8.0	46.5		0	1.5	17,160	13,925	0	7,275	225		93.3	44.2	1.7	9.2	31.7	
2010		145.5			4.5								22,460						10.0
	11-15		4.0	27.5		0	16.5	19,440	17,465	0	22,050	450		123.1	130.1	4.1	91.7	25.5	
2015		169.0			21.0								17,425						27.0
	16-20		9.5	29.5		0.4	30.0	21,955	20,275	7,300	36,450	1,050		138.2	220.3	16.4	172.5	43.2	
2020		189.0			50.6								7,500						48.0

Year	Period	Fuel Cycle Activities (tHM/a) Converter Reactors Fabrication UOX	Fabrication MOX	Reprocessing UOX	Reprocessing MOX	FBRs (Core + Blanket) Fabric	FBRs (Core + Blanket) Reproc
1975							
	1976-80	310	0	0	0	0	0
1980							
	81-85	390	0	0	0	0	0
1985							
	86-90	615	0	0	0	10	0
1990							
	91-95	805	35	375	0	10	10
1995							
	96-00	1,705	45	375	0	30	10
2000							
	2001-05	2,430	90	715	35	65	40
2005							
	06-10	3,220	210	1,455	45	105	75
2010							
	11-15	3,720	170	4,410	90	575	280
2015							
	16-20	4,100	290	7,290	210	1,365	835
2020							

Notes

1. See key to Table 9.20.
2. See notes to Table 9.A.9.
3. Pu recycle in PWR
   - 19.0 kg Puf/tMOX discharged
   - losses : 2% in reprocessing + 8% from Pu241 decay
   - 17.1 kg Puf/tMOX reprocessed.

TABLE 9.A.11. Britain: Fast Breeder Path

Year	Period	CR Capacity (GW) Capac	CR Capacity (GW) Close	CR Capacity (GW) Const	FBR Capacity (GW) Capac	FBR Capacity (GW) Close	FBR Capacity (GW) Const	Converter reactor Fuel (tHM) Fuel Charged	Fuel Disch	Spent F Import	Spent F Reprocessed UOX	Spent F Reprocessed MOX	Spent F Inventory	Fissile Plutonium (tPu_f) In Disch CR Fuel	From CR Reproc	Net from FBRs	To FBR Fuel	To CR MOX	Sep Pu Inventory
1975		5.5			0								0						0
	1976-80		0	2.5		0	0	1,435	1,045	0	0	0		5.0	0	0	0	0	
1980		8.0			0								1,045						0
	81-85		0	4.5		0	0.3	2,295	1,600	0	0	0		7.6	0	0	1.0	0	
1985		12.5			0.3								2,645						(-1.0)
	86-90		0	1.0		0	0	2,115	1,960	0	500	0		9.4	2.3	0.1	0.8	0	
1990		13.5			0.3								4,105						0.6
	91-95		0	1.0		0	0	2,265	2,110	0	1,000	0		10.1	4.6	0.2	0	0	
1995		14.5			0.3								5,215						5.4
	96-00		0	8.0		0	1.2	4,125	2,885	0	1,000	0		13.8	4.6	0.2	4.2	0	
2000		22.5			1.5								7,100						6.0
	2001-05		5.5	12.0		0	1.5	5,805	4,800	0	2,000	0		22.9	9.2	0.7	8.4	0	
2005		29.0			3.0								9,900						7.5
	06-10		2.5	11.0		0	1.5	6,805	5,485	0	4,000	0		26.2	18.4	1.7	9.2	0	
2010		37.5			4.5								11,385						18.4
	11-15		4.5	8.0		0.3	4.8	7,160	6,620	0	6,000	0		31.6	27.5	3.5	29.4	0	
2015		41.0			9.0								12,005						20.0
	16-20		1.0	8.5		0	7.5	8,120	6,955	0	8,000	0		33.2	36.7	8.2	43.6	0	
2020		48.5			16.5								10,960						21.3

Year	Period	Fuel Cycle Activities (tHM/a) Converter Reactors Fabrication UOX	Fabrication MOX	Reprocessing UOX	Reprocessing MOX	FBRs (Core + Blanket) Fabric	FBRs (Core + Blanket) Reproc
1975							
	1976-80	285	0	0	0	0	0
1980							
	81-85	460	0	0	0	5	0
1985							
	86-90	425	0	100	0	10	10
1990							
	91-95	455	0	200	0	10	10
1995							
	96-00	825	0	200	0	30	10
2000							
	2001-05	1,160	0	400	0	65	40
2005							
	06-10	1,360	0	800	0	105	75
2010							
	11-15	1,430	0	1,200	0	250	165
2015							
	16-20	1,625	0	1,600	0	435	300
2020							

Notes

1. See key to Table 9.20.
2. Converter reactor : AGR
   - 4.78 kg Pu_f/tHM discharged
   - losses : 2% in reprocessing + 2% from Pu241 decay
   - 4.59 kg Pu_f/tHM reprocessed.

TABLE 9.A.12. Britain: Fast Breeder/Plutonium Recycle Path

Year	Period	CR Capacity (GW) Capac	CR Capacity (GW) Close	CR Capacity (GW) Const	FBR Capacity (GW) Capac	FBR Capacity (GW) Close	FBR Capacity (GW) Const	Converter reactor Fuel (tHM) Fuel Charged	Fuel Disch	Spent F Import	Spent F Reprocessed UOX	Spent F Reprocessed MOX	Spent F Inventory	Fissile Plutonium (tPu_f) In Disch CR Fuel	From CR Reproc	Net from FBRs	To FBR Fuel	To CR MOX	Sep Pu Inventory
1975		5.5			0								0						0
	1976-80		0	2.5		0	0	1,435	1,045	0	0	0		5.0	0	0	0	0	
1980		8.0			0								1,045						0
	81-85		0	4.5		0	0.3	2,295	1,600	0	0	0		7.6	0	0	1.0	0	
1985		12.5			0.3								2,645						(-1.0)
	86-90		0	1.0		0	0	2,115	1,960	0	500	0		9.4	2.3	0.1	0.8	0	
1990		13.5			0.3								4,105						0.6
	91-95		0	1.0		0	0	2,265	2,110	740	5,755	0		10.1	27.5	0.2	0	22.3	
1995		14.5			0.3								1,200						6.0
	96-00		0	8.0		0	1.2	4,125	2,885	2,335	5,220	0		13.8	27.5	0.2	4.2	23.5	
2000		22.5			1.5								1,200						6.0
	2001-05		5.5	12.0		0	1.5	5,805	4,800	1,800	5,665	735		22.9	40.3	0.7	8.4	29.6	
2005		29.0			3.0								1,400						9.0
	06-10		2.5	11.0		0	1.5	6,805	5,485	2,935	7,245	775		26.2	49.9	1.7	9.2	40.4	
2010		37.5			4.5								1,800						11.0
	11-15		4.5	8.0		0.3	4.8	7,160	6,620	8,085	12,330	975		31.6	84.2	3.5	29.4	51.3	
2015		41.0			9.0								3,200						18.0
	16-20		1.0	8.5		0	7.5	8,120	6,955	6,785	12,400	1,335		33.2	88.2	8.2	43.6	50.8	
2020		48.5			16.5								3,200						20.0

Year	Period	Fuel Cycle Activities (tHM/a) Converter Reactors Fabrication UOX	Fabrication MOX	Reprocessing UOX	Reprocessing MOX	FBRs (Core + Blanket) Fabric	FBRs (Core + Blanket) Reproc
1975							
1980	1976-80	285	0	0	0	0	0
1985	81-85	460	0	0	0	5	0
1990	86-90	425	0	100	0	10	10
1995	91-95	455	(150)	1,200	0	10	10
2000	96-00	825	(155)	1,200	0	30	10
2005	2001-05	1,160	(195)	1,200	200	65	40
2010	06-10	1,360	(270)	1,590	210	105	75
2015	11-15	1,430	(340)	2,935	265	250	165
2020	16-20	1,625	(340)	2,840	360	435	300

Notes

1. See key to Table 9.20.
2. See notes to Table 9.A.11.
3. In calculating reprocessing activity required for a given Pu yield, 1 kg imported (LWR) spent fuel is assumed to be equivalent to 1.33 kg AGR fuel; average annual figures in the fuel cycle activities section are adjusted accordingly.
4. Pu recycle in unspecified non-British PWRs; spent MOX fuel returned to Britain for reprocessing
   - 19.0 kg Pu_f/tMOX discharged
   - losses: 2% in reprocessing + 8% from Pu^{241} decay
   - 17.1 kg Pu_f/tMOX reprocessed.

TABLE 9.A.13. United States: Fast Breeder Path

Year	Period	CR Capacity (GW) Capac	CR Close	CR Const	FBR Capacity (GW) Capac	FBR Close	FBR Const	Converter reactor Fuel (tHM) Fuel Charged	Fuel Disch	Spent F Import	Spent F Reprocessed UOX	Spent F Reprocessed MOX	Spent F Inventory	Fissile Plutonium (tPu_f) In Disch CR Fuel	From CR Reproc	Net from FBRs	To FBR Fuel	To CR MOX	Sep Pu Inventory
1975		36.0			0								0						0
	1976-80		0	28.0		0	0	7,460	5,110	0	0	0		32.1	0	0	0	0	
1980		64.0			0								5,110						0
	81-85		0	51.0		0	0	13,415	9,130	0	0	0		57.4	0	0	0	0	
1985		115.0			0								14,240						0
	86-90		0	35.0		0	0	17,195	14,255	0	0	0		89.7	0	0	0	0	
1990		150.0			0								28,495						0
	91-95		0	48.5		0	1.5	22,775	18,705	0	0	0		117.6	0	0	5.3	0	
1995		198.5			1.5								47,200						(-5.3.)
	96-00		0	68.5		0	1.5	30,670	24,915	0	3,750	0		156.7	20.8	0.6	9.2	0	
2000		267.0			3.0								68,365						6.9
	2001-05		36.0	84.5		0	1.5	37,890	33,820	0	7,500	0		212.7	41.6	1.7	9.2	0	
2005		315.5			4.5								94,685						41.0
	06-10		28.0	76.5		0	1.5	42,905	38,835	0	15,000	0		244.3	83.3	2.7	9.2	0	
2010		364.0			6.0								118,520						117.8
	11-15		51.0	66.0		0	30.0	45,525	44,265	0	30,000	0		278.4	166.5	6.3	163.5	0	
2015		379.0			36.0								132,785						127.1
	16-20		35.0	52.5		0	52.5	46,645	45,175	0	45,000	0		284.1	249.8	27.1	302.7	0	
2020		396.5			88.5								132,960						101.3

Year	Period	Fuel Cycle Activities (tHM/a) Converter Reactors Fabrication UOX	Fabrication MOX	Reprocessing UOX	Reprocessing MOX	FBRs (Core + Blanket) Fabric	FBRs Reproc
1975							
	1976-80	1,495	0	0	0	0	0
1980							
	81-85	2,685	0	0	0	0	0
1985							
	86-90	3,440	0	0	0	0	0
1990							
	91-95	4,555	0	0	0	30	0
1995							
	96-00	6,135	0	750	0	65	40
2000							
	2001-05	7,580	0	1,500	0	105	75
2005							
	06-10	8,580	0	3,000	0	140	115
2010							
	11-15	9,105	0	6,000	0	990	450
2015							
	16-20	9,330	0	9,000	0	2,370	1,425
2020							

Notes

1. See key to Table 9.20.
2. Converter reactor : PWR
   - 6.29 kg Pu_f/tHM discharged
   - losses : 2% in reprocessing + 10% from Pu241 decay
   - 5.5 $kgPu_f$/tHM reprocessed.

TABLE 9.A.14. United States: Fast Breeder/Plutonium Recycle Path

Year	Period	CR Capacity (GW) Capac	CR Capacity (GW) Close	CR Capacity (GW) Const	FBR Capacity (GW) Capac	FBR Capacity (GW) Close	FBR Capacity (GW) Const	Converter reactor Fuel (tHM) Fuel Charged	Fuel Disch	Spent F Import	Spent F Reprocessed UOX	Spent F Reprocessed MOX	Spent F Inventory	Fissile Plutonium (tPu_f) In Disch CR Fuel	From CR Reproc	Net from FBRs	To FBR Fuel	To CR MOX	Sep Pu Inventory
1975		36.0			0								0						0
	1976-80		0	28.0		0	0	7,460	5,110	0	0	0		32.1	0	0	0	0	
1980		64.0			0								5,110						0
	81-85		0	51.0		0	0	13,415	9,130	0	0	0		57.4	0	0	0	0	
1985		115.0			0								14,240						0
	86-90		0	35.0		0	0	17,195	14,255	0	0	0		89.7	0	0	0	0	
1990		150.0			0								28,495						0
	91-95		0	48.5		0	1.5	22,775	18,705	0	0	0		117.6	0	0	5.3	0	
1995		198.5			1.5								47,200						(-5.3)
	96-00		0	68.5		0	1.5	30,670	24,915	0	3,750	0		156.7	20.8	0.6	9.2	1.9	
2000		267.0			3.0								68,365						5.0
	2001-05		36.0	84.5		0	1.5	37,890	33,820	0	7,500	0		213.5	41.6	1.7	9.2	30.1	
2005		315.5			4.5								94,685						9.0
	06-10		28.0	76.5		0	1.5	42,905	38,835	0	22,435	65		256.9	125.6	2.7	9.2	102.1	
2010		364.0			6.0								111,020						26.0
	11-15		51.0	66.0		0	30.0	45,525	44,265	0	36,500	1,000		321.4	219.7	6.3	163.5	43.5	
2015		379.0			36.0								117,785						45.0
	16-20		35.0	52.5		0	52.5	46,645	45,175	0	49,100	3,400		302.4	330.6	27.1	302.7	28.0	
2020		396.5			88.5								110,480						72.0

Year	Period	Fuel Cycle Activities (tHM/a) Converter Reactors Fabrication UOX	Converter Reactors Fabrication MOX	Converter Reactors Reprocessing UOX	Converter Reactors Reprocessing MOX	FBRs (Core + Blanket) Fabric	FBRs (Core + Blanket) Reproc
1975							
1980	1976-80	1,495	0	0	0	0	0
1985	81-85	2,685	0	0	0	0	0
1990	86-90	3,440	0	0	0	0	0
1995	91-95	4,555	0	0	0	30	0
2000	96-00	6,120	15	750	0	65	40
2005	2001-05	7,380	200	1,500	0	105	75
2010	06-10	7,900	680	4,485	15	140	115
2015	11-15	8,815	290	7,300	200	990	450
2020	16-20	9,145	185	9,820	680	2,370	1,425

Notes

1. See key to Table 9.20.
2. See notes to Table 9.A.13.
3. Pu recycle in PWR
   - 19.0 kg Pu_f/tMOX discharged
   - losses : 2% in reprocessing + 8% from Pu241 decay
   - 17.1 kg Pu_f/tMOX reprocessed.

TABLE 9.A.15. Soviet Union: Fast Breeder Path

Year	Period	CR Capacity (GW) Capac	Close	Const	FBR Capacity (GW) Capac	Close	Const	Converter reactor Fuel (tHM) Fuel Charged	Fuel Disch	Spent F Import	Spent F Reprocessed UOX	MOX	Spent F Inventory	Fissile Plutonium (tPu_f) In Disch CR Fuel	From CR Reproc	Net from FBRs	To FBR Fuel	To CR MOX	Sep Pu Inventory
1975		5.5			0								0						0
	1976-80		0	6.0		0	0	1,350	845	0	0	0		5.3	0	0	0	0	
1980		11.5			0								845						0
	81-85		0	7.5		0	1.0	2,210	1,580	0	1,875	0		9.9	11.4	0	3.5	0	
1985		19.0			1.0								550						7.9
	86-90		0	17.0		0	0	4,210	2,780	0	1,875	0		17.5	11.4	0.4	2.6	0	
1990		36.0			1.0								1,455						17.1
	91-95		0	29.5		0	1.5	7,645	5,165	0	3,750	0		32.5	22.7	0.7	5.3	0	
1995		65.5			2.5								2,870						35.2
	96-00		0	55.5		0	1.5	14,135	9,470	0	7,500	0		59.6	45.4	1.3	9.2	0	
2000		121.0			4.0								4,840						72.7
	2001-05		5.5	59.0		0	1.5	20,425	15,935	0	15,000	0		100.2	90.8	2.4	9.2	0	
2005		174.5			5.5								5,775						156.7
	06-10		6.0	59.0		0	27.0	26,405	21,955	0	22,500	0		138.1	136.2	5.7	147.5	0	
2010		227.5			32.5								5,230						151.1
	11-15		7.5	47.5		1.0	46.0	30,860	27,500	0	30,000	0		173.0	181.5	25.6	265.8	0	
2015		267.5			77.5								2,730						92.4
	16-20		17.0	65.5		0	46.5	36,910	32,840	0	30,000	0		206.5	181.5	59.8	283.3	0	
2020		316.0			124.0								5,570						50.4

Year	Period	Fuel Cycle Activities (tHM/a) Converter Reactors Fabrication UOX	MOX	Reprocessing UOX	MOX	FBRs (Core + Blanket) Fabric	Reproc
1975							
	1976-80	270	0	0	0	0	0
1980							
	81-85	440	0	375	0	20	0
1985							
	86-90	840	0	375	0	25	25
1990							
	91-95	1,530	0	750	0	55	25
1995							
	96-00	2,825	0	1,500	0	90	65
2000							
	2001-05	4,085	0	3,000	0	130	100
2005							
	06-10	5,280	0	4,500	0	895	410
2010							
	11-15	6,170	0	6,000	0	2,100	1,290
2015							
	16-20	7,380	0	6,000	0	3,240	2,400
2020							

Notes

1. See key to Table 9.20.
2. Converter reactor : PWR
   - 6.29 kg Pu_f/tHM discharged
   - losses : 2% in reprocessing + 2% from Pu241 decay
   - 6.05 kg Pu_f/tHM reprocessed.

TABLE 9.A.16. Soviet Union: Fast Breeder/Plutonium Recycle Path

Year	Period	CR Capacity (GW) Capac	CR Capacity (GW) Close	CR Capacity (GW) Const	FBR Capacity (GW) Capac	FBR Capacity (GW) Close	FBR Capacity (GW) Const	Converter reactor Fuel (tHM) Fuel Charged	Fuel Disch	Spent F Import	Spent F Reprocessed UOX	Spent F Reprocessed MOX	Spent F Inventory	Fissile Plutonium (tPu_f) In Disch CR Fuel	From CR Reproc	Net from FBRs	To FBR Fuel	To CR MOX	Sep Pu Inventory
1975		5.5			0								0						0
	1976-80		0	6.0		0	0	1,350	845	0	0	0		5.3	0	0	0	0	
1980		11.5			0								845						0
	81-85		0	7.5		0	1.0	2,210	1,580	0	1,875	0		9.9	11.4	0	3.5	0	
1985		19.0			1.0								550						7.9
	86-90		0	17.0		0	0	4,210	2,780	0	1,875	0		17.5	11.4	0.4	2.6	14.1	
1990		36.0			1.0								1,455						3.0
	91-95		0	29.5		0	1.5	7,645	5,165	2,380	7,500	0		38.4	45.4	0.7	5.3	33.8	
1995		65.5			2.5								1,500						10.0
	96-00		0	55.5		0	1.5	14,135	9,470	2,530	10,780	470		73.8	73.3	1.3	9.2	60.4	
2000		121.0			4.0								2,250						15.0
	2001-05		5.5	59.0		0	1.5	20,425	15,935	4,315	17,625	1,125		125.5	125.9	2.4	9.2	108.1	
2005		174.5			5.5								3,750						26.0
	06-10		6.0	59.0		0	27.0	26,405	21,955	10,295	27,990	2,010		183.4	203.7	5.7	147.5	45.9	
2010		227.5			32.5								6,000						42.0
	11-15		7.5	47.5		1.0	46.0	30,860	27,500	11,500	33,900	3,600		192.3	266.7	25.6	265.8	9.5	
2015		267.5			77.5								7,500						59.0
	16-20		17.0	65.5		0	46.5	36,910	32,840	4,660	35,970	1,530		210.5	243.8	59.8	283.3	18.3	
2020		316.0			124.0								7,500						61.0

Year	Period	Fuel Cycle Activities (tHM/a) Converter Reactors Fabrication UOX	Fabrication MOX	Reprocessing UOX	Reprocessing MOX	FBRs (Core + Blanket) Fabric	FBRs (Core + Blanket) Reproc
1975							
1980	1976-80	270	0	0	0	0	0
1985	81-85	440	0	375	0	20	0
1990	86-90	745	95	375	0	25	25
1995	91-95	1,305	225	1,500	0	55	25
2000	96-00	2,425	400	2,155	95	90	65
2005	2001-05	3,365	720	3,525	225	130	100
2010	06-10	4,975	305	5,600	400	895	410
2015	11-15	6,105	65	6,780	720	2,100	1,290
2020	16-20	7,260	120	7,195	305	3,240	2,400

Notes

1. See key to Table 9.20.
2. See notes to Table 9.A.15.
3. Pu recycle in PWR
   - 19.0 kg Pu_f/tMOX discharged
   - losses : 2% in reprocessing + 8% from Pu241 decay
   - 17.1 kg Pu_f/tMOX reprocessed.

TABLE 9.A.17. Brazil, China, East Germany, Italy, Spain: Fast Breeder Path

Year	Period	CR Capacity (GW) Capac	CR Close	CR Const	FBR Capacity (GW) Capac	FBR Close	FBR Const	Converter reactor Fuel (tHM) Fuel Charged	Fuel Disch	Spent F Import	Spent F Reprocessed UOX	Spent F Reprocessed MOX	Spent F Inventory	Fissile Plutonium (tPu_f) In Disch CR Fuel	From CR Reproc	Net from FBRs	To FBR Fuel	To CR MOX	Sep Pu Inventory
1975		2.5			0								0						0
	1976-80		0	3.0		0	0	645	395	0	0	0		2.5	0	0	0	0	
1980		5.5			0								395						0
	81-85		0	10.0		0	0	1,840	1,000	0	0	0		6.3	0	0	0	0	
1985		15.5			0								1,395						0
	86-90		0	10.5		0	0	3,025	2,140	0	0	0		13.5	0	0	0	0	
1990		26.0			0								3,535						0
	91-95		0	15.0		0	0	4,750	3.490	0	0	0		22.0	0	0	0	0	
1995		41.0			0								7,025						0
	96-00		0	44.0		0	0	9,975	6,280	0	0	0		39.5	0	0	0	0	
2000		85.0			0								13,305						0
	2001-05		2.5	46.5		0	0	15,060	11,360	0	0	0		71.4	0	0	0	0	
2005		129.0			0								24,665						0
	06-10		3.0	51.5		0	7.5	20,580	16,505	0	9,375	0		103.8	52.0	0	26.3	0	
2010		177.5			7.5								31,795						25.7
	11-15		10.0	62.5		0	7.5	26,900	22,490	0	9,375	0		141.5	52.0	3.2	45.8	0	
2015		230.0			15.0								44,910						35.1
	16-20		10.5	83.0		0	7.5	35,270	29,180	0	18,750	0		183.5	104.0	8.4	45.8	0	
2020		302.5			22.5								55,340						101.7

Year	Period	Fuel Cycle Activities (tHM/a) Converter Reactors Fabrication UOX	Fabrication MOX	Reprocessing UOX	Reprocessing MOX	FBRs (Core + Blanket) Fabric	FBRs Reproc
1975							
1980	1976-80	130	0	0	0	0	0
1985	81-85	370	0	0	0	0	0
1990	86-90	605	0	0	0	0	0
1995	91-95	950	0	0	0	0	0
2000	96-00	1,995	0	0	0	0	0
2005	2001-05	3,010	0	0	0	0	0
2010	06-10	4,115	0	1,875	0	150	0
2015	11-15	5,380	0	1,875	0	325	200
2020	16-20	7,055	0	3,750	0	525	375

Notes

1. See key to Table 9.20.
2. Converter reactor : PWR
   - 6.29 kg Puf/tHM discharged
   - losses : 2% in reprocessing + 10% from Pu241 decay
   - 5.55 kg Puf/tHM reprocessed.

TABLE 9.A.18. Brazil, China, East Germany, Italy, Spain: Fast Breeder/Plutonium Recycle Path

Year	Period	CR Capacity (GW) Capac	CR Close	CR Const	FBR Capacity (GW) Capac	FBR Close	FBR Const	Converter reactor Fuel (tHM) Fuel Charged	Fuel Disch	Spent F Import	Spent F Reprocessed UOX	Spent F Reprocessed MOX	Spent F Inventory	Fissile Plutonium (tPu_f) In Disch CR Fuel	From CR Reproc	Net from FBRs	To FBR Fuel	To CR MOX	Sep Pu Inventory
1975		2.5			0								0						0
	1976-80		0	3.0		0	0	645	395	0	0	0		2.5	0	0	0	0	
1980		5.5			0								395						0
	81-85		0	10.0		0	0	1,840	1,000	0	0	0		6.3	0	0	0	0	
1985		15.5			0								1,395						0
	86-90		0	10.5		0	0	3,025	2,140	0	0	0		13.5	0	0	0	0	
1990		26.0			0								3,535						0
	91-95		0	15.0		0	0	4,750	3,490	0	0	0		22.0	0	0	0	0	
1995		41.0			0								7,025						0
	96-00		0	44.0		0	0	9,975	6,280	0	0	0		39.5	0	0	0	0	
2000		85.0			0								13,305						0
	2001-05		2.5	46.5		0	0	15,060	11,360	0	0	0		71.4	0	0	0	0	
2005		129.0			0								24,665						0
	06-10		3.0	51.5		0	7.5	20,580	16,505	0	9,375	0		103.8	52.0	0	26.3	14.7	
2010		177.5			7.5								31,795						11.0
	11-15		10.0	62.5		0	7.5	26,900	22,490	0	9,375	0		147.7	52.0	3.2	45.8	9.4	
2015		230.0			15.0								44,910						11.0
	16-20		10.5	83.0		0	7.5	35,270	29,180	0	18,260	490		187.5	109.7	8.4	45.8	59.3	
2020		302.5			22.5								55,340						24.0

Year	Period	Fuel Cycle Activities (tHM/a) Converter Reactors Fabrication UOX	Fabrication MOX	Reprocessing UOX	Reprocessing MOX	FBRs (Core + Blanket) Fabric	FBRs Reproc
1975							
	1976-80	130	0	0	0	0	0
1980							
	81-85	370	0	0	0	0	0
1985							
	86-90	605	0	0	0	0	0
1990							
	91-95	950	0	0	0	0	0
1995							
	96-00	1,995	0	0	0	0	0
2000							
	2001-05	3,010	0	0	0	0	0
2005							
	06-10	4,015	100	1,875	0	150	0
2010							
	11-15	5,315	65	1,875	0	325	200
2015							
	16-20	6,660	395	3,650	100	525	375
2020							

Notes

1. See key to Table 9.20.
2. See notes to Table 9.A.17.
3. Pu recycle in PWR
   - 19.0 kg Pu_f/tMOX discharged
   - losses : 2% in reprocessing + 8% from Pu241 decay
   - 17.1 kg Pu_f/tMOX reprocessed.

TABLE 9.A.19. Czechoslovakia, Poland, Sweden: Fast Breeder Path

Year	Period	CR Capacity (GW) Capac	Close	Const	FBR Capacity (GW) Capac	Close	Const	Converter reactor Fuel (tHM) Fuel Charged	Fuel Disch	Spent F Import	Spent F Reprocessed UOX	Spent F Reprocessed MOX	Spent F Inventory	Fissile Plutonium (tPu_f) In Disch CR Fuel	From CR Reproc	Net from FBRs	To FBR Fuel	To CR MOX	Sep Pu Inventory
1975		3.5			0								0						0
	1976-80		0	1.0		0	0	515	430	0	0	0		2.7	0	0	0	0	
1980		4.5			0								430						0
	81-85		0	5.0		0	0	1,115	695	0	0	0		4.4	0	0	0	0	
1985		9.5			0								1,125						0
	86-90		0	3.0		0	0	1,435	1,180	0	0	0		7.4	0	0	0	0	
1990		12.5			0								2,305						0
	91-95		0	6.5		0	0	2,200	1,650	0	0	0		10.4	0	0	0	0	
1995		19.0			0								3,955						0
	96-00		0	16.5		0	0	4,150	2,765	0	0	0		17.4	0	0	0	0	
2000		35.5			0								6,720						0
	2001-05		3.5	18.0		0	0	5,950	4,730	0	0	0		29.8	0	0	0	0	
2005		50.0			0								11,450						0
	06-10		1.0	21.0		0	0	8,115	6,435	0	0	0		40.5	0	0	0	0	
2010		70.0			0								17,885						0
	11-15		5.0	19.5		0	4.5	9,910	8,690	0	5,625	0		54.7	31.2	0	15.8	0	
2015		84.5			4.5								20,950						15.4
	16-20		3.0	24.0		0	4.5	12,285	10,480	0	5,625	0		65.9	31.2	1.9	27.5	0	
2020		105.5			9.0								25,805						21.0

Year	Period	Fuel Cycle Activities (tHM/a) Converter Reactors Fabrication UOX	Fabrication MOX	Reprocessing UOX	Reprocessing MOX	FBRs (Core + Blanket) Fabric	FBRs Reproc
1975							
1980	1976-80	105	0	0	0	0	0
1985	81-85	225	0	0	0	0	0
1990	86-90	285	0	0	0	0	0
1995	91-95	440	0	0	0	0	0
2000	96-00	830	0	0	0	0	0
2005	2001-05	1,190	0	0	0	0	0
2010	06-10	1,625	0	0	0	0	0
2015	11-15	1,980	0	1,125	0	90	0
2020	16-20	2,455	0	1,125	0	195	120

Notes

1. See key to Table 9.20.
2. Converter reactor : PWR
   - 6.29 kg Pu_f/tHM discharged
   - losses : 2% in reprocessing + 10% from Pu241 decay
   - 5.55 kg Pu_f/tHM reprocessed.

TABLE 9.A.20. Czechoslovakia, Poland, Sweden: Fast Breeder/Plutonium Recycle Path

Year	Period	CR Capacity (GW) Capac	Close	Const	FBR Capacity (GW) Capac	Close	Const	Converter reactor Fuel (tHM) Fuel Charged	Fuel Disch	Spent F Import	Spent F Reprocessed UOX	Spent F Reprocessed MOX	Spent F Inventory	Fissile Plutonium (tPu_f) In Disch CR Fuel	From CR Reproc	Net from FBRs	To FBR Fuel	To CR MOX	Sep Pu Inventory
1975		3.5			0								0						0
	1976-80		0	1.0		0	0	515	430	0	0	0		2.7	0	0	0	0	
1980		4.5			0								430						0
	81-85		0	5.0		0	0	1,115	695	0	0	0		4.4	0	0	0	0	
1985		9.5			0								1,125						0
	86-90		0	3.0		0	0	1,435	1,180	0	0	0		7.4	0	0	0	0	
1990		12.5			0								2,305						0
	91-95		0	6.5		0	0	2,200	1,650	0	0	0		10.4	0	0	0	0	
1995		19.0			0								3,955						0
	96-00		0	16.5		0	0	4,150	2,765	0	0	0		17.4	0	0	0	0	
2000		35.5			0								6,720						0
	2001-05		3.5	18.0		0	0	5,950	4,730	0	0	0		29.8	0	0	0	0	
2005		50.0			0								11,450						0
	06-10		1.0	21.0		0	0	8,115	6,435	0	0	0		40.5	0	0	0	0	
2010		70.0			0								17,885						0
	11-15		5.0	19.5		0	4.5	9,910	8,690	0	5,625	0		54.7	31.2	0	15.8	8.4	
2015		84.5			4.5								20,950						7.0
	16-20		3.0	24.0		0	4.5	12,285	10,480	0	5,625	0		69.4	31.2	1.9	27.5	5.6	
2020		105.5			9.0								25,805						7.0

Year	Period	Fuel Cycle Activities (tHM/a): Converter Reactors Fabrication UOX	Fabrication MOX	Reprocessing UOX	Reprocessing MOX	FBRs (Core + Blanket) Fabric	FBRs Reproc
1975							
1980	1976-80	105	0	0	0	0	0
1985	81-85	225	0	0	0	0	0
1990	86-90	285	0	0	0	0	0
1995	91-95	440	0	0	0	0	0
2000	96-00	830	0	0	0	0	0
2005	2001-05	1,190	0	0	0	0	0
2010	06-10	1,625	0	0	0	0	0
2015	11-15	1,925	55	1,125	0	90	0
2020	16-20	2,420	35	1,125	0	195	120

Notes

1. See key to Table 9.20.
2. See notes to Table 9.A.19.
3. Pu recycle in PWR
   - 19.0 kg Pu_f/tMOX discharged.

TABLE 9.A.21. India: Fast Breeder Path

Year	Period	CR Capacity (GW) Capac	Close	Const	FBR Capacity (GW) Capac	Close	Const	Converter reactor Fuel (tHM) Fuel Charged	Fuel Disch	Spent F Import	Spent F Reprocessed UOX	Spent F Reprocessed MOX	Spent F Inventory	Fissile Plutonium (tPu_f) In Disch CR Fuel	From CR Reproc	Net from FBRs	To FBR Fuel	To CR MOX	Sep Pu Inventory
1975		0.5			0								0						0
	1976-80		0	0.5		0	0	485	420	0	0	0		1.2	0	0	0	0	
1980		1.0			0								420						0
	81-85		0	0.5		0	0	765	700	0	0	0		1.9	0	0	0	0	
1985		1.5			0								1,120						0
	86-90		0	2.5		0	0	1,875	1,550	0	0	0		4.3	0	0	0	0	
1990		4.0			0								2,670						0
	91-95		0	2.0		0	0	3,060	2,795	0	0	0		7.7	0	0	0	0	
1995		6.0			0								5,465						0
	96-00		0	5.0		0	0	5,420	4,770	0	0	0		13.1	0	0	0	0	
2000		11.0			0								10,235						0
	2001-05		0.5	6.5		0	0	8,660	7,880	0	0	0		21.7	0	0	0	0	
2005		17.0			0								18,115						0
	06-10		0.5	8.5		0	0	12,830	11,790	0	0	0		32.4	0	0	0	0	
2010		25.0			0								29,905						0
	11-15		0.5	9.0		0	1.5	17,485	16,380	0	3,750	0		45.0	9.6	0	5.3	0	
2015		33.5			1.5								42,535						4.3
	16-20		2.5	14.0		0	1.5	23,630	22,135	0	3,750	0		60.9	9.6	0.6	9.2	0	
2020		45.0			3.0								60,920						5.3

Year	Period	Fuel Cycle Activities (tHM/a) Converter Reactors Fabrication UOX	Fabrication MOX	Reprocessing UOX	Reprocessing MOX	FBRs (Core + Blanket) Fabric	FBRs (Core + Blanket) Reproc
1975							
	1976-80	100	0	0	0	0	0
1980							
	81-85	150	0	0	0	0	0
1985							
	86-90	375	0	0	0	0	0
1990							
	91-95	610	0	0	0	0	0
1995							
	96-00	1,085	0	0	0	0	0
2000							
	2001-05	1,730	0	0	0	0	0
2005							
	06-10	2,565	0	0	0	0	0
2010							
	11-15	3,500	0	750	0	30	0
2015							
	16-20	4,725	0	750	0	65	40
2020							

Notes

1. See key to Table 9.20.
2. Converter reactor : HWR
   - 2.75 kg Pu_f/tHM discharged
   - losses : 2% in reprocessing + 5% from Pu241 decay
   - 2.55 kg Pu_f/tHM reprocessed.

TABLE 9.A.22. Rest of the World (Non-FBR Countries)

Year	Period	CR Capacity (GW) Capac	Close	Const	FBR Capacity (GW) Capac	Close	Const	Converter reactor Fuel (tHM) Fuel Charged	Fuel Disch	Spent F Import	Spent F Reprocessed UOX	MOX	Spent F Inventory	Fissile Plutonium (tPu_f) In Disch CR Fuel	From CR Reproc	Net from FBRs	To FBR Fuel	To CR MOX	Sep Pu Inventory
1975		3.5											0						
	1976-80		0	4.0				1,155	800	0				3.9					
1980		7.5											800						
	81-85		0	19.0				4,055	2,375	0				11.5					
1985		26.5											3,175						
	86-90		0	7.0				5,215	4,595	0				23.0					
1990		33.5											7,770						
	91-95		0	27.5				9,420	6,980	0				34.5					
1995		61.0											14,750						
	96-00		0	65.0				19,415	13,655	0				67.3					
2000		126.0											28,405						
	2001-05		3.5	77.0				31,090	24,575	0				122.2					
2005		199.5											52,980						
	06-10		4.0	94.0				45,135	37,160	0				185.2					
2010		289.5											90,140						
	11-15		19.0	126.0				62,680	53,205	0				266.2					
2015		396.5											143,345						
	16-20		7.0	145.5				83,535	71,265	0				356.0					
2020		535.0											214,610						

Year	Period	Fuel Cycle Activities (tHM/a): Converter Reactors Fabrication UOX	Fabrication MOX	Reprocessing UOX	Reprocessing MOX	FBRs (Core + Blanket) Fabric	FBRs Reproc
1975							
	1976-80	230	0				
1980							
	81-85	810	0				
1985							
	86-90	1,045	0				
1990							
	91-95	1,885	0				
1995							
	96-00	3,885	0				
2000							
	2001-05	6,220	0				
2005							
	06-10	9,025	0				
2010							
	11-15	12,535	0				
2015							
	16-20	16,705	0				
2020							

Notes

1. See key to Table 9.20.
2. Converter reactor : 90% LWR + 10% HWR.

10

Report of the International Consultative Group on Nuclear Energy

ENERGY AND NUCLEAR POWER

Today, as users of energy, we all face unprecedented responsibilities and uncertainties. Over the next few decades, it is overwhelmingly probable that global petroleum production will level off and begin to fall, the price of fossil fuels will go on rising, and the balance between energy supply and demand will become even more precarious. One result is that the world's energy system will be increasingly vulnerable to error, accident, or deliberate disruption. In that uncertain and delicately balanced future, the survival of our societies, as we know them, may depend on our separate and joint ability to limit energy consumption and expand energy production.

In the past, economic growth and development have always been accompanied by rising energy consumption. The ways in which they are connected to each other are not fully understood. Nor is the extent to which public policy can influence the connection between them. A great deal can certainly be done to use energy more efficiently. But more than that will be needed, if only because most conservation measures require large financial investments, are slow to produce results, and, even then, can have only a limited effect. Meanwhile, population growth and the natural desire for higher living standards, especially in the less developed countries where more than three-quarters of the world's people use less than a fifth of the world's energy, will continue to force up energy demand.

Despite higher prices and general economic recession, the world's annual consumption of primary energy has increased since 1972 by over 20 percent: an amount equivalent to about two and a half times the current annual oil production of Saudi Arabia. But oil production alone can no longer be increased fast enough to meet that rising demand. Moreover, as recent events have shown, oil supplies can be affected by often unpredictable political cir-

Date of origin, 17 January 1980.

cumstances. However much is done to use energy more efficiently, therefore, we must expect the coming decades to be a period in which a growing proportion of demand will have to be satisfied by means other than petroleum – or not be satisfied at all. As the balance between supply and demand becomes more delicate, we must also expect increasingly fierce international competition for the energy sources which are available.

The dangers of either unsatisfied demand for energy or an all-out competition for available energy supplies are self-evident. The impact on individual countries would of course vary, simply because countries differ in their access to energy sources, their dependence on them, and their ability to pay for them. In general, however, energy shortages, inevitably accompanied by sharply rising energy prices, would be bound to reduce industrial output, increase unemployment, and depress living standards. The effects would be particularly severe in developing countries. Far from securing the larger share of world resources they seek, those countries would face the prospect of energy shortages and costs presenting yet another barrier to development programs and perpetuating unemployment, poverty, and hunger. But almost all the developed industrial countries would face real hardships as well.

Contemplating the possibility of unsatisfied energy demand, individual countries, fearing for their national welfare and security, would be tempted to use every asset to their separate advantage. Those with large reserves of essential raw materials and fuels, many of which are in the developing world, would be inclined to retain them or to raise their prices. Those possessing advanced technologies, including energy technologies, most of which are in the developed world, would tend to treat that asset in the same way. Those with larger financial resources would be driven to compete for the energy sources still available, preempting supplies and driving up prices. In such a fiercely competitive atmosphere, the chances of establishing a reasonable relationship between developing countries with raw materials and developed countries with technology, each essential to the other's development, would be minimal. Instead, feelings of insecurity would be accentuated and the risk of international conflict, including even military conflict, greatly increased.

The danger of fiercer international competition and conflict over energy supplies, provoked by the fear of unsatisfied demand, is obviously of universal concern. That only underlines the extent to which energy itself has become an inescapably international issue. The policies of individual states, even if intended to affect only the management of energy within their own borders, have an international impact. In facing a future of growing energy demand and increasing uncertainty about energy supplies and prices, all states must thus accept that global energy interdependence is a harsh reality.

Although we must expect energy demand to increase and traditional energy sources to become harder to obtain at an acceptable price, we do not know exactly how much energy the world will need in twenty, thirty, or forty years' time, or how quickly additional supplies can be made available. Indeed, facts about the extent and cost of future energy supplies are increasingly difficult to establish. One necessary conclusion is that long-term energy policies have to be designed not merely to match a specified supply to a defined demand, since we can be sure of neither, but also to create a capacity for managing uncertainty. It is important, therefore, that a sufficient range of energy supply options should be developed in time for a rational choice to be made between them as the pattern of future demand develops.

The options we need to manage future uncertainty should include the best available techniques for producing, converting, and using energy. The longer-term potential of nuclear fission energy, which is one of the supply sources already available to us, must be seen in that context.

Today, nuclear fission power equivalent to about 2 percent of world primary energy supply provides only about 7 percent of our electricity (although the proportion reaches over 25 percent in some countries). Projections of future nuclear power growth vary. However, programs already in hand mean that its contribution will grow considerably during the 1980s. A typical estimate would suggest that, by 1985, it will represent more than 5 percent of primary energy and generate some 17 percent of the world's electricity. By 2000, given the considerable scope for its further expansion, the role of nuclear power is likely to be larger still, and by 2020 it could be called upon to meet a fifth of all the world's primary energy needs. Already, therefore, it represents a significant part of the response to rising energy demand, while, for the uncertain long-term future, its expanded use constitutes one of the substantial energy supply options which can and should be placed at the world's disposal.

The option of expanding nuclear power supply will not be available for the long term unless its development is carefully sustained during the intervening period. In the past few years, the slower growth of electricity demand, a generally unfriendly economic climate, and a variety of delays and difficulties in building, licensing, and operating nuclear reactors have combined to inhibit nuclear power programs. Economic constraints seem likely to bear heavily on nuclear investment in some countries, especially by private utilities, during the next few years as well. Timely development of nuclear options is particularly difficult in such a period. Just for that reason, it is all the more important that governments and industries should take deliberate steps to maintain a rational balance of research and industrial capacity and an adequate supply of experienced scientists and trained technicians, so that the option of future nuclear power expansion is not closed off.

Expansion for its own sake, for the wrong reasons or at the wrong time, is obviously to be avoided; planning for energy uncertainty requires the freedom not to pursue, as well as to pursue, any of the options available. Growth of energy demand, albeit at an uncertain rate, together with the danger of fiercer competition for energy supplies is, however, to be expected. With that in mind, we believe that the continued development of nuclear power is needed as an important contribution to retaining a capacity for meeting the world's future energy needs.

ESTABLISHING NUCLEAR OPTIONS

Current programs will result in the wider commercial deployment of nuclear fission reactors similar to those already in service, fueled with natural or low-enriched uranium but capable of extracting hardly 1 percent of the energy potentially available in uranium itself. Decisions to install these so-called thermal reactors in particular countries reflect a number of different considerations: the relative cost of generating electricity by nuclear, as against other, means; the existence of an electricity grid able to absorb generating units of the relevant size; the technological capacity to support a nuclear program; the availability of investment capital; the desire to reduce dependence on scarce or imported fuels. Especially in the developing world, those considerations limit the number of countries in which nuclear power is likely to be used before the end of this century.

In the longer term, the potential deployment of today's thermal reactors is more generally limited by the fact that economically recoverable uranium, like oil, is a finite commodity. Improved methods of uranium enrichment, fuel design, and operation could somewhat extend the utility of current reactor types. Although its economic attractions are a matter of dispute, some contribution to uranium conservation might also be made by recycling plutonium and uranium separated from spent fuel into thermal reactors of present types. Because their intrinsic efficiency in using uranium is limited, however, these reactors cannot be expected to satisfy an indefinitely rising demand for nuclear power, into the twenty-first century. More efficient forms of uranium fueled thermal reactor are therefore worthy of development, as a valuable way of extending the viability of the present fuel cycle. But even that could not be more than an interim means of limiting uranium consumption. For the long term, new and more advanced uranium-conserving technologies must be prepared.

If further development proved it feasible, a substantial saving of uranium might be achieved in the longer term by using thorium in thermal reactors. Ultimately, however, the way to escape from the constraint of uranium availability in the twenty-first century is seen to be to use fast breeder reactors: that is, reactors which "breed" more fissile plutonium than they consume fissile uranium, and which thus, in effect, extract more of the energy potential from uranium itself.

Given uncertainties about the long-term energy future, it is impossible to predict the rate of commercial deployment of fast breeder reactors in different areas of the world. In any case, the total number likely to be deployed commercially during the early stages could not have any larger an effect on world uranium consumption over the next thirty to forty years than might the use of more efficient thermal reactors. With the longer term in mind, however, we are convinced of the need to provide an option to deploy commercial fast breeder reactors. We are also convinced that the option will be available only if a major effort is made throughout the remainder of the present century to establish it, by further development and by successfully operating commercial-scale prototypes.

That effort must continue to involve substantial economic support and political commitment on the part of individual governments, with the main burden being carried in countries within the limited group, including France, Britain, West Germany, Japan, the Soviet Union, and the United States, which is in the forefront of fast breeder reactor technology. However, given the scope and complexity of the work still needed, few countries could reasonably bear the financial burden of a completely independent fast breeder reactor program, especially if it meant giving up development of other energy supply options. An even wider measure of international cooperation than already exists should also, therefore, be sought.

Because fast breeder reactors both produce and consume plutonium, their deployment entails reprocessing spent reactor fuel, to separate plutonium (and depleted uranium) for further use. There are other arguments for civil reprocessing as well, carrying different weights in different countries and circumstances: to concentrate nuclear waste into an environmentally acceptable form; to limit long-term access to plutonium; to deal with spent fuel which is particularly difficult to store, such as that from gas-graphite reactors; to separate plutonium for recycling into thermal reactors. It is, however, the potential commercial deployment of fast breeder reactors, and the need to have sufficient capacity in place to separate plutonium in time to meet their needs, which constitutes the most generally persuasive reason for developing reprocessing as a routine stage in the nuclear fuel cycle.

In parallel with a program to establish the fast breeder reactor option commercially, work must therefore continue on demonstrating the commercial feasibility of large-scale reprocessing of spent fuel from both thermal and breeder reactors. While allowance must also be made for the weight attached to other arguments in particular national circumstances, the rate at which commercial reprocessing capacity is actually expanded should depend principally upon the rate at which it proves necessary to deploy fast breeder reactors. We believe, however, that a deliberate program to develop and demonstrate commercial-scale reprocessing is needed now, as part of the preparation of nuclear options for the future.

NUCLEAR SAFETY AND THE PUBLIC INTEREST

Whatever technology may permit or economics demand, the future of nuclear power depends on earning and retaining a sufficient measure of public acceptance. Especially in most of the Western industrial countries, therefore, it depends on reducing public concern over the possibility of nuclear accidents, the problems of nuclear waste, and the health effects of low-level radiation. Ultimately, it also depends on people being convinced that the wider use of nuclear power is compatible with the sort of society in which they want to live.

The question is not whether the civil use of nuclear energy entails risks. It does. Every stage in the fuel cycle, from uranium mining to waste disposal, carries risks to those who engage in it and to the community at large. So, however, do all other activities connected with the supply of energy: mining, transporting, and burning coal, for example, or carrying liquid natural gas, or establishing large hydroelectric dams, or producing and fabricating solar installations. Risks of accident, injury, disease, and death cannot be eliminated from any of those activities, any more than from flying, driving, or eating.

While their relative significance varies from country to country, the real questions about nuclear safety are quite different. Are nuclear facilities designed and operated with due regard for the public interest? Do national institutions provide adequate information about nuclear and other risks and the means of containing them? Can nuclear risks and the risks of providing energy by other means be assessed accurately enough to compare them fairly with each other? How should nuclear risks be weighed against the alternative risks associated with the limits on future energy supply if nuclear power were not available? It is on answers to real questions such as these, rather than on any mere identification of risks in the abstract, that the future of nuclear power should depend.

Nuclear power is sometimes identified as a prominent example of the technical complexity and institutional centralization against which sentiment in at least some countries has become stronger. Some, taking an extreme view, have even accused nuclear power of leading toward necessarily authoritarian state control. There is no simple way of responding to such assertions. Nuclear power is a complex, costly, and large-scale technology, which requires central control if stringent standards are to be maintained. It must be shown in practice, however, that the central controls needed to protect the public interest can be combined with responsiveness and accountability. The institutions responsible for nuclear power must therefore take unusual pains to demonstrate their sensitivity to reasonable public fears or opinions and to establish and maintain an unblemished reputation for open and honest public communication.

Some of the hazards associated with nuclear power are said or thought to be of a special kind. That is particularly true of radiation, statements about the possible long-term effects of which arouse widespread fear. Yet radiation hazards exist as a normal feature of daily life, in industry, in medical treatment, or in the natural environment, and would so exist if nuclear fission had never been discovered. Partly for that reason, it has taken less time to assess them than it has to come to grips with the long-term health hazards of many other activities, such as the production of coal, tin, asbestos, or toxic chemicals, the smoking of tobacco, or the consumption of alcohol. The assessment shows clearly that the radiation hazards of nuclear power can, in principle, be held within publicly acceptable limits, especially when measured against the hazards associated with alternative energy sources.

To say that nuclear risks can be held within acceptable limits is not to argue for complacency. What has to be demonstrated to the public is that the standards of public and occupational safety which are attainable in principle will also be enforced and maintained in practice. In general, nuclear power has a record of public and occupational safety which compares favorably with that of other industries, including other energy industries. In particular cases, the implementation and enforcement of safety criteria could nevertheless be improved. That, for example, is one lesson of the accident in March 1979 at the Three Mile Island nuclear power station near Harrisburg, in the United States, where the fact that the accident happened and developed as it did is testimony to the need, in that particular case, to improve the national application of design and operating standards.

Another area needing continued attention is that of nuclear waste management. Much of the nuclear waste now in existence has come from military programs. As the amount of electricity generated by nuclear means increases, however, a growing volume of spent nuclear fuel, containing plutonium, depleted uranium, and highly radioactive fission products, will require careful handling and treatment.

While interim storage procedures for spent fuel are well established, techniques for disposing finally of highly active waste products separated in reprocessing, or of spent fuel itself if it is not to be reprocessed, are still being developed, tested, or demonstrated. Because it will be some years before any large amount of nuclear power waste has to be disposed of permanently, that does not imply a need to delay other nuclear operations, such as reactor construction or licensing. It does imply that governments, supported by industry, should accelerate the testing and demonstration of effective and acceptable methods for disposing finally of reprocessed or unreprocessed nuclear waste, so that the public may have a reasonable assurance that safe provision will be available as the quantity of waste from nuclear power increases.

In addressing the real questions about nuclear safety, it is easy to lose a sense of proportion. Numerous public inquiries into nuclear programs have explored the public and occupational hazards of nuclear power, to an extent unparalleled in the case of comparable industries. Their general conclusion has always been that nuclear risks can practicably be held to a level universally acceptable in other connections. Perhaps inevitably, that conclusion has received less public attention than the catalog of the risks themselves. Yet there is no risk-free way of maintaining and increasing energy supplies. Standards used to compare the alternative ways of doing so must therefore be even-handed, and judgment must not be distorted by applying one set of standards to nuclear power and a different set to other energy sources. Governments have a duty to ensure that their citizens are equally well informed about the risks associated with all forms of energy supply.

Governments also have an international responsibility in regard to nuclear safety. A major nuclear accident may cause damage beyond national borders, and it is in any case likely to increase public concern in other countries. Conversely, work on nuclear safety or waste management in one country may be relevant to other countries as well. There is a clear need, therefore, for governments to intensify their cooperation in these areas, directly or through international agencies, to share information and experience freely, and to do all they can to assist countries with less developed nuclear power industries in achieving high standards of safety. Relevant information should be provided promptly to other countries in case of a nuclear accident, and other countries should be ready to respond, if practicable, to requests for assistance, bilaterally or under the auspices of the International Atomic Energy Agency (IAEA).

We believe that public concern over nuclear safety can be dissipated only by demonstrating convincingly that the risks inherent in producing nuclear power are very low and, in any case, no greater than the risks of alternatives, and that those responsible for nuclear power are firmly committed to, and capable of, protecting the public safety. Government and industry have a shared duty to provide that demonstration. At the same time, the extent to which nuclear power is publicly accepted depends upon a wider range of political and social considerations than we have here presented. Rational assessment of that wider range imposes responsibilities not only on industry and government but also on all groups which influence or contribute to public attitudes.

NUCLEAR TRADE AND NUCLEAR PROLIFERATION

The main impediment to international nuclear trade and collaboration is the fear of nuclear weapons proliferating. Many would regard the almost unimaginable destructive power of the nuclear weapons already deployed as an even graver risk to international security in general. It is fears of further proliferation, however, which have come more specifically to hamper the peaceful nuclear relations between states. Governments may disagree about the management of that problem. There is nevertheless widespread international understanding that a world in which more states control nuclear weapons, or where more nuclear weapons are deployed, would be an even more dangerous one for all.

Weapons proliferation is not an inevitable, or even a probable, result of expanded civil nuclear energy use. Indeed, no nuclear weapons force has yet depended for its existence on civil nuclear power development. A country embarking on nuclear armament might well find that a self-sufficient civil nuclear power industry could shorten the time needed to produce material for weapons. But easier and cheaper routes to weapons acquisition are likely to lie through the establishment, in the first place, of separate facilities committed to military use. Moreover, having a substantial nuclear power industry which depends on international trade may deter a government from risking that portion of its national energy supply by embarking on a nuclear weapons program in defiance of its trading partners.

The risk of weapons proliferation would continue to exist even if nuclear power did not. Particular stages of civil nuclear development, undertaken in particular circumstances, may nevertheless increase the fear that weapons proliferation will occur in future. The stages in question are especially those involving enrichment or reprocessing facilities capable of producing the high-enriched uranium or separating the plutonium usable in weapons. To the extent that they are seen to be needed for civil purposes, establishing such facilities, provided they are under international safeguards, cannot reasonably be seen as a deliberate step toward acquiring weapons. The fear that they will prove to be such a step may arise, however, if their exclusively civil purpose is not made immediately credible.

Countries operating or planning to operate power reactors requiring low-enriched uranium fuel must be expected to want assured access to enrichment services on equitable terms. In the aggregate, the enrichment plants already in operation or scheduled to come into service have ample capacity to satisfy world requirements until close to the end of the century. Thus, the balance between total demand and total supply, although it may not be the only consideration, implies no urgent need for additional countries to undertake enrichment, especially in view of its technical difficulty and cost. That will be relevant, however, only if all countries in need of enriched uranium have solid grounds for believing that enrichment services will, in fact, be available to them internationally on acceptable terms.

Confidence on that score has been shaken in the past by unilateral changes in the terms on which enrichment services were provided by the United States. Some of that effect has been offset by the subsequent emergence of additional suppliers, in Western Europe and the Soviet Union, since a diversification of supply sources is itself one form of supply assurance. International confidence will be fully rebuilt, however, only if all suppliers of enrichment services provide credible evidence that all international contracts will be fulfilled on the terms originally agreed.

A thoroughly credible assurance of international supply would tend to weaken the argument for additional countries to establish civil enrichment plants. Some might nevertheless consider exercising their sovereign right to do so, for reasons of technological development, supply security, or economic advantage. However, because high-enriched uranium probably offers the shortest route to weapons production, they would have to take account of the fact that the international need for a convincing demonstration of their exclusively peaceful purpose would be all the greater if there were no longer an obvious energy security argument for independent enrichment.

Plutonium separated from spent fuel in a reprocessing plant offers another possible route to weapons manufacture. One important difference between the cases of enrichment and reprocessing plants, however, is that reprocessing is not also immediately essential to operating most current power reactors, the spent fuel from which can instead be held in storage. Countries looking to the future may nevertheless want to be sure of having access, if needed, to the option of commercial reprocessing. Their desire to ensure such access by building their own reprocessing plants will be all the stronger if they see any reason to fear that adequate reprocessing services may not be available internationally, on acceptable terms, when they need them. Confidence in international supply is therefore as important potentially in the reprocessing case as it is immediately in the case of enrichment.

Because separated plutonium, like high-enriched uranium, is potentially usable in weapons, and thus "sensitive," all countries involved in its civil production have reason to see that it is separated, held, and used in circumstances which minimize fears of its diversion to a military program. International safeguards represent the first and most important step in that direction. But the burden on a safeguards system is bound to become heavier to the extent that the number of reprocessing plants in the world increases. So, for the countries concerned, is the burden of demonstrating that all such plants have an exclusively civil purpose. As in the case of uranium enrichment, therefore, minimizing the fear of possible proliferation argues for matching the number of facilities separating or handling plutonium under safeguards to the scale of nuclear power needs, including the need for all countries to have dependable access to the services and materials required for energy supply.

Although the problems of horizontal and vertical proliferation exist independently, fears of proliferation are liable to grow as the expansion of nuclear power creates pressure for additional countries to undertake enrichment or reprocessing. Those fears can be avoided or reduced only by providing convincing evidence on two counts: that civil nuclear programs involving "sensitive" facilities or materials will remain exclusively peaceful, and that the international supply of nuclear materials, services, and technology will be maintained on agreed and predictable terms. Averting the fear and risk that proliferation may stem from the expanded use of nuclear power depends not on technical contrivances or unilateral restrictions, which must be either insufficient or counterproductive, but on constructing that international bargain of confidence.

Confidence in international nuclear supply cannot be based on rhetoric alone. It can be established only by the consistent performance of suppliers, complemented rather than countermanded by their national legislation, and reinforced as necessary by formal guarantees, but demonstrated, above all, in practice.

As to confidence in the exclusively civil purpose of nuclear power programs, the primary means of establishing that is, and should remain, the even-handed application of safeguards under the aegis of the IAEA. As nuclear power use expands, that safeguards system has to be further developed and more broadly applied. It can also usefully be reinforced by formal non-proliferation commitments and guarantees. The process must necessarily, however, be voluntary. Countries which have already accepted IAEA safeguards, or have subscribed to the Non-Proliferation Treaty or other relevant treaties, have done so freely, on the grounds that their national interests were thus served. States will equally make and maintain such commitments in the future not as a result of international pressure but because they are convinced that their national security will most effectively be assured by doing so.

One way confidence in both international supply and peaceful purpose might be further reinforced is by developing arrangements to conduct "sensitive" processes or hold "sensitive" materials not only under safeguards but also under multinational auspices. The IAEA has taken a lead in studying possible multinational nuclear fuel cycle arrangements, implementation of which might reduce fears of both proliferation and unilateral restrictions on international trade. Its efforts to explore their feasibility deserve general support. In particular, provided conditions for deposit and release were clearly defined and agreed, a commitment to place in international custody separated plutonium not needed for immediate civil use could help to alleviate fears aroused by the construction of additional reprocessing plants.

Multinational arrangements will never be easy to establish or administer, and they will have a useful role to play only if countries with interests in both non-proliferation and nuclear power determine, through negotiation, that they will serve those two interests at once. They urgently deserve fuller international consideration, however, as one possible means of countering some of the fears associated with the expansion of nuclear power.

Whatever is done to combat fears of proliferation must be done in a spirit of international cooperation rather than confrontation. Apart from the wider damage to international relations which confrontation may cause, it is likely to prove counterproductive by serving only to encourage additional countries to develop self-sufficient national nuclear fuel cycle industries, despite their high cost. Cooperation, in turn, must be founded on reciprocal restraint. Any attempt to engage in political or economic extortion by exploiting particular strength—whether strength derived from ownership of resources, control of technology, or a capacity to disturb international stability—is thus strongly antithetical to it. Provided there is reciprocal restraint, however, we are convinced that international cooperation can create more and stronger barriers to nuclear proliferation than international competition, confrontation, or conflict, and that the immediate goal should therefore be to establish the "bargain of confidence" we have here described.

CONDITIONS FOR THE FUTURE

To summarize the major conclusions of our work together, we believe that, if nuclear power is to be available to meet an increasing fraction of the world's future energy needs, at least five conditions will have to be satisfied.

The first condition is that nuclear power, despite the difficulty of the short-term climate, will have to be systematically developed, without interruption or undue delay.

The possibility of unsatisfied energy demand and the danger of intense competition for energy supplies represent serious risks to security, development, and welfare. Nuclear power has an important part to play in containing those risks, by helping to meet the increasing demand for energy. Governments and industries must ensure that it is preserved and prepared as a substantial energy supply option, not only for the next decades but also for the long-term future.

The second condition is that nuclear power must earn and retain public acceptance.

The political and social issues bearing on acceptance extend beyond the nuclear energy field. Within that field, however, the issue of nuclear safety is particularly relevant. Government and industry again share a duty to demonstrate that the public and occupational hazards of nuclear power, in relation to the risks of alternatives, are acceptably small, that those responsible for nuclear power are committed to protecting the public safety, and that safety standards attainable in principle will also be maintained in practice. In that connection, testing and demonstration of effective and acceptable methods for disposing finally of nuclear waste must be accelerated, and international cooperation in both waste management and nuclear safety must be intensified.

The third condition is that technologies for using uranium more efficiently must be developed and tested as soon as possible, with both the coming decades and the twenty-first century in mind.

In the first instance, that means providing ways to make better use of uranium in existing and improved types of thermal reactors. Although the proper timetable for their deployment cannot yet be determined, it also means demonstrating the commercial feasibility of fast breeder reactors as an option for the longer term, as well as the feasibility of the commercial-scale reprocessing necessary to that option. Future decisions about whether or when to deploy new uranium-conserving technologies commercially have to be taken in the light of particular and changing circumstances, but development of the options themselves cannot safely be delayed.

The fourth condition is that the fear of nuclear weapons proliferation resulting from an expansion of nuclear power must be further reduced.

A convincing demonstration is needed that fuel cycle activities, including enrichment and reprocessing in particular, will be matched to nuclear power needs and have an exclusively peaceful purpose. That cannot be achieved by technical contrivance or unilateral action based on strength. It can be achieved only by self-restraint and by voluntary application of the evolving system of IAEA safeguards, reinforced by formal non-proliferation commitments and possibly by new multinational arrangements.

The fifth condition, closely linked to the fourth, is that countries depending on nuclear technology, services, or materials to ensure their energy supply must be convinced of continued international access to them, under safeguards, on acceptable terms.

Unless all are so convinced, their need for security of energy supply will drive more countries to seek a degree of nuclear self-sufficiency, at excessive cost, which may exacerbate fears of proliferation. Assurance of international supply might again be reinforced by multinational arrangements and can certainly be strengthened by credible guarantees. But the assurance itself can come only from practical experience of reliable performance in international nuclear trade and cooperation.

It is the international aspects of nuclear power which have particularly concerned us. The recent erosion of confidence in international nuclear relations must be repaired. There must, in fact, be re-created a sense of global responsibility, on the part of all states, strong enough to support practical arrangements for sharing both nuclear resources and technology internationally on mutually acceptable terms. The only way of doing that is to construct a new "bargain of confidence," in which credible guarantees of peaceful purpose will be balanced by firm assurances of access to nuclear services, materials, and technology. Establishing that bargain depends on resolving existing disagreements by compromise. It is in that spirit that we have subscribed to this report.

Ian Smart, Chairman
Adviser on International Energy Affairs

Miguel Barandiarán Alcorta
Subdirector Técnico de Iberduero SA
Spain

Russell W. Fox
Ambassador-at-Large for Australia for Nuclear Non-Proliferation and Safeguards

Paulo Nogueira Batista
Career Ambassador
President of Nuclebras
Brazil

Karl-Heinz Beckurts
Chairman of the Board
Kernforschungsanlage Jülich
Federal Republic of Germany

Juan Eibenschutz
Director General de Energía
Secretariat de Patrimonio y Fomento Industrial
Mexico

Akbar Etemad
Former President
Atomic Energy Organization of Iran

David Fischer
Assistant Director General for External Relations
International Atomic Energy Agency

Bertrand Goldschmidt
Governor from France to the International Atomic Energy Agency

Ryukichi Imai
General Manager for Engineering
Japan Atomic Power Company

Jon Jennekens
President
Atomic Energy Control Board
Canada

Måns Lönnroth
Secretariat for Futures Studies
Sweden

Robert Press
Cabinet Office
United Kingdom

Mason Willrich
Vice-President for Corporate Planning
Pacific Gas and Electric Company
United States

Myron Kratzer (Rapporteur)
Consultant, International Energy Associates Ltd.

Glossary and Abbreviations

Note: Entries are alphabetized by term, not by abbreviation.

Abbreviation	*Term*	*Definition*
AGR	Advanced gas-cooled reactor	See under *reactor.*
Am	Americium	Element of atomic number 95.
Am241	*Americium 241*	*Radioactive isotope* formed by *decay* of *plutonium 241.*
a	Annum, year	
A/E	Architect/engineer	Enterprise responsible for engineering design of a *reactor.*
AEC	Atomic Energy Commission (United States)	
AECB	Atomic Energy Control Board (Canada)	
AECL	Atomic Energy of Canada, Ltd.	
AFR	Away from reactor	Describing storage facility for *spent fuel* not at site of a *reactor.*
BWR	Boiling water reactor	See under *reactor.*
	Breeder reactor	See under *reactor.*

Abbreviation	*Term*	*Definition*
Btu	British thermal unit	Unit of heat (1 Btu = 1,055 *joules*).
	Burnup	Measure of the time period in relation to power level during which a stated amount of *fuel* is subjected to *irradiation* in a *reactor*.
CANDU	Canadian deuterium uranium (reactor)	See under *reactor*.
	Capacity factor	See under *load factor*.
CEGB	Central Electricity Generating Board (Britain)	
CPE	Centrally planned economy	Any of the following states: Soviet Union, Poland, Romania, East Germany, Czechoslovakia, Hungary, Bulgaria, Yugoslavia, Albania, China, Vietnam, North Korea, Kampuchea, Laos, Mongolia.
	Centrifuge	See under *enrichment*.
	Civex	See under *reprocessing*.
	Closed cycle	See under *fuel cycle*.
CEA	Commissariat à l'Energie Atomique (France)	
	Conversion	Conventionally, the process of converting an oxide form of *uranium* into a fluoride compound (*uranium hexafluoride*) in preparation for *enrichment*.
CR	Converter reactor	See under *reactor*.
	Coprocessing	See under *reprocessing*.
	Criticality	Point in time at which a *reactor* achieves a self-sustaining *fission* chain reaction.

Abbreviation	*Term*	*Definition*
Comecon	Council for Mutual Economic Assistance	Economic organization linking the Soviet Union, East European states, and Mongolia.
	Decay	Process by which atomic nuclei of a *radioactive isotope* decay spontaneously to form another isotope.
	Degraded uranium	See under *uranium.*
	Denatured plutonium	See under *plutonium.*
	Depleted uranium	See under *uranium.*
DWK	Deutsche Gesellschaft für Wiederaufarbeitung von Kernbrennstoffen (West Germany)	
	Diffusion	See under *enrichment.*
$	Dollar	United States dollar ($1 = 100 cents = 1,000 mills).
ENDC	Eighteen Nation Disarmament Committee	
EdF	Electricité de France	
EPRDA	Electric Power Resources Development and Adjustment Council (Japan)	
ERDA	Energy Research and Development Administration (United States)	
	Enriched uranium	See under *uranium.*
	Enrichment	Process of increasing the proportion of the *fissile uranium 235 isotope* in *uranium* by any of a variety of techniques involving separative work.
	Centrifuge, gas centrifuge	Enrichment method based on subjecting *uranium hexafluoride* gas to centrifugal force in a rotating cylinder.

Abbreviation	*Term*	*Definition*
	Diffusion, gaseous diffusion	Enrichment method based on repeated passage of *uranium hexafluoride* gas through porous membranes.
	Toll enrichment	Enterprise in which the *uranium* to be enriched is owned by the customer of the service, rather than by the enricher.
EAR	Estimated additional resources	Geological resources (of *uranium*) in identified deposits or related formations but not included in *reasonably assured resources.*
	Eurochemic	Multinational consortium of West European states, under the auspices of the European Nuclear Energy Agency of the *Organization for European Economic Cooperation* (later the *Nuclear Energy Agency* of the *Organization for Economic Cooperation and Development*), set up to build, own and operate a *reprocessing* plant in Belgium.
	Eurodif	Multinational consortium established by France, Italy, Spain, Belgium, and Iran to invest in *enrichment* capacity in France.
Euratom	European Atomic Energy Community	One of three component institutions (with the European Economic Community and the European Coal and Steel

Abbreviation	*Term*	*Definition*
		Community) of the European Communities.
EJ	Exajoule	See under *joule.*
	Fabrication, fuel fabrication	Process of preparing *fuel elements* for use in a *reactor.*
	Refabrication	Fuel fabrication process dealing with *recycled fuel* material in a *closed cycle* version of the *fuel cycle.*
FBR	Fast breeder reactor	See under *reactor.*
	Fast reactor	See under *reactor.*
	Fertile (material)	Material in which *irradiation,* predominantly by high-energy ("fast") neutrons, breeds new *fissile material* (as *plutonium 239* is bred in fertile *uranium 238*).
	Fissile (material)	See under *fission.*
Pu_f	Fissile plutonium	See under *plutonium.*
	Fission	Nuclear reaction in which atomic nuclei split into roughly equal parts, releasing energy.
	Fissile (material)	Material in which atomic nuclei are predominantly susceptible to fission induced by low-energy ("thermal") neutrons, for example, *uranium 235, plutonium 239.*
	Fissionable (material)	Material in which atomic nuclei are susceptible to fission.
	Fission product	Atomic nuclei produced by fission of larger nuclei.
	Fissionable (material)	See under *fission.*
	Fission product	See under *fission.*
	Fuel, reactor fuel	Material containing *fissile*

Abbreviation	*Term*	*Definition*
		material within which *fission* occurs in the core of a *reactor.*
	Fuel element	Fabricated component of reactor fuel containing *fissile material.*
	Spent fuel	Fuel material withdrawn from a *reactor* after *irradiation* on reaching the intended *burnup.*
	Fuel cycle	Sequence of processes involving *reactor fuel* materials, including "front end" processes preceding the loading of fuel into a *reactor* and "back end" processes dealing with *spent fuel* after withdrawal from a reactor.
	Closed cycle	Version of fuel cycle in which *spent fuel* is subjected to *reprocessing* and *recycle.*
	Once-through cycle	Version of fuel cycle in which *spent fuel* is stored indefinitely without *reprocessing* and *recycle.*
	Fuel element	See under *fuel.*
	Fuel fabrication	See under *fabrication.*
	Full-scope safeguards	See under *safeguards.*
	Fusion	Nuclear reaction in which atomic nuclei combine to form a heavier nucleus, releasing energy.
	Gas centrifuge	See under *enrichment.*
GCR	Gas-cooled reactor	See under *reactor.*
	Gaseous diffusion	See under *enrichment.*
	Gas-graphite reactor	See under *reactor.*

Abbreviation	*Term*	*Definition*
GE	General Electric Corporation (United States)	
GESMO	Generic Environmental Statement on Mixed Oxide Fuels	Study of the use of *mixed oxide fuel* in *light water reactors,* undertaken from 1973 by the *Nuclear Regulatory Commission* in the United States.
G	Giga	10^9, billion.
Gtce	Gigaton of coal equivalent	See under *ton of coal equivalent.*
Gtoe	Gigaton of oil equivalent	See under *ton of oil equivalent.*
GW	Gigawatt	See under *watt.*
GWh	Gigawatt hour	See under *watt hour.*
g	Gram	Unit of weight (metric) (1 g = 10^{-3} *kilograms* = 0.002 *pounds*).
GDP	Gross domestic product	
GNP	Gross national product	
	Half-life	Time for the rate of radiation emission from a *radioactive isotope* to decline from its initial maximum to half that value.
HM	Heavy metal	Collective term for heavy metallic elements, commonly of atomic numbers ≥ 90, including *thorium, uranium, plutonium.*
	Heavy water	Deuterium oxide (D_2O): a compound of deuterium or "heavy hydrogen" (^2_1H), forming 0.02% of ordinary ("light") water, used as a moderator and/or coolant in some types of *reactor.*

Abbreviation	*Term*	*Definition*
HWR	Heavy water reactor	See under *reactor*.
HEU	High-enriched (highly enriched) uranium	See under *uranium*.
HLW	High level waste	See under *waste*.
HTR	High temperature reactor	See under *reactor*.
	Hot cell	Facility designed to permit remote handling of highly *radioactive* material.
IAEA	International Atomic Energy Agency	
IEA	International Energy Agency	
INFCE	International Nuclear Fuel Cycle Evaluation	Multinational review, from 1977 to 1980, of requirements and options for different versions and stages of the *fuel cycle*, with particular reference to considerations of nuclear weapons proliferation.
	Irradiation	Process of subjecting materials to a flux of radiation (especially of neutrons) released by *fission* in a *reactor*, resulting in, for example, further fission in *fissile material* or neutron capture in *fertile material*.
	Isotope	One form of an element, the atomic nuclei of whose isotopes all contain the same number of protons (and thus have the same atomic number) but different numbers of neutrons (and thus have different mass numbers); for example, all

Abbreviation	*Term*	*Definition*
		isotopes of *uranium* have 92 protons in the nucleus, but the *uranium 235* isotope has 143 (235 − 92) neutrons, whereas the *uranium 238* isotope has 146.
J	Joule	SI unit of energy and work (1 J = 10^7 ergs = 278 × 10^{-6} *watt hours* = 948 × 10^{-6} *British thermal units* = 34.1 × 10^{-12} *tons of coal equivalent* = 22.7 × 10^{-12} *tons of oil equivalent*).
EJ	*Exajoule*	10^{18} joules (1 EJ = 278 *terawatt hours* = 34.1 *million tons of coal equivalent* = 22.7 *million tons of oil equivalent*).
k	Kilo	10^3, thousand.
kg	Kilogram	Unit of weight (metric) (1 kg = 10^{-3} *tons* = 2.204 *pounds*).
kW	Kilowatt	See under *watt.*
kWh	Kilowatt hour	See under *watt hour.*
KWU	Kraftwerk Union (West Germany)	
LDC	Less developed country	Synonym of "developing country," "Third World."
LWR	Light water reactor	See under *reactor.*
LMFBR	Liquid metal fast breeder reactor	See under *reactor.*
L_f	Load factor, capacity factor	Average proportion of designed capacity of a *reactor* or process plant actually in use over a given time period; for

Abbreviation	*Term*	*Definition*
		example, a 100 *megawatt* reactor operating at a 65% load factor over 100 hours produces 6,500 *megawatt hours* of electricity.
LEU	Low-enriched uranium	See under *uranium.*
LLW	Low-level waste	See under *waste.*
	Magnox	See under *reactor.*
MLW	Medium-level waste	See under *waste.*
M	Mega, million	10^6.
MW	Megawatt	See under *watt.*
MWd	Megawatt day	See under *watt day.*
MWh	Megawatt hour	See under *watt hour.*
m	Meter	Unit of length (metric) (1 m = 39.37 inches).
Mtce	Million tons of coal equivalent	See under *ton of coal equivalent.*
Mtoe	Million tons of oil equivalent	See under *ton of oil equivalent.*
MOX	Mixed oxide	*Reactor fuel* material containing oxides of different elements, commonly of *uranium* (UO_2) and *plutonium* (PuO_2).
Unat	Natural uranium	See under *uranium.*
NNWS	Non-nuclear-weapon state	Any state not qualified as a *nuclear-weapon state.*
NPT	Non-Proliferation Treaty	Treaty on the Non-Proliferation of Nuclear Weapons; opened for signature 1 July 1968, entered into force 5 March 1970.
NEA	Nuclear Energy Agency	Agency of the *Organization for Economic Cooperation and Development.*
NPC	Nuclear Power Company (Britain)	

Abbreviation	*Term*	*Definition*
	Nuclear reactor	See under *reactor.*
NRC	Nuclear Regulatory Commission (United States)	
NSSS	Nuclear steam supply system	The principal components of a *reactor,* including nuclear components, from which steam is derived to drive a turbine.
NWS	Nuclear-weapon state	Defined in the *Non-Proliferation Treaty* as a state having manufactured and exploded a nuclear explosive device before 1 January 1967 (i.e., the United States, the Soviet Union, Britain, France, or China).
	Once-through cycle	See under *fuel cycle.*
OECD	Organization for Economic Cooperation and Development	
OEEC	Organization for European Economic Cooperation	
OAPEC	Organization of Arab Petroleum Exporting Countries	
OPEC	Organization of Petroleum Exporting Countries	
Pu	Plutonium	Element of atomic number 94; extremely rare in nature and predominantly created by *irradiation* of *uranium* (especially *uranium 238*) in a *reactor.*
	Denatured plutonium	Plutonium in which *fissile isotopes* are heavily diluted by blending with other isotopes.

Abbreviation	*Term*	*Definition*
Pu_f	*Fissile plutonium*	*Plutonium 239* and *plutonium 241 isotopes.*
Pu239	*Plutonium 239*	*Fissile isotope* of plutonium.
Pu241	*Plutonium 241*	*Fissile isotope* of plutonium.
lb	Pound	Unit of weight (avoirdupois) (1 lb = 0.4536 *kilograms*).
	Power reactor	See under *reactor.*
PWR	Pressurized water reactor	See under *reactor.*
	Purex (process)	See under *reprocessing.*
	Radioactive	See under *radioactivity.*
	Radioactive waste	See under *waste.*
	Radioactivity	Spontaneous emission of radiation from the atomic nucleus of a radioactive *isotope* in the form of alpha or beta particles or gamma rays, resulting in progressive *decay,* at a rate expressed by the *half-life* of successive radioactive isotopes, until a stable (nonradioactive) nucleus remains.
	Reactor, nuclear reactor	Controllable assembly using nuclear *fission* to produce energy, whose basic components are a core containing *fissile material,* a moderator (in the case of a *thermal reactor*) to maximize the availability of low-energy ("thermal") neutrons, and a coolant to remove energy in the form of heat from the core.
AGR	*Advanced gas-cooled reactor*	British version of *gas-cooled reactor,* fueled

Abbreviation	*Term*	*Definition*
		with *low-enriched uranium.*
BWR	*Boiling water reactor*	Version of *light water reactor,* in which the coolant is allowed to boil.
	Breeder reactor	Reactor in which the rate at which *fissile material* is created by *irradiation* of *fertile material* exceeds that at which fissile material is consumed by *fission.*
CANDU	*Canadian deuterium uranium (reactor)*	Canadian version of *heavy water reactor,* in which *heavy water* is used as both moderator and coolant.
CR	*Converter reactor*	Reactor in which the rate at which *fissile material* is consumed and converted into thermal energy by *fission* exceeds that at which fissile material is created by *irradiation.*
FBR	*Fast breeder reactor*	*Breeder reactor* in which *fission* is caused predominantly by high-energy ("fast") neutrons.
	Fast reactor	Reactor, without a moderator, in which *fission* is caused predominantly by high-energy ("fast") neutrons.
GCR	*Gas-cooled reactor*	*Thermal reactor* type in which the coolant is a gas such as carbon dioxide or helium and the moderator is commonly graphite.
	Gas-graphite reactor	*Thermal reactor* with gas coolant and graphite moderator (see *gas-cooled reactor*).

Abbreviation	Term	Definition
HWR	*Heavy water reactor*	*Thermal reactor* type in which deuterium oxide (D_2O) or *heavy water* forms the moderator, commonly (but not necessarily) fueled with *natural uranium.*
HTR	*High temperature reactor*	*Thermal reactor* type operating at particularly high temperature, with a gas coolant (e.g., helium) and a graphite moderator, fueled with *high-enriched uranium* and possibly also with *thorium* and/or *uranium 233.*
LWR	*Light water reactor*	*Thermal reactor* type in which ordinary ("light") water forms both moderator and coolant, commonly fueled with *low-enriched uranium.*
LMFBR	*Liquid metal fast breeder reactor*	Version of *fast breeder reactor* in which the coolant is liquid sodium metal, commonly fueled with a *mixed oxide* of *uranium* and *plutonium.*
	Magnox (reactor)	Early British version of *gas-cooled reactor,* fueled with *natural uranium* contained in Magnox (magnesium oxide).
	Power reactor	Reactor designed primarily for generating electricity.
PWR	*Pressurized water reactor*	Version of *light water reactor,* in which the coolant is pressurized to prevent boiling.
	Research reactor	Reactor designed for research use, rather than

Abbreviation	Term	Definition
		primarily for generating electricity.
	Thermal reactor	Reactor in which *fission* is caused predominantly by low-energy ("thermal") neutrons.
	Reactor fuel	See under *fuel.*
RBU	Reaktor Brennelement Union (West Germany)	
RAR	Reasonably assured resources	Geological resources (of *uranium*) in identified deposits, assessed and classified as extractable at a stated cost.
	Recycle	Process of returning *fissile material* (and/or *fertile material*) recovered by *reprocessing spent fuel* for use in new *fuel elements* in a *closed cycle* version of the *fuel cycle.*
	Refabrication	See under *fabrication.*
	Reprocessing	Chemical process of disaggregating (dissolving) *spent fuel* into its component parts (e.g., *uranium, plutonium,* and *waste fission products*).
	Civex (process)	Proposed reprocessing method to minimize extraction of *plutonium* in discrete form.
	Coprocessing	Reprocessing method yielding recovered *uranium* and *plutonium* combined as a single stream, rather than as separate products.
	Purex (process)	Conventional method of reprocessing by solvent extraction.

Abbreviation	*Term*	*Definition*
R&D	Research and development	
	Research reactor	See under *reactor*.
RWE	Rheinisch-Westfälisches Elektrizitätswerk (West Germany)	
	Safeguards	System of materials accounting and inspection to verify observance of a commitment to restrict the use made of nuclear materials, facilities, or technology, administered by national authorities or, in the case of international safeguards, by the *International Atomic Energy Agency* or a regional agency (e.g., *Euratom*).
	Full-scope safeguards	International (*International Atomic Energy Agency*) safeguards applied to all *fissionable material* in all peaceful nuclear activities under the control of a given country, especially of a *non-nuclear-weapon state* party to the *Non-Proliferation Treaty*.
SWU	Separative work unit	Unit of work performed in the *enrichment* of *uranium*, conventionally understood as being expressed in equivalent *kilogram* units (kgSWU).
	Serena (agreement)	A joint company for licensing information on the *fast breeder reactor*, established in 1977 by France and West Germany

Abbreviation	*Term*	*Definition*
		on behalf of themselves and their collaborators, Italy, Belgium, and the Netherlands.
SSEB	South of Scotland Electricity Board (Britain)	
	Spent fuel	See under *fuel.*
SALT	Strategic Arms Limitation Talks	
TVA	Tennessee Valley Authority (United States)	
T	Tera	10^{12}.
TW	Terawatt	See under *watt.*
TWh	Terawatt hour	See under *watt hour.*
	Thermal reactor	See under *reactor.*
	Thermonuclear (reaction)	Nuclear reaction, especially *fusion,* occurring only at extremely high temperatures ($\sim 10^8$ degrees centigrade).
Th	Thorium	Element of atomic number 90.
Th232	*Thorium 232*	The predominant *isotope* in thorium, capable of being used as a *fertile material* for conversion by *irradiation* into *uranium 233.*
	Tlatelolco, Treaty of	Treaty for the Prohibition of Nuclear Weapons in Latin America; opened for signature 14 February 1967.
	Toll enrichment	See under *enrichment.*
t	Ton (tonne)	Unit of weight (metric) (1 t = 0.984 long tons = 1.102 short tons = 2,204 pounds). N.B. *Unless otherwise stated, all tons used in this volume are*

Abbreviation	Term	Definition
		metric tons of 1,000 kilograms.
tce	Ton of coal equivalent	Primary energy calorifically equivalent to 1 *ton* of standard coal (1 tce = 0.67 *tons of oil equivalent* = 29.3×10^9 *joules*).
Gtce	*Gigaton of coal equivalent*	10^9 tons of coal equivalent.
Mtce	*Million tons of coal equivalent*	10^6 tons of coal equivalent.
toe	Ton of oil equivalent	Primary energy calorifically equivalent to 1 *ton* of standard oil (1 toe = 1.5 *tons of coal equivalent* = 44×10^9 *joules*).
Gtoe	*Gigaton of oil equivalent*	10^9 tons of oil equivalent.
Mtoe	*Million tons of oil equivalent*	10^6 tons of oil equivalent.
	United Reprocessors	Multinational arrangement for collaboration on *reprocessing* marketing and technology between enterprises in Britain, France, and West Germany.
USDOE	United States Department of Energy	
U	Uranium	Element of atomic number 92.
	Degraded uranium	Uranium in which the proportion of the *uranium 235 isotope* has been reduced by *fission.*
	Depleted uranium	Uranium in which the proportion of the *uranium 235 isotope* has been reduced below the level in *natural uranium.*

Abbreviation	*Term*	*Definition*
	Enriched uranium	Uranium in which the proportion of the *uranium 235 isotope* has been increased to above the level (0.71%) in *natural uranium.*
HEU	*High-enriched (highly enriched) uranium*	Uranium containing ≥ 20% of *uranium 235.*
LEU	*Low-enriched uranium*	Uranium containing > 0.71% but < 20% of *uranium 235.*
Unat	*Natural uranium*	Uranium of the isotopic composition found in nature (99.285% *uranium 238,* 0.71% *uranium 235,* 0.005% uranium 234), ore containing which is commonly measured in terms of the content of the prevalent uranium oxide (U_3O_8).
U233	*Uranium 233*	*Fissile isotope* extremely rare in nature but created by *irradiation* of *thorium 232* in a *reactor.*
U235	*Uranium 235*	*Fissile isotope* present in *natural uranium.*
U236	*Uranium 236*	Nonnatural *isotope,* created by *irradiation,* whose neutron absorption propensity qualifies it as an adulterant for *enrichment* purposes.
U238	*Uranium 238*	The preponderant *isotope* in *natural uranium,* capable of being used as a *fertile material* for conversion by *irradiation* into *plutonium.*

Abbreviation	*Term*	*Definition*
UF_6	*Uranium hexafluoride*	Compound of uranium commonly produced by *conversion* for use, in gaseous form, as feed for *enrichment.*
UOX	*Uranium oxide*	Oxide compounds of uranium, commonly as UO_2, used in *reactor fuel.*
	Urenco	Tripartite industrial and commercial *enrichment* consortium established by Britain, West Germany, and the Netherlands.
	Waste, radioactive waste	Waste materials from any part of the *fuel cycle* that include *radioactive* atomic nuclei, commonly classified in terms of the relative intensity of their *radioactivity.*
HLW	*High-level waste*	Waste with a high level of *radioactivity.*
LLW	*Low-level waste*	Waste with a low level of *radioactivity.*
MLW	*Medium-level waste*	Waste with an intermediate level of *radioactivity.*
W	Watt	SI unit of power (1 W = 1 *joule*/second). N.B. *Reactors and reactor capacity can be rated either in terms of the thermal power level attained within the reactor, measured in watts thermal (Wth), or in terms of electrical power output, measured in watts electrical (We); in this volume, all values are on the latter basis, and all*

Abbreviation	*Term*	*Definition*
		references to reactor and capacity ratings are to be read as in watts electrical (We).
GW	*Gigawatt*	10^9 watts.
kW	*Kilowatt*	10^3 watts.
MW	*Megawatt*	10^6 watts.
TW	*Terawatt*	10^{12} watts.
Wd	Watt day	Unit of power output in relation to time, used to measure *burnup* of *fuel* in a *reactor* (1 Wd = 86.4 × 10^3 *joules*).
MWd	*Megawatt day*	10^6 watt days.
Wh	Watt hour	Unit of power output in relation to time, used to measure electricity supplied (1 Wh = 3.6 × 10^3 *joules*).
GWh	*Gigawatt hour*	10^9 watt hours.
kWh	*Kilowatt hour*	10^3 watt hours.
MWh	*Megawatt hour*	10^6 watt hours.
TWh	*Terawatt hour*	10^{12} watt hours.
	Weapons grade (material)	Material including *fissile material,* of a composition particularly suitable for use in nuclear weapons.
WEC	World Energy Conference	
WOCA	World outside centrally planned economies area	World excluding the *centrally planned economies.*

Authors

KARL-HEINZ BECKURTS is senior vice-president and a member of the Managing Board of Siemens AG and, during the ICGNE study, was president of the Kernforschungsanlage Jülich.

THOMAS J. CONNOLLY is professor and associate chairman of mechanical engineering at Stanford University and, during the ICGNE study (1978–79), was on sabbatical leave at the Kernforschungsanlage Jülich.

DAVID FISCHER is a special adviser to the IAEA and, during the ICGNE study, was assistant director general for external relations at the IAEA.

RUSSELL W. FOX is a judge of the Federal Court of Australia and, during the ICGNE study, was Australian ambassador-at-large for nuclear non-proliferation and safeguards.

BERTRAND GOLDSCHMIDT is retired but, during the ICGNE study, was director of international relations at the Commissariat à l'Energie Atomique in France and a member (later chairman) of the Board of Governors of the IAEA.

ULF HANSEN is professor of power engineering at the University of Essen and, during the ICGNE study, was deputy head of the Systems Analysis and Technology Evaluation (STE) division at the Kernforschungsanlage Jülich.

RYUKICHI IMAI is Japanese ambassador to Kuwait and, during the ICGNE study, was general manager for engineering at the Japan Atomic Power Company.

WOLFGANG JAEK is a mechanical engineer and a research fellow in the Systems Analysis and Technology Evaluation (STE) division at the Kernforschungsanlage Jülich.

MUNIR AHMAD KHAN is a nuclear engineer who has been a member of the Board of Governors of the IAEA and, since 1972, the chairman of the Pakistan Atomic Energy Commission.

MYRON B. KRATZER is a senior consultant at International Energy Associates Limited in Washington, D.C.

RICHARD K. LESTER is assistant professor of nuclear engineering at Massachusetts Institute of Technology and, during the ICGNE study (1977–78), was a visiting research fellow in the International Relations Division of the Rockefeller Foundation.

MÅNS LÖNNROTH is a member of the Swedish government's Secretariat for Futures Studies in Stockholm and of the Department of Technology and Social Change at the University of Linköping.

ROBERT PRESS is a part-time adviser to the Cabinet Office in London.

IAN SMART is an independent adviser on international energy affairs and, during the ICGNE study (1977–78), was director of studies and deputy director of the Royal Institute of International Affairs.

WILLIAM WALKER is a senior fellow in the Science Policy Research Unit at the University of Sussex and, during the ICGNE study, was a research fellow at the Royal Institute of International Affairs.

MASON WILLRICH is vice-president for corporate planning at the Pacific Gas and Electric Company and, during the ICGNE study (1977–79), was director for international relations at the Rockefeller Foundation.

International Consultative Group on Nuclear Energy

Members since
1 November 1977

*Mr. Miguel Barandiarán Alcorta
Subdirector Técnico de Iberduero SA
Spain

*Mr. Paulo Nogueira Batista
Career Ambassador
President of Nuclebras
Brazil

*Professor Dr. Karl-Heinz Beckurts
Chairman of the Board
Kernforschungsanlage Jülich GmbH
Federal Republic of Germany

*Contralmirante Carlos Castro Madero
Presidente
Comisión Nacional de Energía Atómica
Argentina

*Ing. Juan Eibenschutz
Director General de Energía
Secretariat de Patrimonio y Fomento Industrial
Mexico

*Dr. Akbar Etemad
Former President
Atomic Energy Organization of Iran

*Mr. David Fischer
Assistant Director General for External Relations
International Atomic Energy Agency

Lord Flowers FRS
Rector
Imperial College of Science and Technology
United Kingdom

*HE The Hon. Mr. Justice Russell W. Fox
Ambassador-at-Large for Australia for Nuclear Non-Proliferation and Safeguards

*Dr. Bertrand Goldschmidt
Governor from France to the International Atomic Energy Agency

*Dr. Ryukichi Imai
General Manager for Engineering
Japan Atomic Power Company

*Mr. Jon Jennekens
President
Atomic Energy Control Board
Canada

*Mr. Munir Ahmad Khan
Chairman
Pakistan Atomic Energy Commission

H. E. Ambassador Edvard Kljun
Yugoslavia

*Mr. Måns Lönnroth
Secretariat for Futures Studies
Sweden

* Members of ICGNE in January 1980

*Dr. Robert Press CB CBE
Cabinet Office
United Kingdom

Dr. Homi Sethna
Government of India
Atomic Energy Commission

*Mr. Ian Smart
Adviser on International Energy Affairs
United Kingdom

*Dr. Mason Willrich
Vice-President for Corporate Planning
Pacific Gas and Electric Company
United States

Secretariat

Mr. Myron B. Kratzer (Rapporteur)
Consultant, International Energy Associates Ltd.
United States

Dr. Richard K. Lester (Research Associate)
Department of Nuclear Engineering
Massachusetts Institute of Technology
United States

Mr. William Walker (Secretary)
Royal Institute of International Affairs
United Kingdom

Index

The Johns Hopkins University Press

World Nuclear Energy: Toward a Bargain of Confidence

This book was composed in Times Roman text and Helios display type by Capitol Communication Systems. It was printed on S. D. Warren's 50-lb. Sebago Eggshell paper and bound in Holliston Roxite A cloth by Universal Lithographers.

59 0462879 9

Introduction to VHDL